Proceedings of the International Symposium on
FRACTURE MECHANICS

Pergamon Titles of Related Interest

Carlsson   MECHANICAL BEHAVIOR OF MATERIALS IV

Hearn   MECHANICS OF MATERIALS

Kachaniwsky   IMPACT ON PRODUCTIVITY OF NON-FERROUS
METALLURGICAL PROCESSES

Macmillan   QUALITY AND PROCESS CONTROL IN REDUCTION AND
CASTING OF ALUMINIUM AND OTHER LIGHT METALS

Rigaud   ADVANCES IN REFRACTORIES FOR THE METALLURGICAL
INDUSTRIES

Ruddle   ACCELERATED COOLING OF ROLLED STEEL

Salter   GOLD METALLURGY

Smith   FATIGUE CRACK GROWTH — 30 YEARS PROGRESS

Tait   FRACTURE & FRACTURE MECHANICS, CASE STUDIES

Valluri   ADVANCES IN FRACTURE RESEARCH

Related Journals

(Free sample copies available upon request)

ACTA METALLURGICA
CANADIAN METALLURGICAL QUARTERLY
SCRIPTA METALLURGICA

Proceedings of the International Symposium on

FRACTURE MECHANICS

Winnipeg, Canada
August 23–26, 1987

**Co-Sponsored by
the Basic Sciences and Materials
Engineering Sections of METSOC and
the Canadian Committee for Research
on the Strength and Fracture of Materials**

Vol. 6 Proceedings of the Metallurgical Society
of the Canadian Institute of Mining and Metallurgy

Edited by

W. R. TYSON
PMRL/CANMET, Ottawa, Ontario, Canada

B. MUKHERJEE
Ontario Hydro Research, Toronto, Canada

PERGAMON PRESS

NEW YORK · OXFORD · BEIJING · FRANKFURT
SÃO PAULO · SYDNEY · TOKYO · TORONTO

U.S.A.	Pergamon Press, Maxwell House, Fairview Park, Elmsford, New York 10523, U.S.A.
U.K.	Pergamon Press, Headington Hill Hall, Oxford OX3 0BW, England
PEOPLE'S REPUBLIC OF CHINA	Pergamon Press, Room 4037, Qianmen Hotel, Beijing, People's Republic of China
FEDERAL REPUBLIC OF GERMANY	Pergamon Press, Hammerweg 6, D-6242 Kronberg, Federal Republic of Germany
BRAZIL	Pergamon Editora, Rua Eça de Queiros, 346, CEP 04011, Paraiso, São Paulo, Brazil
AUSTRALIA	Pergamon Press Australia, P.O. Box 544, Potts Point, N.S.W. 2011, Australia
JAPAN	Pergamon Press, 8th Floor, Matsuoka Central Building, 1-7-1 Nishishinjuku, Shinjuku-ku, Tokyo 160, Japan
CANADA	Pergamon Press Canada, Suite No. 271, 253 College Street, Toronto, Ontario, Canada M5T 1R5

Copyright © 1988 Canadian Institute of Mining and Metallurgy

First edition 1988

Library of Congress Cataloging in Publication Data

International Symposium on Fracture Mechanics (1987: Winnipeg, Man.)
Proceedings of the International Symposium on Fracture Mechanics, Winnipeg, Canada, August 23–26, 1987 edited by W. R. Tyson, B. Mukherjee.
p. cm. — (Proceedings of the Metallurgical Society of the Canadian Institute of Mining and Metallurgy; vol. 6)
"Co-sponsored by the Basic Sciences and Materials Engineering Sections of METSOC and the Canadian Committee for Research on the Strength and Fracture of Materials."
Includes index.
1. Fracture mechanics — Congresses. I. Tyson, W. R.
II. Mukherjee, B. III. Metallurgical Society of CIM. Basic Sciences Section. IV. Metallurgical Society of CIM. Materials Engineering Section. V. Canadian Committee for Research on the Strength and Fracture of Materials. VI. Title. VII. Series.
TA409.I56   1987   620.1'126—dc19   88–2504

British Library Cataloguing in Publication Data

International Symposium on Fracture Mechanics (1987: Winnipeg, Man.)
Proceedings of the International Symposium on Fracture Mechanics, Winnipeg, Canada, August 23–26 1987.
1. Materials. Fracture. Mechanics
I. Title II. Tyson, W. R. III. Mukherjee, B.
IV. Series
620.1'126
ISBN 0–08–035764–4

Printed in Great Britain by A. Wheaton & Co. Ltd., Exeter

INTRODUCTION

This volume contains the proceedings of a Symposium held as a part of the 26th Annual Conference of Metallurgists held in Winnipeg, Manitoba, 23-26 August 1987. The Symposium was jointly sponsored by the Basic Sciences and Materials Engineering Sections of the Metallurgical Society of the Canadian Institute of Mining and Metallurgy, and the Canadian Committee for Research on the Strength and Fracture of Materials.

The Symposium, patterned after the Conference on "Fracture Control in Engineering Structures" (Can. Met. Quart. 19 #1, Jan. - Mar. 1980) held at a previous Conference of Metallurgists in Sudbury, Ontario, in August 1979, was opened with a tutorial session of invited reviews of the state of the art in key areas of fracture research followed by four sessions of presentations of current work. The topics covered a broad range of materials and methods of evaluation, reflecting the pervasive interest in avoiding fracture and fatigue in traditional and novel materials.

We would like to thank the sponsoring organizations for making this Symposium possible, and our employers and colleagues at Ontario Hydro and CANMET for invaluable support, both administrative and technical, in preparing for the Symposium and for processing the manuscripts. Finally, we would like to thank the participants, and organizers of the individual sessions - Graham Bellamy, Bob Coote, Jacques Masounave, and Pat Nicholson - for making the Symposium a success.

W.R. Tyson
PMRL/CANMET
Ottawa

B. Mukherjee
Ontario Hydro Research
Toronto

Symposium Co-chairmen

<u>Page</u>

TUTORIAL

MICROMECHANISMS OF FRACTURE

J.D. Embury and F. Zok
Materials Science and Engineering
McMaster University, JHE-358
Hamilton, Ontario L8S 4L7
CANADA

ABSTRACT

Various forms of damage may accumulate in engineering materials during monotonic deformation. The fracture mode is governed by which form of damage first reaches a critical level. Here, the various types of damage and fracture are reviewed. Metallographic evidence of the damage in a range of ferrous and non-ferrous materials under various stress states is presented and related to the fracture modes. Fracture mechanism maps are used to show the competition between failure modes, particularly for tests performed under superimposed pressure.

KEYWORDS

Fracture mechanism maps; damage accumulation; microcracks; ductile fracture; localized shear failure; delamination.

INTRODUCTION

When stresses are applied to a material it is essential to be able to predict whether the response will involve elastic or plastic flow or some form of fracture. However, such predictions are difficult because the competition between these modes of response depends on the detailed microstructure, the prevailing state of stress and the thermal-mechanical history of the material.

Thus in the past decades material scientists have pursued two well defined approaches to the description of fracture processes which can be approximated as macroscopic and microscopic in concept. **The macroscopic approach, termed fracture mechanics,** asks what the resistance to fracture is if the material or structure contains a pre-existing flaw of known size. In this approach we can characterize resistance to failure in a quantitative manner by designating a critical value of a parameter such as stress intensity factor or energy release rate.

1

In addition, a microstructural approach can be pursued which asks what processes can become locally favorable relative to continued elastic or plastic flow. These processes can include a wide variety of failure mechanisms such as brittle fracture or cleavage, decohesion of grain boundaries, ductile failure by void growth, delamination, or localized shear.

The description of the competition between these local events, and continued plastic flow involves a complex array of factors such as the scale of the microstructure, local chemistry of interfaces, stress state and strain path. These descriptions are essentially empirical and depend on the use of continuum descriptions to indicate the levels of stress and strain in the material or the growth of voids, and the use of criteria such as critical stress levels which are difficult to verify or to calculate from first principles.

Despite these shortcomings the microscopic approach allows a new dimension of the fracture process to be explored, mainly the concept of **damage accumulation prior to failure.** This is a concept of wide applicability including compressive failure of granular materials, ductile failure in metals, wear processes, failure in forming operations and the degradation of composites. Thus the approach taken in the current work will be to consider **damage processes in terms of nucleation and growth events,** and to consider both the stress state and microstructural dependence of these events. It is important to emphasize the link between damage accumulation and fracture mechanics in that in many situations failure occurs by the local accumulation of damage into a critical sized flaw which initiates failure rather than by growth of a pre-existing defect.

At this point it is of value to put the damage accumulation process into a practical engineering context. Consider first the classical situation of the increase in fracture resistance of structural steels with temperature as shown by the K_{Ic} data in Fig. 1. A consideration of the stresses at the crack tip indicates that the local stresses to cause cleavage failure are of the order of 2,000 to 2,500 MPa. However, the event which nucleates cleavage appears to be heterogenous, and involves the injection of cracks from carbides as shown in Fig. 2 (a). Thus the control of the ductile brittle transition involves both

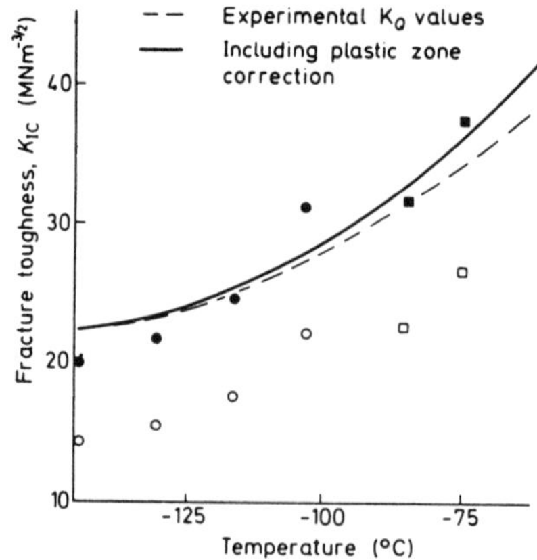

Fig. 1   Variation in fracture toughness with temperature for a high nitrogen steel (after Knott, 1973).

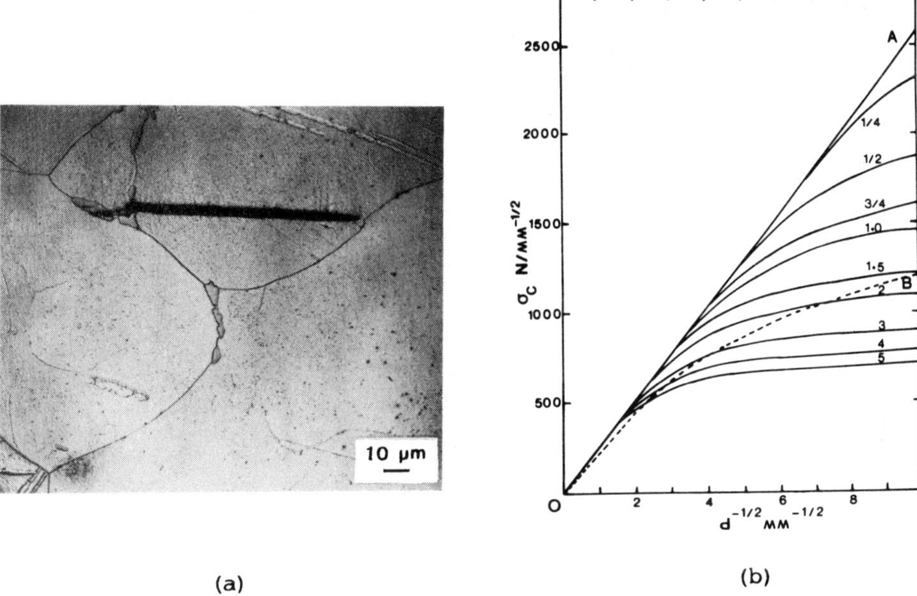

(a)   (b)

Fig.2   (a) Optical micrograph of a stable microcrack in the ferrite matrix
           originating from a grain boundary carbide (after Timbres, 1970).   (b)
           The effect of grain size, d, and carbide size (in μm) on the cleavage
           fracture strength of steel (after Petch, 1986).

the control of grain size, and the scale of the carbides which inject cleavage
nuclei as shown in Figure 2 (b). In the above case damage is in the form of
microcracks and their spatial organization into a cleavage front. This process
can be observed directly by conducting bend tests in the SEM to determine
whether macroscopic crack growth occurs by cleavage events or by void growth
(Luo et al., 1985). It is worth noting that current models of heterogenous
cleavage are static in form. It is possible that the condition is a dynamic one
which involves the velocity of the crack which can be injected from the carbide
into the ferrite. Kinetic criteria of this type are now being developed   by
Hack et al. (1987).

A second important aspect to consider in the use of structural materials is that
as the flow stress of a material is increased, the range of failure modes which
can occur increases rapidly due both to the high local stresses which can result
in decohesion events, and to the occurrence of localized strain due to the
difficulty of maintaining work hardening capacity at high stress levels. These
are shown schematically in Fig. 3. It can be seen that the higher the level of
the flow stress the lower the useful strains which can be imposed prior to the
attainment of a failure mode. However, in terms of materials development, it
may be possible to avoid some failure modes by the use of nanometer scale
materials, amorphous solids and ceramics.

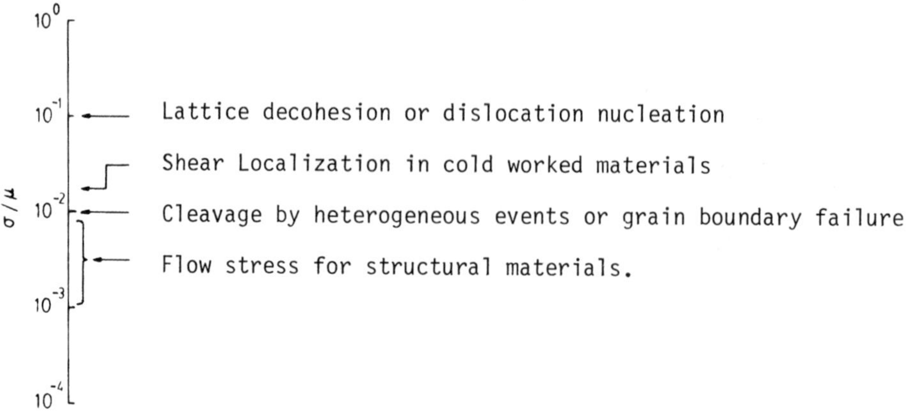

Fig. 3 Schematic drawing showing the increase in available fracture modes at high stress levels.

MODES OF FAILURE

For clarity let us consider specific modes of failure and proceed to indicate how they relate to damage accumulation processes. In order to accomplish this let us consider the pressure dependence of flow and fracture which can be explored using axisymmetric stress states of the type illustrated in Fig. 4. By varying the superimposed pressure, or by introducing notches in the tensile samples, we can attain a wide range of stress states over which fracture can be studied.

The major failure modes found in monotonic loading are cleavage fracture, ductile fracture by void growth, brittle intergranular fracture and localized shear. Examples of these modes for a variety of materials are shown in the fractographic evidence in Fig. 5.

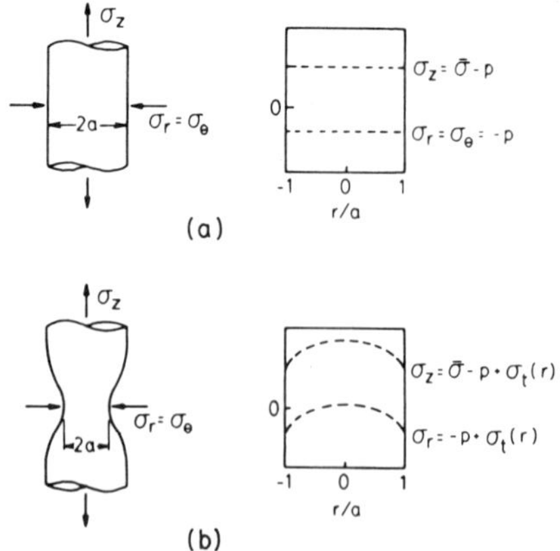

Fig. 4. Schematic drawings showing tensile tests under superimposed hydrostatic pressure, p, and the corresponding stress distributions (a) prior to necking and (b) after necking (or in pre-notched samples). Note the additional hydrostatic tension, σ_t, which results from the necked (or notched) geometry.

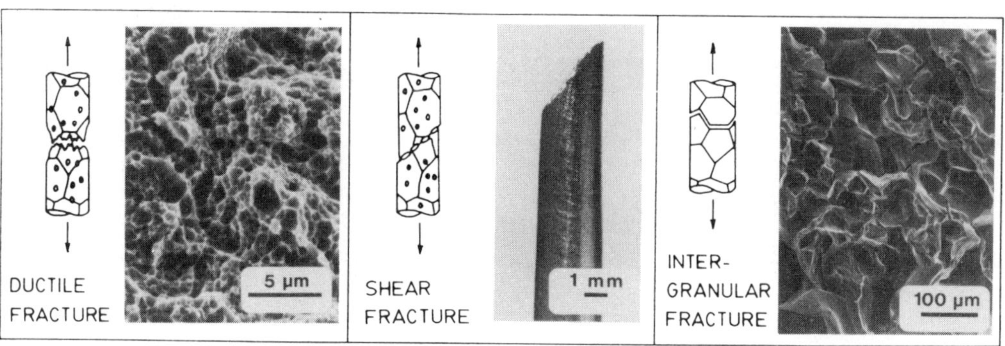

Fig. 5.　Schematic drawings and fractographic evidence of various fracture
modes at room temperature.　Examples are (from left to right):
fracture surface of a 4340 quenched and tempered steel,
macrophotograph of a directionally solidified Al–Ni eutectic alloy,
and the fracture surface of a brittle Cu–0.02% Bi alloy.

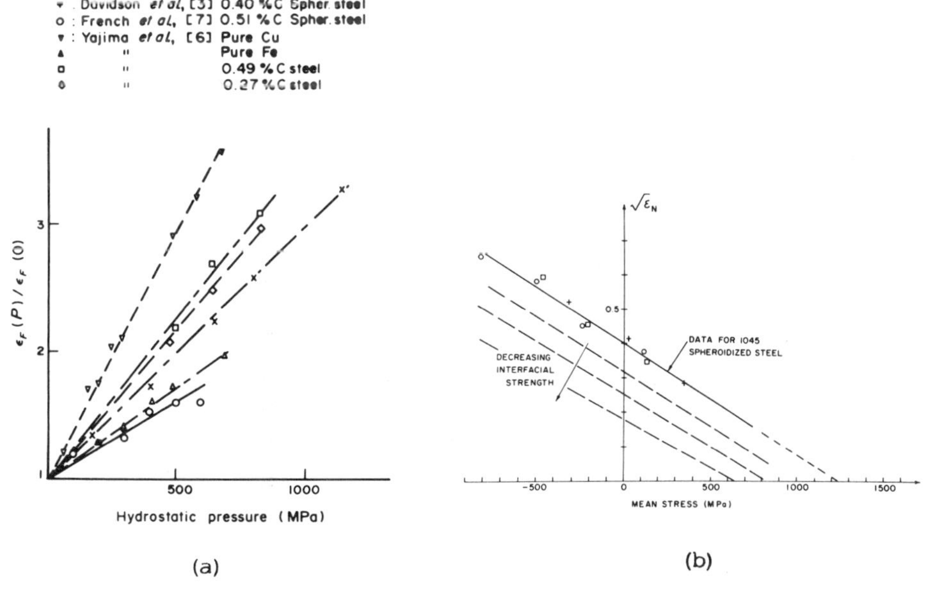

(a)　　　　　(b)

Fig. 6.　Diagrams showing the effects of superimposed pressure on (a) the ratio
of the fracture strain under pressure, $\epsilon_F(P)$, to the fracture strain
at atmospheric pressure, $\epsilon_F(0)$, for a variety of materials, and (b)
the void nucleation strain ϵ_N in a spheroidized steel.　Extrapolation
of the data in (b) to $\epsilon_N = 0$ gives an estimate of the interfacial
strength (after Brownrigg et al., 1983).

By conducting a series of tests at different imposed pressures, we can determine both the pressure dependence of the overall ductility and the pressure dependence of damage initiation and growth processes as illustrated in Fig. 6 and 7. These types of diagrams are of commercial relevance because in a number of materials containing high volume fractions of second phase particles the nucleation event determines the ductility. Thus measurement of the pressure dependence of the nucleation event yields an approximate value for the critical stress for interfacial decohesion for various types of particles as shown in Fig. 6 (b). This type of data may be of value in optimizing features such as particle size and distribution, interfacial chemistry and particle volume fraction in both conventional materials and materials fabricated by techniques such as powder compaction.

By considering the pressure dependence of various events, a map or diagram of the competitive fracture processes can be developed. These are termed fracture or failure mechanism maps, and the quantitative description of their development is given in abbreviated form in the Appendix. We now consider how damage processes lead to fracture in various materials and how fracture maps can be used to depict the competition between these processes.

Cleavage and Brittle Intergranular Fracture

At low temperatures and high hydrostatic tensile stresses, many crystalline solids fail in a brittle manner. Cracks may nucleate through a slip or twinning process or by injection from second phases, and propagate either by transgranular cleavage or by an intergranular path.

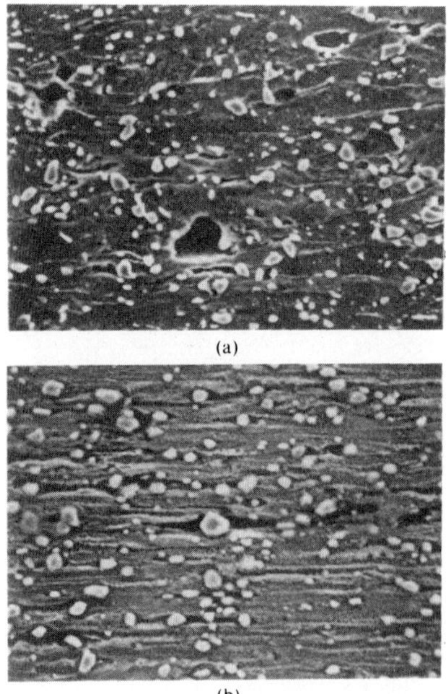

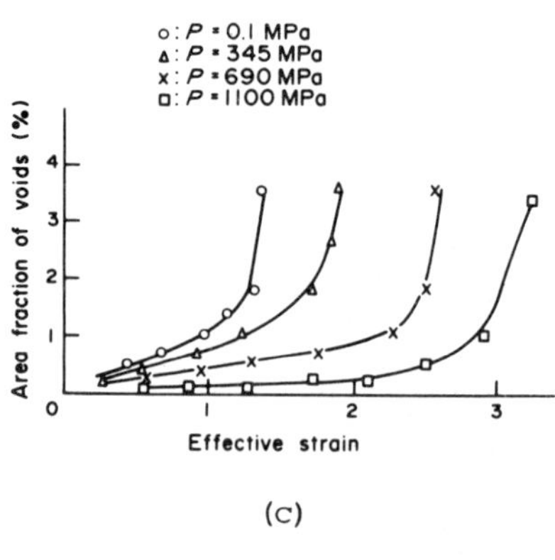

Fig. 7.　　SEM micrographs showing the damage due to void growth in a 1045 spheroidized steel at (a) atmospheric pressure and (b) a superimposed pressure of 1100 MPa. (c) A diagram showing the effect of pressure on the development of void damage with plastic strain (after Brownrigg et al., 1983).

Figure 8 shows the fracture surface of a Fe-1.2% P alloy which failed by a mixture of cleavage and intergranular fracture with negligible macroscopic ductility. A longitudinal section through the sample (Fig. 9 (a)) did not reveal any subsurface microcracks suggesting that fracture occurred immediately after crack nucleation. When tested under a superimposed pressure of 350 MPa the material failed at a strain of 0.1. Here there were many subsurface microcracks (Fig. 9 (b)) indicating that the pressure suppressed the propagation of the first nucleated crack. At yet higher pressures (>600 MPa), the material exhibited much higher ductility as shown by the pronounced necking in Fig. 9 (c).

The progressive change in the macroscopic ductility and the fracture behaviour shows the role of hydrostatic pressure in suppressing the growth and linkage of brittle microcracks. The behaviour can be described by the following simple model. At low pressures, fracture occurs when the first microcrack forms: this generally occurs at the yield point. At higher pressures the cracks are

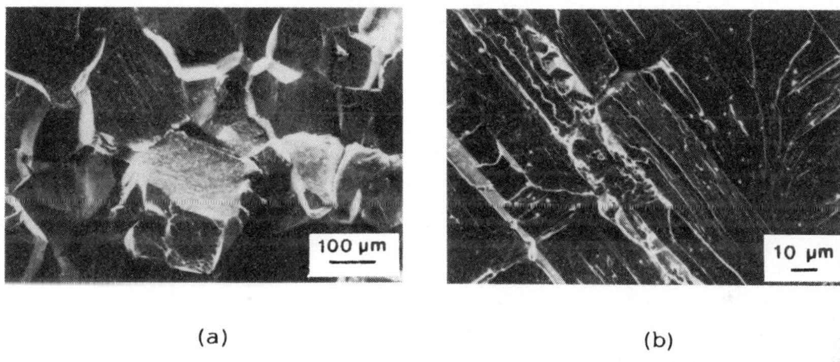

(a)   (b)

Fig. 8.   SEM micrographs of the fracture surface of an Fe-P alloy showing (a) grain boundary facets and (b) cleavage facets (after Teirlinck, 1983).

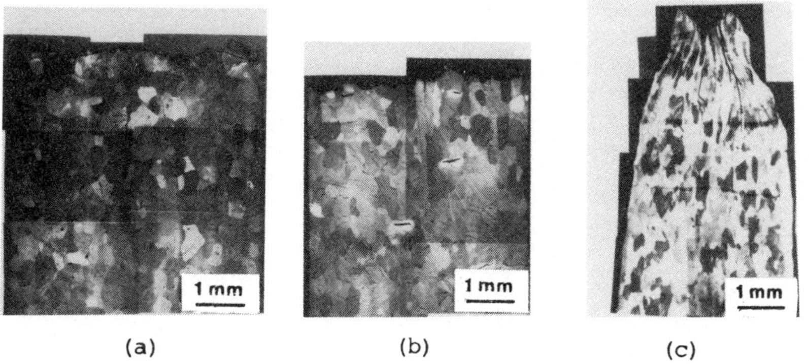

(a)   (b)   (c)

Fig. 9.   Longitudinal sections through the fractured Fe-P samples tested under pressures of (a) 0.1 MPa,   (b) 350 MPa, and (c) 870 MPa (after Teirlinck, 1983).

initially stable. But, as the stress is increased, the cracks grow and link until at some point a critical macrocrack is formed. The critical length of this defect can be estimated from a fracture mechanics approach.

Thus it is clear that in order to model brittle intergranular failure cognizance must be taken of the variation of microcrack density with strain and stress state. The interaction of the microcracks, and the formation of a critical defect by microcrack accumulation is a difficult problem, but is likely to be of great value in understanding the fracture of both metals and ceramics.

Ductile Fracture

Most metals and alloys fail by ductile fracture at room temperature. This mode involves the nucleation of voids at second phase particles, the growth of the voids by plastic flow of the matrix, and finally the coalescence of the voids into a macroscopic defect which results in fracture. Void nucleation and growth can be viewed as damage accumulation processes: when the damage reaches a critical level, the material fails.

Damage in ductile materials can be characterized by the area or volume fraction of voids, or by some average void dimension. The latter idea is consistent with Brown and Embury's (1973) simple geometric model for ductile fracture in which the void dimension in the tensile direction much reach a critical value dependent only on the interparticle spacing. Thus we can develop a simple fracture model in which voids grow from some initial size (= diameter of the particle) to a critical value independent of any superimposed pressure. Furthermore, we can describe the damage level at any strain by the ratio of the current void dimension in the tensile direction to the critical value of this dimension at fracture. This damage parameter therefore takes on a value of zero before void nucleation, and a value of one at fracture. Intermediate values indicate the extent of void development, and thus the degree of damage at any point during deformation.

Figure 10 shows a fracture mechanism map for a 1045 spheroidized steel in $\bar{\sigma} - \sigma_m$ space. The map shows a large regime of ductile fracture, with cleavage and shear fracture predominating at high and low mean stresses, respectively. Damage contours are also drawn on the map to show void development during deformation. The models used to construct the maps are briefly outlined in the Appendix and in more detail in Teirlinck et al. (1987).

Although the models used here to describe damage and fracture are rather simple, they do rationalize the various changes in fracture mode which are observed experimentally for a range of materials. They also predict the pressure-dependence of fracture strain for ductile materials in reasonable agreement with experimental measurements. The model for ductile fracture does not, however, account for the increase in void density with strain, the interaction of voids during void growth, and the effects of particle clustering which may lead to the formation of a critical defect prematurely.

Studying ductile fracture in materials which are microstructurally relatively simple (such as the spheroidized steels) provides some fundamental knowledge from which we can begin to understand the behaviour of more complex engineering materials under various stress states. One such group of materials which has been studied here is a series of 4340 quenched and tempered steels with various yield strengths. The material was austenitized at 900 $^{\circ}$C for 30 minutes, oil quenched, and tempered for 1 hour at the designated temperatures. Figure 11 shows the effect of temper temperature and pressure on fracture strain and fracture stress. The materials exhibit essentially the same plastic behaviour

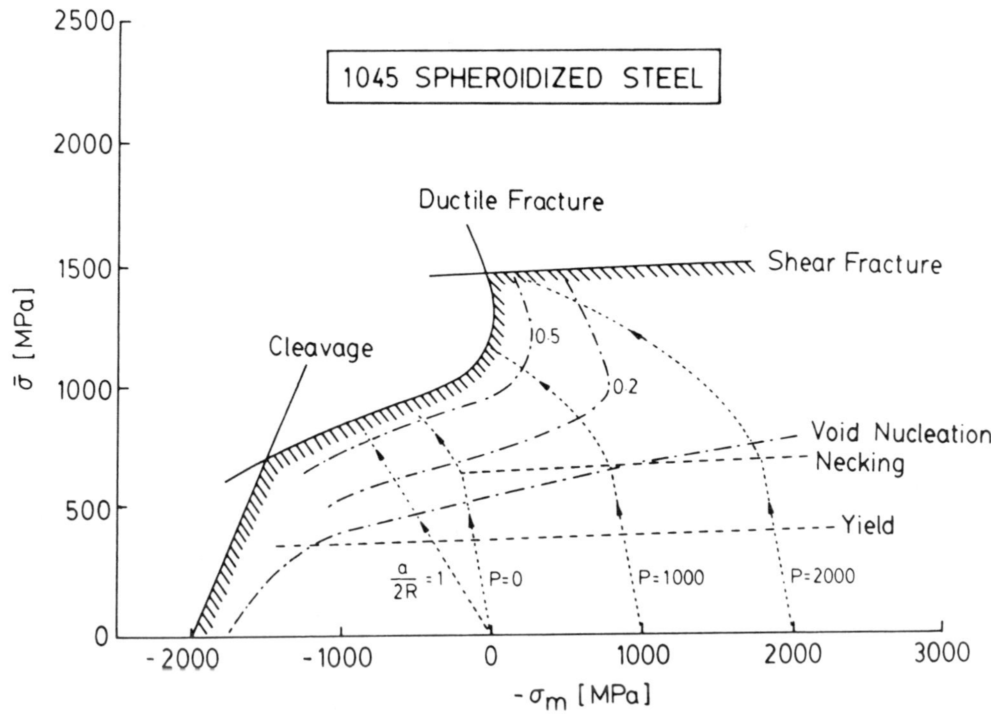

Fig. 10   A fracture mechanism map for a 1045 spheroidized steel showing fracture loci, contours of constant damage, and the stress trajectories for tensile tests conducted on cylindrical samples at various pressures (p = 0, 1000, 2000 MPa) and in pre-notched samples (a/2R = 1).

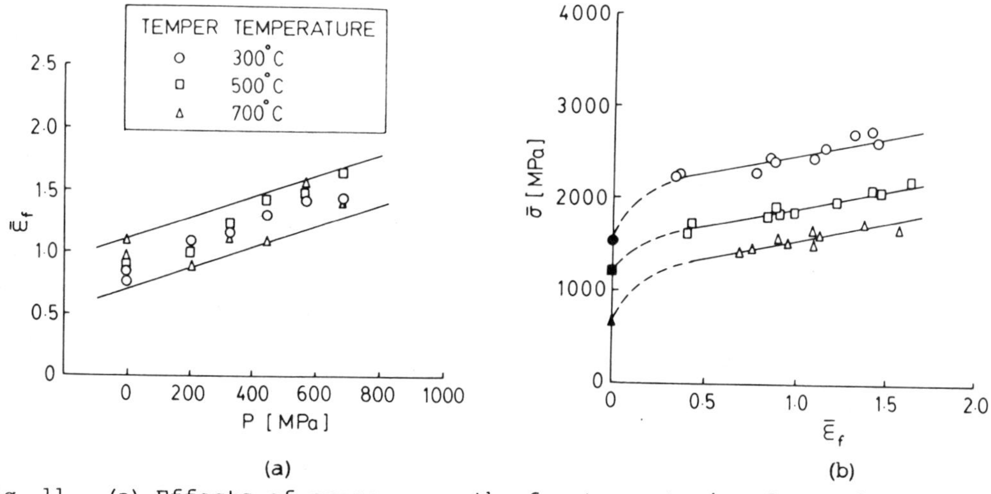

(a)   (b)

Fig. 11   (a) Effects of pressure on the fracture strain of a series of 4340 quenched and tempered steels. (b) Flow curves for the 4340 steels calculated from the fracture conditions at various pressures and in pre-notched samples. The solid symbols correspond to the yield points.

apart from differences in yield stress due to the temper temperatures.
Furthermore, their fracture strains increase with pressure at a rate which is
independent of the intial yield strength.

Under all stress states the 4340 steels exhibited one of two types of fracture
typified by the fracture surfaces in Fig. 12. One was the typical cup-cone
fracture with a dimpled central region bounded by shear lips. The other type
consisted of a small central dimpled region, a large region containing radially
oriented cracks and narrow shear lips. The delamination in the latter fracture
type appears to be due to a grain boundary tearing mechanism (Fig. 13), and has
been frequently observed in other high strength steels (Hero et al., 1975).

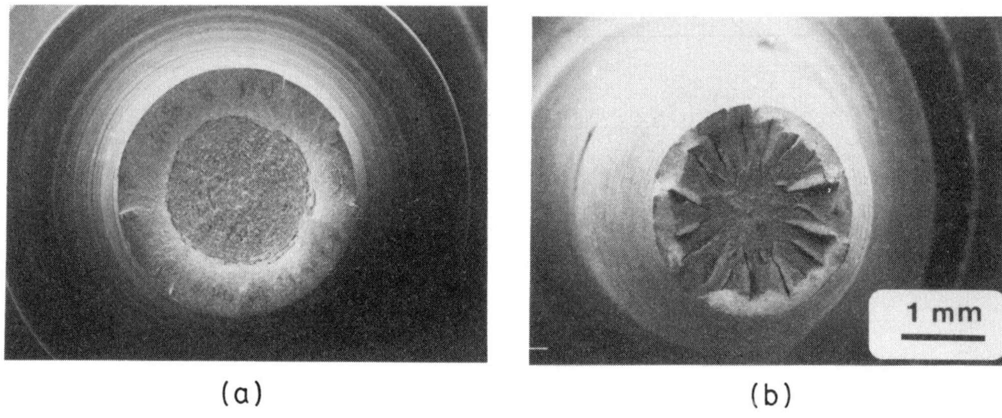

(a)   (b)

Fig. 12   SEM micrographs of the fracture surfaces of 4340 steels showing (a)
           classical ductile fracture and (b) ductile fracture accompanied by
           delamination.

(a)   (b)

Fig. 13   SEM micrographs showing (a) delamination at the fracture surface and
           (b) elongated ferrite grains on the delamination surface.

The delamination in the 4340 steels was not simply correlated with the yield strength or the superimposed pressure. Figure 14 does, however, show a trend when the fracture strains are plotted against the maximum (most tensile) normal stress, σ_N, acting on the tangential planes at fracture. The normal stress has a component due to the superimposed pressure, and one due to the hydrostatic tension arising from a neck or notch, the latter being calculated using Bridgman's (1952) analysis. It appears that delamination occurs when a critical combination of plastic strain and normal stress is attained. The behaviour may be explained qualitatively using a damage model in the following way. We can crudely characterize the resistance to delamination by a critical tensile stress which depends on the local level of damage. The damage may be in the form of voids formed at inclusions or grain boundary carbides. As the strain is increased the damage level increases, and thus the critical stress for delamination decreases. If the critical damage level is not attained prior to fracture, then the material fails in the classical ductile manner. Otherwise, delamination occurs prior to failure, and consequently we observe the large radial cracks on the fracture surface. It is difficult to quantitatively describe the damage development leading to delamination since it will depend in a complex way on the current level of plastic strain as well as the stress state.

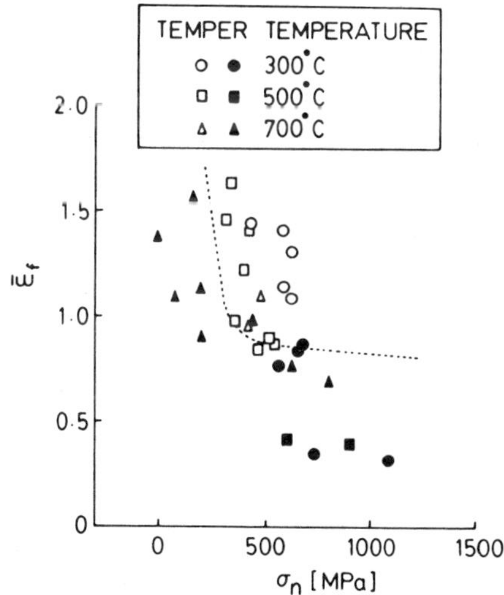

Fig. 14   A diagram showing the effects of fracture strain and maximum normal stress on the occurence of delamination in 4340 steels. Here the open symbols correspond to the fractured samples which exhibited delamination.

The competition between continued plastic flow, ductile fracture and delamination in the 4340 steels is shown in a fracture map in Fig. 15. The map was constructed from the experimental measurements and the fractographic observations. The lines for delamination are simply schematic and are intended to show whether or not delamination occurs prior to failure.The stress trajectories correspond to unnotched samples tested at atmospheric pressure. At the highest yield stress, delamination occurs for pressures slightly greater than atmospheric. At the intermediate yield stress, delamination occurs at

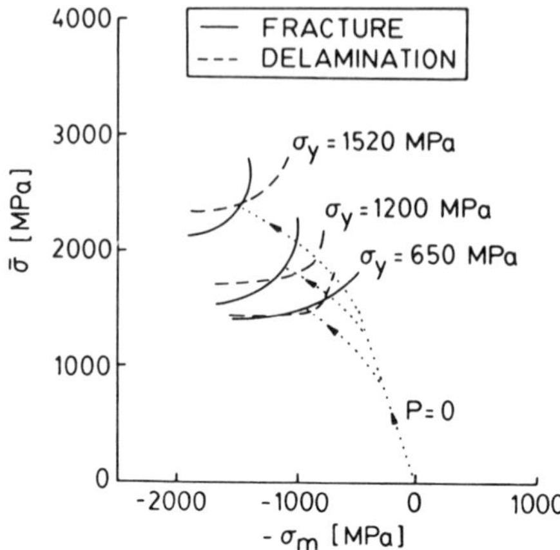

Fig. 15   A simplified fracture map for 4340 steels of various yield strengths showing the competition between fracture and delamination.

pressures greater than and including atmospheric pressure, but not in pre-notched samples. At the lowest yield stress, delamination occurs only near atmospheric pressure, and can be suppressed by a small superimposed pressure or by introducing a notch in the sample.

Further study is required to understand the conditions under which delamination occurs, and the interaction or competition between delamination and ductile fracture.

Localized Shear Fracture

In some cases where the stress levels are high (σ/μ= 10^{-2} to 10^{-1}), deformation can become localized into a narrow band of material. If the work hardening rate of the material is insufficient to offset the additional local stress, the material will fail catastrophically by the accumulation of damage within the band. It should be noted that localized shear failure involves two simultaneous problems, viz. a) the localization of plastic flow into macroscopic non-crystallographic bands (Korbel et al., 1986), and b) the nucleation and growth of voids within the shear band due to the **high local strain** levels.

Figure 16 shows a fractured tensile sample of a 7075 Al alloy which has undergone localized shear failure. The macroscopic strain at fracture was only 0.1 although the elongated voids on the fracture surface indicate that the local strains were much greater than this value (probably of order 1.0). The onset of shear localization in this alloy is independent of the superimposed pressure up to 1100 MPa suggesting that the onset is triggered by a structural softening and **not** by damage accumulation. Once formed, however, the development of the shear band is sensitive to the pressure. At low pressures (<500 MPa), fracture occurs within the shear band almost immediately after the onset of localization. At high pressures (>500 MPa) the original shear band is stablized while conjugate bands are formed on intersecting planes at 90° to one another (Fig. 17(a)). The load then drops incrementally with extension until the cross-section has been completely thinned (Fig. 17 (b)).

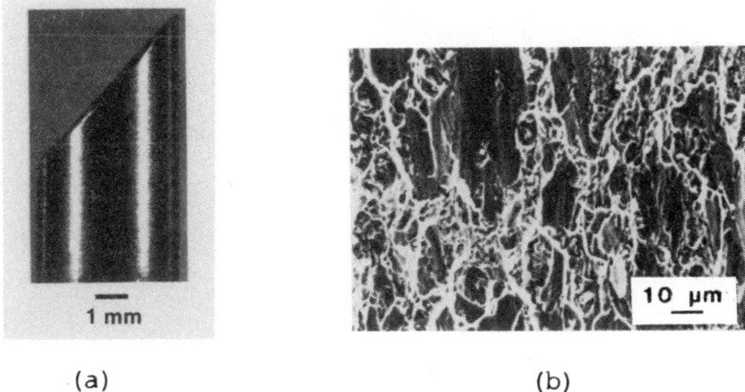

(a) (b)

Fig. 16   Shear fracture in a 7075 Al alloy tested under a pressure of 350 MPa. (a) Macrophotograph of the fractured sample.   (b) Fracture surface showing the void sheeting mechanism (after Teirlinck, 1983).

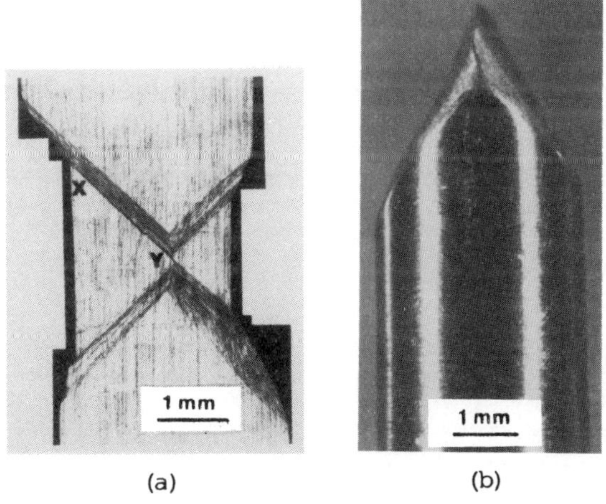

(a) (b)

Fig. 17   Shear band formation leading to plastic rupture in the 7075 alloy tested under a pressure of 1100 MPa.   (a)   Longitudinal section through sample prior to failure.   (b) Macrophotograph of sample after failure (after Teirlinck, 1983).

These observations clearly reflect the influence of hydrostatic pressure in the development of dilational damage within the shear bands. At low pressures, shear localization leads to a rapid accumulation of damage in the form of elongated voids. The damage results in a geometric softening which reduces the work hardening rate sufficiently to preclude plastic flow elsewhere in the material. Once started, the damage accumulation process becomes antocatalitic and ultimately leads to failure with only a small amount of macroscopic ductility. At high pressures, the rate of damage accumulation is substantially

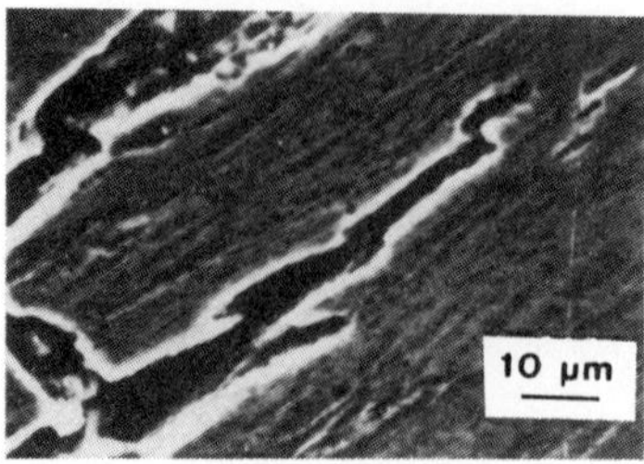

Fig. 18   SEM micrograph showing elongated voids in a shear band in the 7075
alloy tested at 1100 MPa pressure (after Teirlinck, 1983).

lower allowing plastic flow within the conjugate shear bands. Since geometric
softening within the bands is retarded by the superimposed pressure, the
intersecting bands can continue to deform stably until plastic rupture occurs.
Note, however, that even at pressures of 1100 MPa there is substantial
dilational damage within the bands (Fig. 18) although clearly it never reaches
the critical level for catastrophic failure. This result may be of more general
applicability in materials such as amorphous metals or polymers where
superimposed pressure may allow much greater useful ductilities to be obtained.

FUTURE DEVELOPMENTS

In optimizing the selection of engineering materials there is a need to
understand the deformation and fracture behaviour of a wide range of materials
including polymers, ceramics, and composites. Fracture mechanics provides one
aspect of this description. In addition, studies aimed at linking the
micromechanisms of deformation and damage accumulation to fracture mechanics are
required.

Figure 19 shows an example of how the pressure influences the ductility of a
directionally solidified Al-Ni eutectic alloy. Here there is a rapid increase
in fracture strain with pressure up to 350 MPa beyond which there is no effect
of pressure. Further examination of this and other systems may help elucidate
the role of stress state on the fibre-matrix interface and in turn how interface
behaviour affects macroscopic behaviour.

Currently work is in progress to examine the pressure dependence of a wide range
of materials including directionally solidified composites, intermetallics, and
composites of crystalline and amorphous materials.

The simple types of experiments and failure maps described here relate to
modifications of tensile loading. However, the concept of damage accumulation
and its stress state dependence can be applied to a much wider range of problems
including compressive loading (Ashby and Hallam, 1986), wear processes (Chiu,
1986), and fatigue loading. Thus an important future area of research is to
link macroscopic mechanics with damage accumulation processes or damage
mechanics both via the continuum models of Kachanov (1958), Hult (1978), etc.,
and via microstructurally based descriptions such as those given in the current
work. Furthermore, it would be useful to link these approaches to a variety of
important problems which limit the use of conventional materials such as
fracture resistance of notched structures, wear, and cyclic loading.

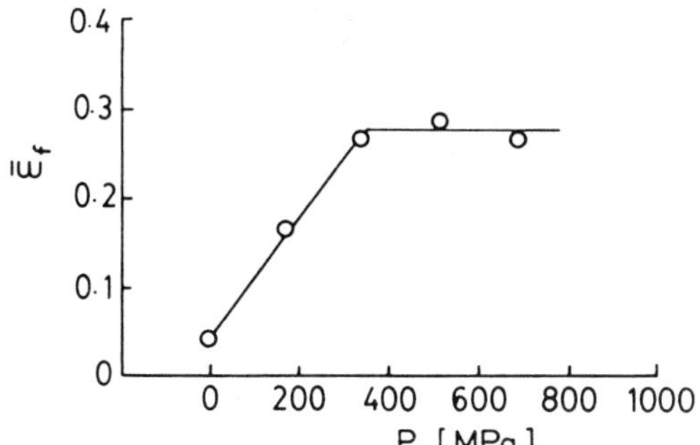

Fig. 19   Variation in fracture strain with pressure for a directionally
           solidified Al - Ni eutectic alloy.

SUMMARY

Various forms of damage may accumulate in engineering materials during monotonic
deformation. The fracture mode will depend on which type of damage reaches a
critical level first. Damage may take the form of microcracks, randomly
distributed voids, voids localized within shear bands, or the delamination of
interfaces.

The rate of damage developments depends on the plastic strain and the stress
state. Here we have shown how damage development can be examined by conducting
simple tension tests at various superimposed pressures. The experimental
observations form the basis for simple physical models which correlate the
micromechanisms of damage with the macroscopic fracture conditions. The damage
and fracture processes can then be depicted by fracture mechanism maps in stress
space.

ACKNOWLEDGEMENTS

The authors are grateful to NSERC for both research support and the provision of
a graduate scholarship. In addition, they wish to express their gratitude to
their present and former colleagues for valuable discussions.

REFERENCES

Ashby, M.F., and S.D. Hallam (1986). Acta Metall., 34, 497.

Bridgman, P.W. (1952). Studies in Large Plastic Flow and Fracture, McGraw-
Hill, New York.

Brown,   L.M. and J.D. Embury (1973).   Proc. 3rd Int. Conf.
Strength of Metals and Alloys, Cambridge, U.K., p. 164.

Brownrigg, A., W.A. Spitzig, O. Richmond, D. Teirlinck and J.D. Embury (1983).
Acta Metall., 31, 1141.

Chiu, C. (1986) M. Eng. Thesis, McMaster University, Hamilton, Ontario, Canada.

Hack, J.E., S.F. Chen, and D.J. Srolowitz. Acta Metall., in press.

Hero, H., J. Evensen and J.D. Embury (1975). Can. Metall. Quarterly, 14 (2), 117.

Hult, J. (1978). In Mechanisms of Deformation and Fracture, K.E. Easterling (Ed.), Pergamon Press Ltd., Oxford, U.K. p. 233.

Kachanov, L.M. (1958). IZV. Akad Nauk. SSR O.T.N. Teckh. Nauk., 8, 26.

Knott, J.F. (1973). Fundamentals of Fracture Mechanics, Butterworths, London, p. 227.

Korbel, A., J.D. Embury, M. Hatherly, P.L. Martin and H.W. Erbsloh (1986). Acta Metall., 34, 1999.

Luo, L.G., A.I. Quarrington and J.D. Embury, (1985). Eng. Fract. Mech., 21, 465.

Petch, N.J. (1986). Acta Metall., 34, 1387.

Rice, J.R. and D.M. Tracey (1969). J. Mech. Phys. Solids, 17, 201.

Teirlinck, D. (1983). Ph.D. Thesis, McMaster University, Hamilton, Canada.

Teirlinck, D., F. Zok, J.D. Embury, and M.F. Ashby, Acta Metall. to be published.

Timbres, D.H. (1970). Ph.D. Thesis, McMaster University, Hamilton, Canada.

APPENDIX

Ductile Fracture

Here the criteria for ductile fracture in notched and unnotched tensile samples under various superimposed pressures are presented in abbreviated form. Complete details of this model are presented elsewhere (Teirlinck et al., 1987).

Ductile fracture is assumed to occur when the size of the voids, r_3, in the tensile direction reaches a critical value $r_{3\ crit}$ which is independent of stress state. The void size is evaluated by integrating Rice and Tracey's (1969) solution for the growth rate of an isolated void from some initial value, r_0, at the nucleation strain, ε_{vn}, to the critical value at fracture.

Void nucleation is assumed to occur when the local stress reaches a critical value, σ_I, sufficient to cause particle-matrix decohesion or particle fracture. This criterion can be written approximately as:

$$\varepsilon_{vn} = \varepsilon_{vn}^0 \left[1 + \frac{P}{\sigma_I}\right]^2$$

where ε_{vn}^0 is the nucleation strain at p = 0.

In unnotched tensile samples the void size at any strain is given by:

$$\frac{r_3}{r_o} = \left[2 \exp \frac{3}{2} \underline{\varepsilon} - 1 \right]^{2/3} \exp \left[0.28 \underline{\varepsilon} + 2.21 (1 + \alpha \ln \alpha - \alpha) - \frac{0.84 \, p}{A (1 - m)} (\bar{\varepsilon}^{1 - m} - \varepsilon_{vn}^{1 - m}) \right]$$

where p is the superimposed pressure; A and m are the power hardening constants in the flow law $\bar{\sigma} = A \bar{\varepsilon}^m$; $\underline{\varepsilon} = \bar{\varepsilon} - \varepsilon_{vn}$; and $\alpha = 1 + 0.38(\varepsilon - m)$. The critical value of r_3 is evaluated by setting $\bar{\varepsilon} = \bar{\varepsilon}_1$ where $\bar{\varepsilon}_1$ is the fracture strain in simple tension with no superimposed pressure. At any other pressure p the fracture strain $\bar{\varepsilon}_2$ is given by:

$$0.44 (\bar{\varepsilon}_1 - \bar{\varepsilon}_2) + 2.21 (\alpha_1 \ln \alpha_1 - \alpha_2 \ln \alpha_2) =$$

$$\frac{-0.84 \, p}{A (1 - m)} (\bar{\varepsilon}_2^{1 - m} - \varepsilon_{vn2}^{1 - m})$$

Similarly for pre-notched tensile samples, the fracture strain $\bar{\varepsilon}_3$ in a sample with a known notch geometry, a/2R, where a is the minimum sample radius and R is the radius of curvature at the root of the notch, is given approximately by:

$$\bar{\varepsilon}_3 = \varepsilon_{vn3} + \frac{1.28 \, \bar{\varepsilon}_1}{1 + D}$$

where $D = 0.56 \sinh \left[1/2 + 3/2 \ln (1 + a/2R) \right]$

Cleavage and Brittle Intergranular Fracture

If the fracture toughness of a material is low then a crack will propagate unstably immediately after nucleation (which generally occurs at the yield point) to cause fracture. The fracture condition is then simply $\bar{\sigma} \geq \sigma_y$.

If the fracture toughness is high or if the superimposed pressure is large, then the crack will not propagate until the tensile stress exceeds the value of:

$$\sigma = p + y \, K_{IC} (\pi c)^{-1/2}$$

where K_{IC} is the fracture toughness, c the semi-crack length and y a constant near unity. This criterion may overestimate the fracture stress if a sufficiently large number of microcracks are formed which interact and link to form a critical defect.

Localized Shear Fracture

The model for shear fracture is a simple geometric one based on void growth within a shear band. Complete details of the model are presented elsewhere (Teirlinck et al., 1987).

When voids grow in a shear band, there is a reduction in the load bearing area which, at constant load, results in an increase in the true shear stress. If the work hardening of the matrix is insufficient to compensate for the increase in stress, then catastrophic failure within the band occurs. Assuming that the number of voids within the band increases proportionally with the strain beyond nucleation, and that the volume of the voids remains roughly constant, the shear fracture criterion can be written as:

$$\left(\frac{\bar{\sigma}^*}{\bar{\sigma}}\right)^{1/m} = \frac{\bar{\sigma}^{1/m} - \sigma_{vn}^{1/m}}{\bar{\sigma}^{*\,1/m} - \sigma_{vn}^{*\,1/m}}$$

where $\bar{\sigma}^*$ and σ_{vn}^* are the shear fracture stress and void nucleation stress at some pressure p^*, and $\bar{\sigma}$ and σ_{vn} are the corresponding values at some other pressure p.

MICROMECHANISMS OF BRITTLE FAILURE IN CERAMICS

David J. Green,
Department of Materials Science and Engineering
The Pennsylvania State University
PA 16802, USA

ABSTRACT
There has been substantial improvements over the last 20 years in the
strength and fracture toughness of ceramics and they are now being
considered for many structural applications, especially at high
temperatures. A key to this progress has been a much better
understanding of the micromechanisms involved in the crack formation
process in brittle materials and in toughening processes. The aim of
this paper is to review the important concepts involved in these
micromechanisms and the way in which they relate to the strength of
ceramics. The approach will be to discuss the various types of
toughening mechanisms and the processes by which critical cracks can
form.

KEYWORDS
Ceramics, toughening mechanisms, strengthening, fracture toughness,
crack formation, failure origins.

INTRODUCTION

Most people have had first hand experience of the brittle failure
process associated with ceramics, such as dropping a plate or breaking a
glass. This failure process generally persists to high temperatures, is
often catastrophic and occurs with little warning. In addition, it often
involves only small amounts of energy and appears to proceed with little
permanent deformation of the material. This lack of toughness and being
able to control brittle failure in ceramics are invariably key
questions in applications that involve structural stresses. It is
critical, therefore, to understand how the brittle fracture process
relates to the structure of the material, especially the micromechanisms
that are involved in the crack propagation process. Although brittle
failure has been a key factor in the use of materials since primitive
man fashioned his first tools, an understanding has only been developed
in this century and for ceramics within the last 20 years. During this
latter period there has been dramatic progress in the development of
high strength ceramics and there are now ceramic materials that have
bridged the gap between traditional brittle ceramics and metals. It is
interesting to note that ceramics are available with toughness values
and strength that are equivalent to cast iron, a material that played
such a key role in the Industrial Revolution.

FRACTURE MECHANICS AND STRENGTHENING

The basis for the understanding of the mechanics and thermodynamics of brittle failure was pioneered by Griffith (Griffith, 1920) and this was later developed into the field of linear elastic fracture mechanics by Irwin (Irwin, 1958). For many brittle materials, this approach assumes there are microscopic defects or flaws in the material and the analysis leads to a simple relationship between strength and toughness, i.e.,

$$\sigma_f = K_{IC}/[Y\sqrt{c}] \tag{1}$$

where σ_f is the fracture stress, c is the size of the critical crack that forms from the defect and propagates as a crack, Y is a geometric constant that depends on the loading and the crack shape and K_{IC} is the critical stress intensity factor. In the current ceramic literature, this latter parameter is generally referred to as the fracture toughness of the material.

One can use Eq. 1 as a guide to the general features that one would utilize to increase the strength of a ceramic material. Clearly one approach would be to increase the fracture toughness value without changing the critical flaw size. It has been found that **toughening mechanisms** are available in ceramics and usually involve manipulating and tailoring the microstructure. The mechanisms that can lead to toughening can be categorized into three types, and these are shown schematically in Fig. 1. *Crack tip interactions* occur when a crack tip interacts directly with an obstacle in the microstructure. For *crack tip shielding*, events are triggered by the high stresses in the crack tip region in a way that reduces these stresses. In the third category, ligaments of some type are left behind the advancing crack tip. This *crack bridging* makes it more difficult for the crack faces to open and hence hinders crack propagation.

A second approach to increase strength would be to determine the source of the defects that lead to the **formation of critical cracks** and then to find ways to eliminate or reduce the size of these flaws. It is important to recognize here that ceramics contain various populations of defects that compete as failure origins and within these populations, the defects lead to critical cracks of various sizes. Thus, the strength of ceramics is best represented as a statistical distribution that reflects these variations in defect type and size. In order to reduce the critical flaw size, there are two basic approaches, i.e., one can improve the techniques by which the material was made *(improved processing)* or one can find techniques to eliminate materials that will have the largest critical cracks. This latter approach generally involves the use of *non-destructive evaluation* to find the most important defects or *proof testing* (stressing all components prior to use) and eliminating those that fail or those that give a signal (e.g., acoustic emission) that the failure process is underway.

The final approach to increasing strength can also be deduced from Eq. 1 but it is more subtle. This approach attempts to utilize macroscopic residual stresses in a material and reduces the applied stress in the vicinity of the fracture origin. In ceramics, surface flaws are often a major flaw population and thus techniques that place the surface in compression, reduces the effective tensile stress acting on these flaws.

I. Crack Tip Interactions

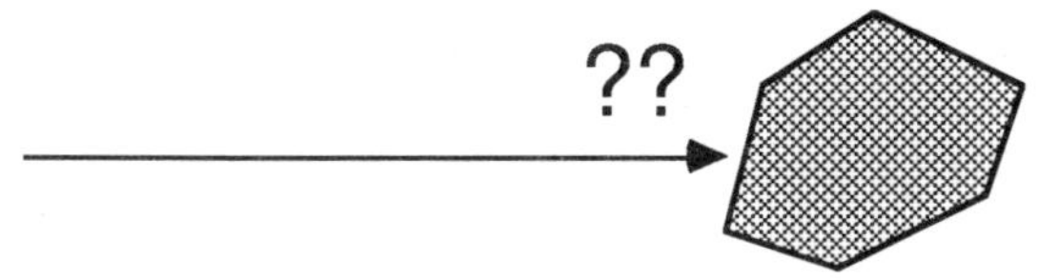

II. Crack Tip Shielding

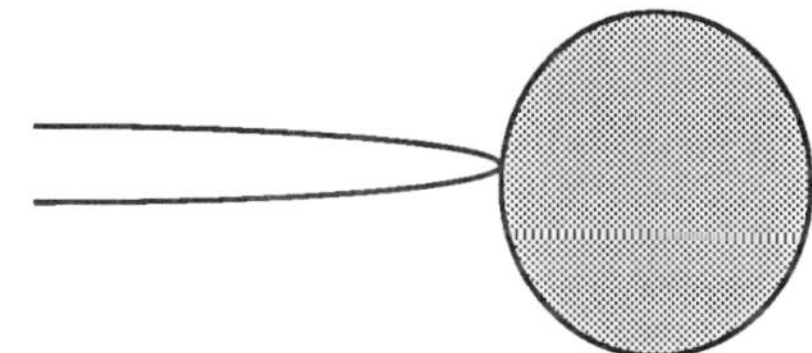

III. Crack Bridging

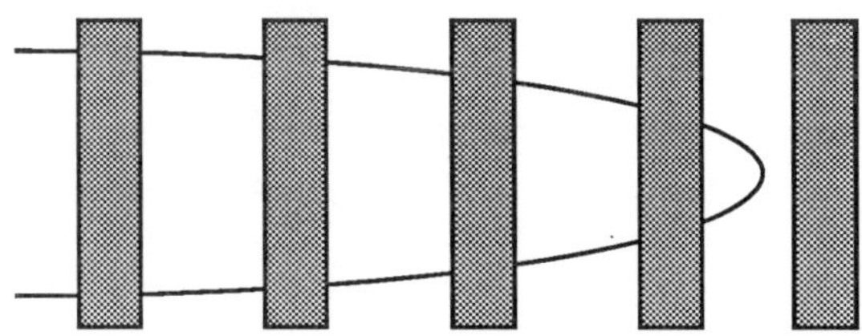

- weak bridging (fiber pull-out)
- strong bridging (metal particles
  or ceramic whiskers)

Fig. 1. Schematic illustrations of general
toughening mechanisms for ceramics.

In glasses, for example, tempering processes that lead to residual surface compression have been utilized for some time.

As indicated in this section two key issues in increasing the strength of ceramics are the toughening mechanisms and the processes by which defects form into critical cracks. Thus, it is the purpose of this paper to discuss the micromechanisms involved in these two processes, although the emphasis will be on the former. This type of understanding is critical in being able to put these techniques for increasing the reliability of ceramics into practice. One area that was not discussed in this section is that in some situations, cracks can propagate at stresses less than that predicted by Eq. 1. This sub-critical crack growth is often associated with stress corrosion or creep processes. Clearly, this can be a key issue in determining the lifetime of a ceramic under stress but is beyond the scope of this paper.

TOUGHENING MECHANISMS

As pointed out in the last section, there are three types of toughening mechanisms that are of current interest in ceramics, viz., crack tip interactions, crack tip shielding and crack bridging, and we will discuss examples of each.

<u>A. Crack Tip Interactions.</u>

The primary aim of this type of mechanism is to place obstacles in the crack path to impede crack motion. These obstacles could be second phase particles, whiskers, fibers or possibly, regions that are simply difficult to cleave. One would expect that stress concentrations or residual stresses associated with such obstacles would play a role in this process. There are two different types of consequences that can occur if the crack motion is impeded by an obstacle. In one case, although the crack is pinned by the obstacle, it can by-pass the obstacle by *bowing* around either side of it, remaining on virtually the same plane (Fig. 2a). In the other case, the crack could attempt to completely avoid the obstacle by *deflecting* out of the crack plane. The deflection of the crack front can be accomplished by tilting of the crack path or twisting of the crack front (Fig. 2b and 2c). In a real situation, a combination of bowing and deflection may occur but let us consider them separately.

1. <u>Crack Bowing</u>. This mechanism has been analyzed theoretically and it has been shown by two different types of calculations, that crack bowing should lead to an increase in fracture toughness (Lange, 1970; Evans 1972). There is considerable direct fractographic evidence for such bowing and this has been summarized previously (Green, 1976). An example of a crack bowing round a void in glass is shown in Fig. 3. In this figure the crack front fringes were produced by ultrasonic fractography (Green, 1976). The theoretical analyses of the crack bowing mechanism (Lange, 1970; Evans 1972) do have some shortcomings in that they both assume the obstacles are impenetrable. One would expect the strength and toughness of such obstacles to be a key issue, in that either the obstacle could fail before the bowing process is complete or the obstacles are left behind as unbroken ligaments behind the crack tip. In this latter case, crack bowing becomes a precursor to crack bridging. In the former case, it is possible to adjust the calculated toughening to account for obstacle strength (Green, 1983a). Two other important aspects that were not considered in the analyses are the influence of local stresses around the obstacle and interaction effects.

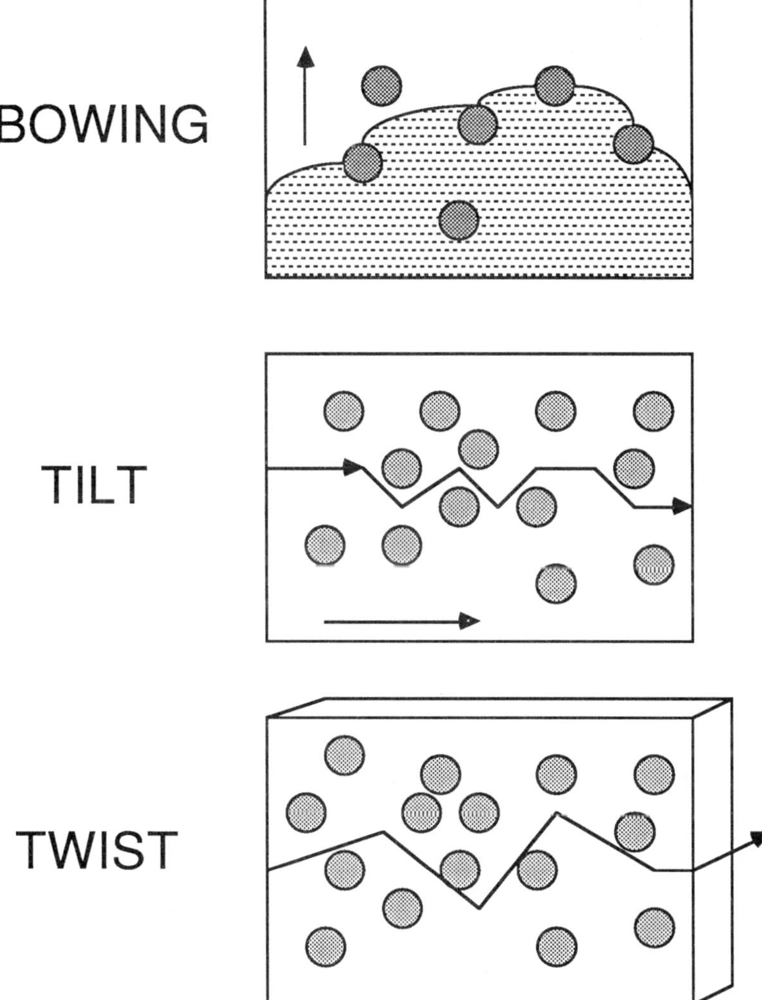

Fig. 2. Schematic illustrations of crack tip interaction mechanisms; a) crack bowing, b) crack deflection (tilt), c) crack deflection (twist).

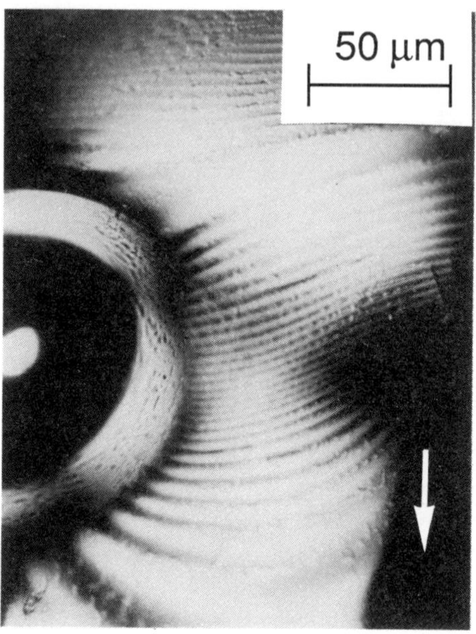

Fig. 3. Bowing of crack front around a void in glass.
Crack front markings produced by ultrasonic
fractography (fringe spacing = 1µs)

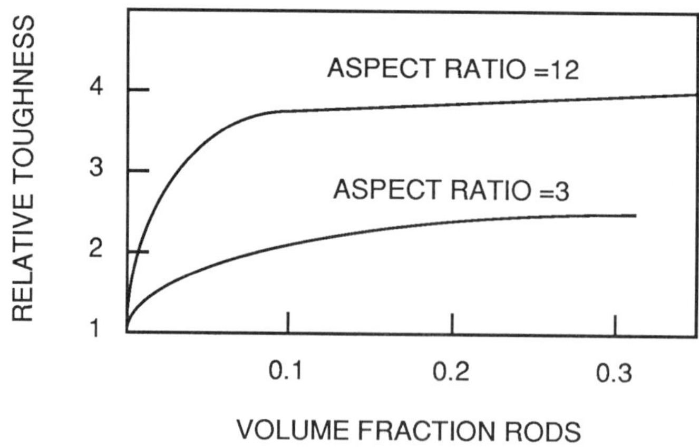

Fig. 4. Effect of rod-shaped particles on toughening
magnitude by crack deflection. After Faber
and Evans (1983a)

2. <u>Crack Deflection.</u> If a crack is deflected out of the plane that is
normal to the applied tensile stress, the crack will no longer be
subjected to the maximum tensile stress and will involve shear
displacements. This mixed mode fracture that occurs on the microscopic
scale is expected to increase the difficulty of crack propagation. Two
types of deflection have been analyzed: tilting of the crack about an
axis parallel to the crack front and twisting about an axis normal to
the crack front. The overall deflection process is manifested as
roughness of the final fracture surface. The reorientation of the crack
plane leads to a reduction of the crack extension force on the deflected
portion (Faber and Evans, 1983a). The crack deflection mechanism was
analyzed by evaluating the stress intensity factors for the various
loading modes on the deflected portion. The various components are then
combined with a mixed mode failure criterion and compared to the crack
resistance of the matrix that surrounds the obstacle. There are several
mixed mode failure criteria and the toughness increase predicted depends
on which criterion is utilized. A fracture mechanics analysis using a
fully coupled failure criterion demonstrated that it is the twist
component that contributes most to the fracture toughness (Faber and
Evans, 1983a). For a random array of obstacles, it has been shown that
the toughening increment depends on the volume fraction and shape of the
particles (Faber and Evans, 1983a). The toughening predicted for rod-
shaped particles is shown schematically in Fig. 4 and predicts that rods
with large aspect ratios impart maximum toughness. This effect is a
result of the increase in twist angle that occurs with increasing aspect
ratio (Faber and Evans, 1983a). For a given volume fraction of
obstacles, shaped as rods, discs or spheres, it has been shown that rods
are the most effective in increasing toughness. As seen in Figure 4, it
was also found that most of the toughening develops for volume fractions
of obstacles < 0.2. An attractive aspect of crack deflection mechanism
is that it is expected to be independent of temperature and particle
size. It should, however, be noted that the fracture mechanics analysis
of crack deflection does not account for local stress fields or local
changes in the crack resistance force as the crack transverses an
obstacle. In the case of localized residual stresses that result from
expansion differences, some temperature sensitivity might be expected
and these stresses may aid in increasing or decreasing the applied crack
extension force. Measurement of fracture toughness values and crack
deflection angles in SiC, Si_3N_4 and a glass-ceramic have been shown to
be consistent with the predictions of the crack deflection mechanism
(Faber and Evans, 1983b, 1983c). Figure 5 is an example of the crack
deflection process that occurs in a glass-ceramic (Faber and Evans,
1983c).

<u>B. Crack Tip Shielding.</u>

There are two mechanisms that have been identified as leading to crack
tip shielding, viz., transformation toughening and stress-induced
microcracking.

<u>1.Transformation Toughening.</u> An elegant approach to the toughening of
ceramics is the use of a stress-induced phase transformation and there
have been several recent reviews on this subject (Evans and Cannon, 1986;
Heuer, 1987; Green, Hannink and Swain, 1988) The most success has been
in the use of zirconia-based materials. Zirconia has a high temperature
tetragonal phase that usually undergoes a martensitic phase
transformation to monoclinic during cooling (~ 1000ºC). In this
direction, the transformation involves a 3 to 5% volume increase and a
simple shear of ~16%. These large transformation strains invoke large

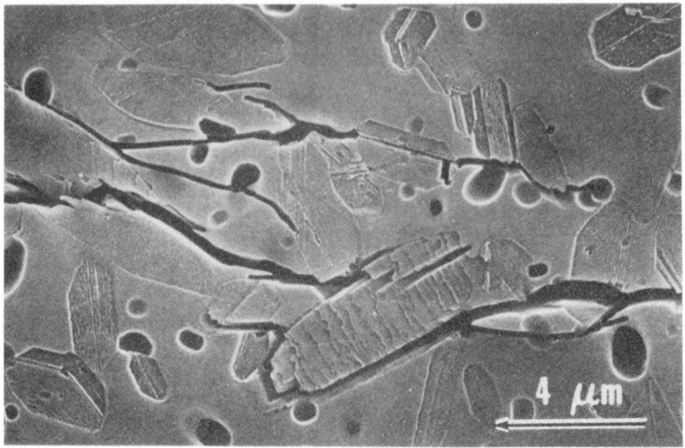

Fig. 5.   Observed crack deflection processes in a glass-ceramic material. Scanning electron micrograph, courtesy of K. T. Faber, Ohio State University.

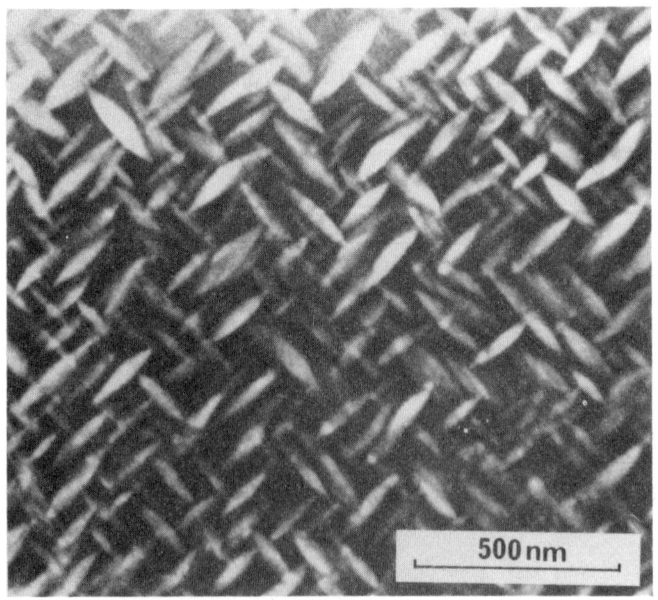

Fig. 6a.   Tetragonal zirconia precipitates in a magnesia partially-stabilized zirconia. Transmission electron micrograph, courtesy R. H. J. Hannink, CSIRO, Australia.

transformation stresses in a material. Thus, when one considers the thermodynamics of the transformation, the strain energy due to the transformation needs to be taken into account. This has the effect that the tetragonal phase, if constrained, can be retained to lower temperatures. In addition, undercooling will also be possible if the monoclinic phase is difficult to nucleate. Indeed, it is possible to retain the tetragonal phase to below room temperature provided the zirconia particle size is below a critical value. The cause of this size effect has been the subject of much debate (Lange and Green, 1981; Ruhle and Heuer, 1984; Heuer and Ruhle, 1985; Chen and Chiao, 1985a, 1985b). It is these materials in which the tetragonal phase is retained to low temperatures that are able to utilize the mechanism of transformation toughening.

There is a wide variety of microstructures that can incorporate the transformation toughening effect (Claussen 1984) but most of the work has been performed on materials in which the zirconia has 1) been precipitated in a cubic zirconia matrix, 2) been formed into a particulate composite (e.g., Al_2O_3/ZrO_2) or 3) been alloyed into a solid solution (e.g., with Y or Ce). Examples of these 3 types of microstructure are shown in Fig. 6. Fracture toughness measurements on these types of materials have shown that fracture toughness values up to ~18 MPa.$\sqrt{m}$ are possible. Thus, a key question is the role of the tetragonal ZrO_2 in this toughening phenomenon and identifying the controlling parameters. In some elegant in-situ observations on the transmission electron microscope (TEM), it was shown that the zirconia phase transformation can be induced by the stress field of a crack (Ruhle and co-workers, 1984). Other types of TEM observations (e.g., Porter and Heuer, 1977; Schoenlein and Heuer, 1983), x-ray diffraction of fracture surfaces (Kosmac, Wagner and Claussen, 1981) and Raman microprobe analysis (Clarke and Adar, 1982) all conclusively support this idea. The basic idea behind the retention of the tetragonal phase and the stress-induced phase transformation is shown schematically in Fig. 7. In the absence of stress, the strain energy involved in the size and shape change of a constrained tetragonal ZrO_2 particle will tend to oppose the transformation but a stress field, such as that illustrated, will help to overcome the constraint.

The next major question is why does this transformation around the crack tip lead to toughening. Figure 8 shows a schematic illustration of a crack surrounded by a transformation zone. If we consider the dilation of the particles within the zone that must occur as it forms, the material around the zone will oppose this increase in zone size. In some ways, the zone in Fig. 8 is analogous to the inclusions in Fig. 7, in that the surrounding material will push back on the transformed material. For the case of the crack, however, the compressive stresses within the zone act to close the crack and hence makes crack propagation more difficult. This reduction in the crack tip stresses reduces the local stress intensity factor and thus, **shields** the crack tip from the applied stresses. Clearly, it is of importance to understand the parameters that control this toughening increment and to determine the magnitude of the toughening produced by the stress-induced phase transformation.

There have been 3 different types of approaches to modelling transformation toughening. One of these (McMeeking and Evans, 1982) considers the influence of the stress field on the stress intensity factor, while the other two (Marshall, Drory and Evans, 1983, Budiansky,

Fig. 6b.   Yttria tetragonal zirconia polycrystal. Scanning
            electron micrograph, courtesy S. Jill Glass, The
            Pensylvania State University.

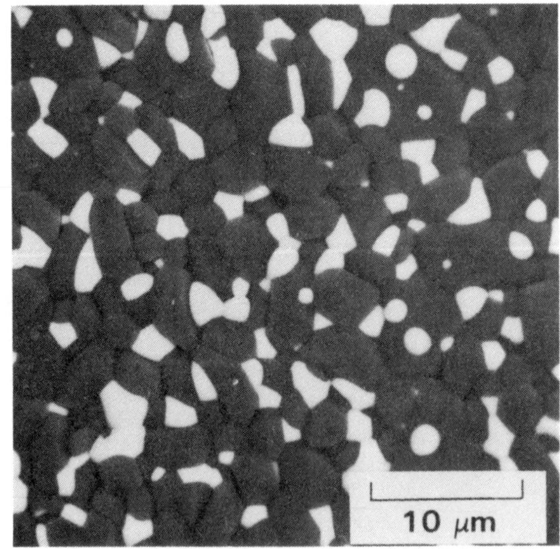

Fig. 6c.   Micrograph of a zirconia toughened alumina composite,
            bright phase is tetragonal zirconia, dark phase is
            alumina. Scanning electron micrograph, back-scattered
            mode.

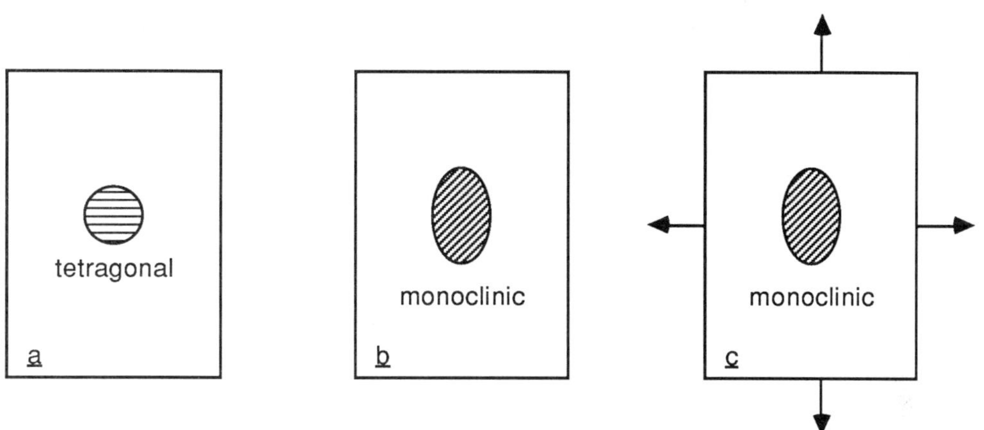

Fig. 7.   Schematic illustration showing effect of constraint on
           zirconia phase transformation and how it can be overcome
           by stress. The constraint of the surrounding material
           opposes the transformation in going from a to b but is
           assisted by a stress in going from a to c.

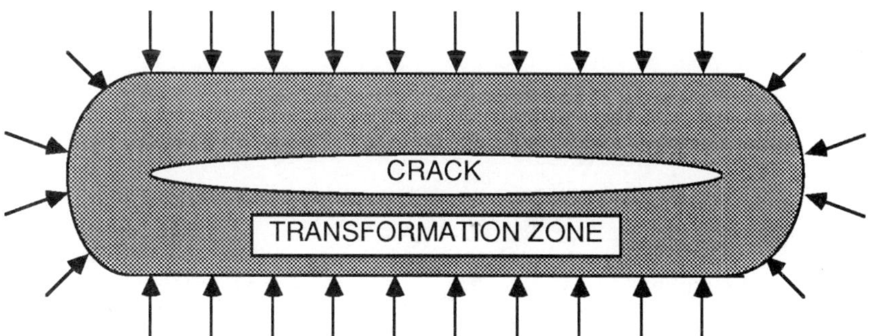

Fig. 8.   Transformation zone around a crack; the dilation within
           the zone is opposed by the surrounding untransformed
           material.

Hutchison and Lambroupolos, 1983) consider a thermodynamics approach. All 3 approaches came to identical conclusions and thus, give a compelling argument about the way the toughness develops and the parameters that control the process. The analyses first consider a crack tip in which there is a frontal transformation zone. The assumption usually made is that the contour of this zone is determined by the mean dilational stress (illustrated in Fig. 9). For this case, it was found that the toughening increment was zero. The result may seem surprising at first but as shown in Fig. 10, particles that transform in the region directly ahead of the crack will give rise to tensile stresses at a point in the crack plane, whereas particles above and below this point will give compressive stresses. The result of the integration is that the net effect is zero. It is only when the zone extends behind the crack tip that the toughening occurs, that is, as the zone extends into the crack wake (Fig. 11). Thus, one expects the toughening to increase as the crack propagates. This type of phenomenon is generally referred to as R curve behavior. For long cracks, it is found that the toughening reaches an asymptotic limit and the toughening (ΔK_{IC}) associated with the limit is given by

$$\Delta K_{IC} = 0.21 E V e^T \sqrt{h} / [1 - \upsilon]$$ (2)

where E and υ are the Young's modulus and Poisson's ratio of the material, e^T is the volumetric transformation strain, V is the volume fraction of material transformed by stress and h is the height of the transformation zone (McMeeking and Evans, 1982). There is a wide variety of assumptions in Eq. 2 and other analyses have considered a wide variety of other possibilities (see Evans and Cannon, 1986 or Green, Hannink and Swain, 1988) but these other approaches only make changes to the numerical constant in Eq. 2.

A useful way of analyzing the toughening is to consider the volumetric stress-strain behavior of an element of material as it enters into the crack tip region and then passes into the crack wake, as shown in Fig. 12. Initially, the element is loaded elastically (UV) and the slope is determined by the bulk modulus of the untransformed material (B). When the element reaches a critical value of the mean stress (σ_m^c), the transformation occurs and a stress drop occurs (VW) with a slope B'. The magnitude of B' depends on the details of the transformation process (Budiansky, Hutchison and Lambroupolos, 1983). Further loading of the element occurs along WT and if the transformed material has the same elastic properties, the slope will again be B. As the element passes into the crack wake unloading occurs towards Y and provided there is no reverse transformation the element will be subjected to compression (σ_m^R) in the remote wake at Y. The toughening increment can be found in a simple fashion as it is simply related to the area under the stress-strain curve (Budiansky, Hutchison and Lambroupolos, 1983). The other point of note here is that even if the transformation reverses, as long as this occurs below W on the line WY, there will still be a finite contribution to the toughening.

It is worth briefly considering the trends in toughness predicted by Eq. 2 and the ways in which it may be optimized. For zirconia-based systems we do not have much control over the dilational strain, e^T but clearly it is useful to maximize V. This also implies that we would like the particles to be of uniform size so that they transform at the same σ_m^c.

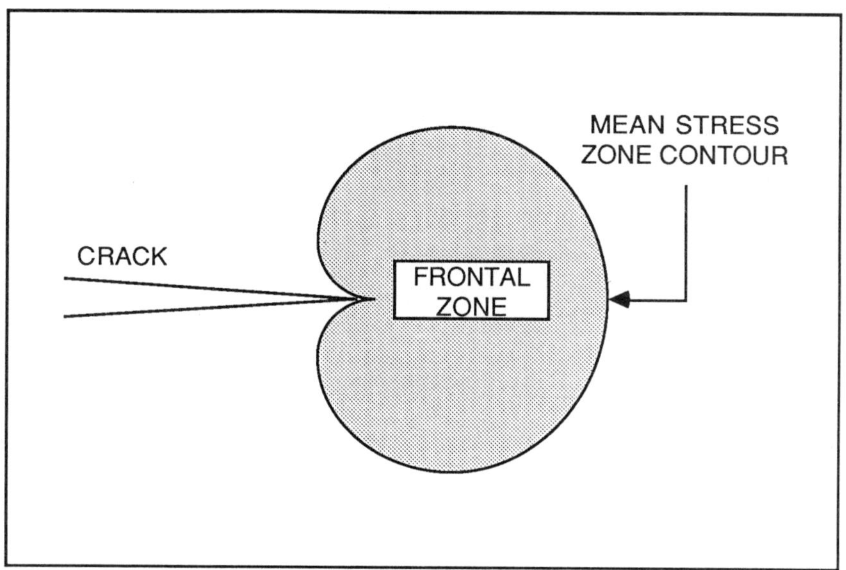

Fig. 9. Schematic illustration showing a frontal zone, in which transformation is nucleated by a critical mean stress.

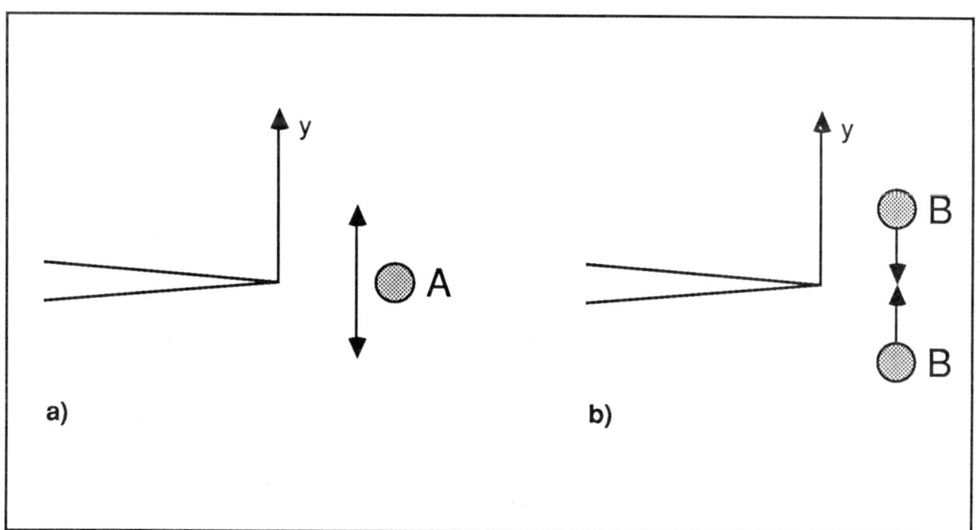

Fig. 10. Effect of transformed region location on crack tip stresses. In a) region A increases stress in a particular direction while in b) regions B decrease stress

     FRACTURE MECHANICS

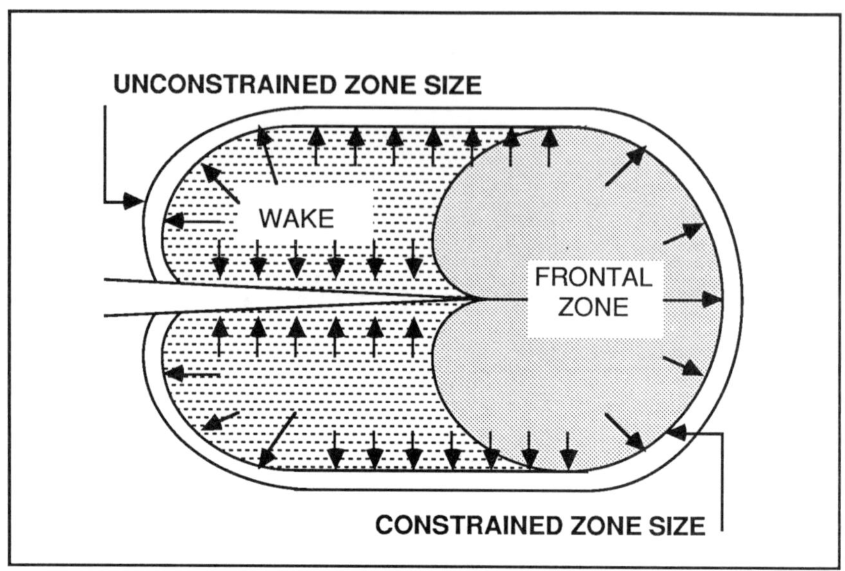

Fig. 11.   Illustration of crack closure stresses that arise as transformed region enters crack wake.

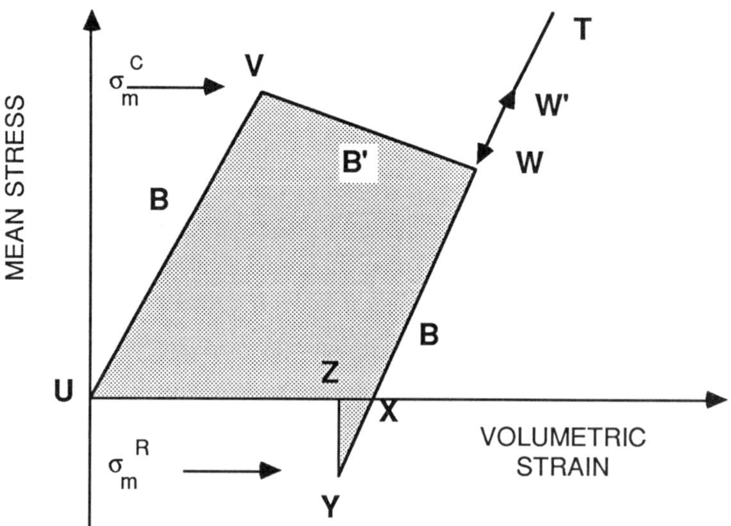

Fig. 12.   Volumetric stress-strain behavior for an element undergoing an irreversible dilatant phase transformation. The toughening is related to the area under the stress-strain curve.

In some cases, increasing V does present difficulties in that it is often more difficult to retain the tetragonal phase and this may be due to particle interaction effects. Increasing E is also expected to be beneficial but this often tends to reduce e^T (McMeeking, 1986) and makes the transformation more difficult to induce at a given particle size. It is generally accepted that h is the most critical parameter in determining the toughness but the parameters that control h are the least quantitatively understood. In general terms, it is clear one wishes to maximize h by making the critical transformation stress as small as possible. One approach is produce materials that contain particles that are just slightly less than the critical size, that is, they are on the verge of transformation. This effect is illustrated in Fig. 13 and shows that small particles are too difficult to transform (small h) and large particles have already transformed (reduces V). In addition Fig. 13 shows that although increasing V can increase toughness, it is also more difficult to retain the tetragonal phase. It is expected that h is also sensitive to temperature in that decreasing temperature should make the transformation easier. This is illustrated in Fig. 14, in which the toughness will depend on the difference between the testing temperature and the normal transformation temperature (T_0). Again the toughness may pass through a maximum because at very low temperatures, the tetragonal phase becomes more difficult to retain. As shown in Fig. 14, it is possible to reduce T_0 by adding solute to the zirconia (increases critical particle size), shifting the toughness curve to lower temperatures. The low toughness at high temperature for transformation-toughened materials implies that their main applications are probably restricted to moderate temperatures. All the general trends discussed in this section have been substantiated qualitatively by experimental results.

Decreasing the critical transformation stress has some important consequences in terms of the relationship between strength and toughness. Importantly, this may mean in some materials, that Eq. 1 is no longer obeyed and that tough materials may invariably have low strengths. This behavior is similar to that observed in metals in the ductile-brittle transition (Evans and Cannon, 1986). In this region, strengths are limited by the yield strength or in this case, the critical transformation stress. This behavior is shown in Fig. 15 (after Swain, 1986), in which the shaded region is that obeyed by Eq. 1. In this region, increasing the fracture toughness (increasing h) for a given critical flaw size leads to a strength improvement but the strength reaches a limit when it approaches the critical transformation stress and is no longer sensitive to the size of the pre-existing flaws. An alternative explanation to the behavior shown in Fig 15 has been put forward (Swain and Rose, 1986; Marshall, 1986) and suggests that as one increases toughness, one samples less of the R curve and that failure occurs at a lower stress. In general, one expects the increased difficulty associated with the propagation of large cracks in a material that undergoes R curve behavior will make the material more tolerant of large cracks.

<u>2. Microcrack Toughening</u>. Ceramics that contain localized residual stresses are known to be capable of microcracking (Green, 1983b). These residual stresses arise in ceramics as a result of phase transformations, thermal expansion anisotropy in single phase materials and thermal expansion or elastic mismatch in multiphase materials. It is also expected that regions of low toughness, such as grain boundaries, would be attractive sites for such cracks. It has been known for some

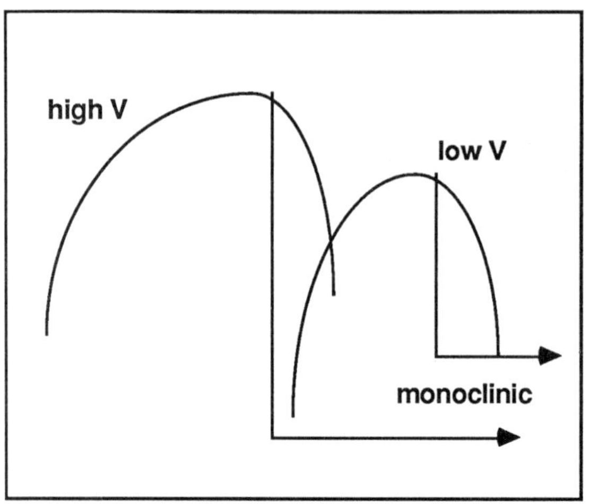

Fig. 13.   Influence of zirconia particle size and transformed
           volume fraction on fracture toughness. Large particles
           cannot be retained as tetragonal phase, while small
           particles need a very high stress to transform.

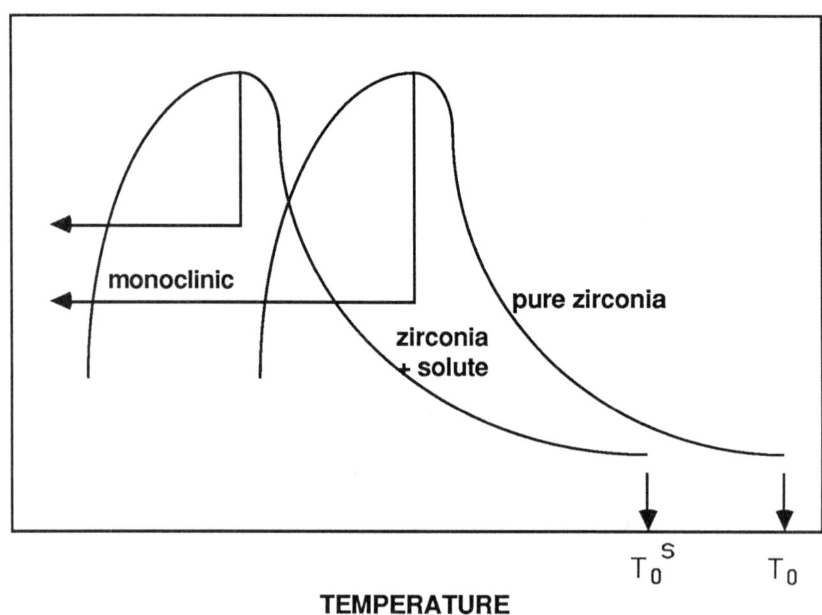

Fig. 14.   Influence of temperature on fracture toughness. The
           toughness increases as temperature is reduced from
           equilibrium temperature but at low temperatures it
           may not be possible to retain the tetragonal phase.

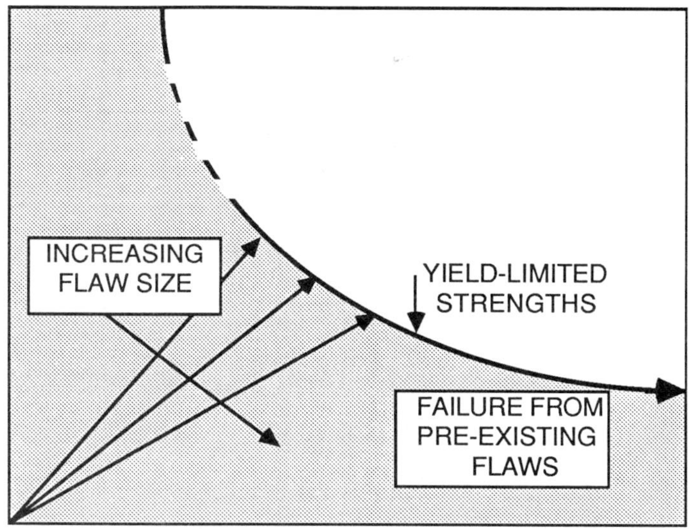

Fig. 15. Failure in ceramics is often dependent on size of
pre-existing flaws (shaded region) but in
transformation-toughened materials a limit can be
reached in which strength is limited by transformation
stress.

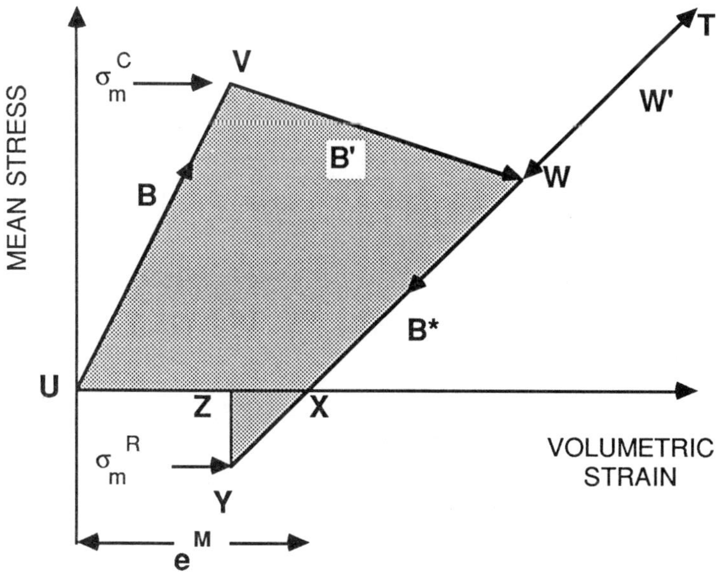

Fig. 16. Volumetric stress-strain behavior for an element
undergoing stress-induced microcracking assuming
there is a volume increase associated with the
microcracking process.

time, that these microcracks can form spontaneously during the fabrication process, provided the grain or particle size is above a critical value. This critical size effect can be analyzed using fracture mechanics and the results for various configurations have been reviewed recently (Green, 1983b) and will be discussed briefly later. Ceramics containing microcracks after fabrication have been associated with good thermal shock resistance (Hasselman, 1969) but it is expected that such materials would have low strengths as the microcracks are likely failure origins. A more attractive proposition, in terms of fracture toughness, is to fabricate materials in which the particle size is below that for spontaneous microcracking but that microcracks could be stress-induced. Calculations that consider microcracks forming under an applied tensile stress indicate that particles down to about one-half of the critical size might be candidates for the formation of stress-induced microcracks (Green, 1983b). In terms of fracture, one would expect that the microcracks would form in a zone around large cracks, much in the same way as a stress-induced phase transformation zone forms. The creation of a microcrack zone around a propagating crack is expected to reduce the stresses near the crack tip, giving rise to **shielding**. An alternative view of the toughening phenomena would be to consider the increased amount of fracture surface that must be associated with the formation of such a zone.

The shielding process associated with a microcrack zone is, in some ways, similar to that involved in transformation toughening. In the latter case, we saw in the last section that dilation of the process zone was a primary factor in the toughening process. The formation of microcracks is generally associated with residual stress fields and when a crack forms in such a field it will give rise to a volume increase as the crack opens. Thus, we would expect that if we can predict the volume increase as a function of the microcrack density, the toughening increment could be predicted simply from the transformation toughening analyses. There are, however, two other effects that are important in the microcrack shielding process and these are different from any effects discussed in transformation toughening. The first of these effects gives rise to additional shielding and is a result of the decrease in the elastic constants that occurs when a material microcracks. This *modulus* effect will itself reduce the stresses in the zone and thus, aids in the shielding process. The other effect is based on the realization that the microcracks must degrade the fracture toughness within the zone. That is, the microcracks have *degraded* the material in the zone and fracture surface is already available when the main crack propagates.

The mean stress-dilational strain behavior of the material in a constrained microcrack zone is shown in Fig. 16. We assume the material is initially uncracked with a bulk modulus B and as with the transformation toughening case, on loading up UV, the material is assumed to microcrack at a critical value of the mean stress. Again, as with the stress-induced transformation, the material will then unload to W with a slope B'. Further loading to T, however, will involve a different slope B*, which is less than B and is therefore, different than the transformation case. Further unloading will also occur down this slope B* until at zero stress (point X) we have a dilation e^M. As the applied stress decreases, we can get further unloading down to point Y and the microcracked material is being residually stressed. It is important to note that even in the absence of the dilation, the decrease in slope from B to B* would give rise to a hysteresis loop and hence is the reason why the modulus effect gives rise to toughening.

This problem has been analyzed (Evans and Faber, 1984) and it was found that for reasonable values of microcrack density, the **toughening increment for a frontal zone** is approximately **zero**. Although this result is the same as the transformation toughening case, the cause is different, in that the degradation of the material by the microcracks is offset by the shielding modulus effect. Thus, again we find that it is the change in zone size and shape that gives rise to the toughening.

We have just determined that the microcrack toughening problem turns out to be virtually analogous to the transformation toughening problem. Thus, the asymptotic toughening increment is expected to be similar to that given by Eq. 2 except that e^T is replaced by e^M. The value for e^M will depend on the details of the microcracking process. For example , for spheres that microcrack under a residual stress σ_R, it has been shown that (Hutchison, 1987)

$$\Delta K_{IC} = 1.12 V \sigma_R h^{1/2} [1 + \upsilon] \tag{3}$$

From the above discussion, it is also expected that R curve behavior should occur in a microcracking material as $\Delta K_{IC} \sim 0$ for a frontal zone but as the microcracks enter the crack wake, dilation toughening will occur (Evans and Faber, 1984). The relationship between h and the critical microcracking stress is, as yet, poorly understood but it will involve the flaw populations present in the residual stress fields and the mechanisms in which they nucleate and form the microcracks.

The overall trends in the fracture toughness increment due to microcracking are very similar to those of transformation toughening. For example, as shown in Fig.17, the toughness is expected to increase with particle size up to the critical particle size for spontaneous microcracking. For larger particle sizes, the material is already microcracked prior to the application of stress and further toughening will only arise if one can increase the microcrack density. The temperature sensitivity could be somewhat different than discussed for transformation toughening, depending on the source of the residual stress. If the microcracks are a result of some type of thermal expansion mismatch, then one would expect that increasing temperature would result in a decrease in toughness, as the magnitude of the residual stress will decrease . On the other hand, if the microcracks were formed by a phase transformation, the toughening effect would be relatively temperature insensitive unless the transformation reverses. Increasing the volume fraction of the microcracks is clearly attractive but again, interaction effects may increase the likelihood of spontaneous transformation and microcrack linking could be a prime source of pre-existing flaws in a material. The factors that control the microcrack zone size are expected to be similar to those that control spontaneous microcracking and include the magnitude of the residual stress, particle size and the size (and shape) of defects within the residually-stressed regions (Green, 1983b). In a similar way to transformation toughening, one would expect that there will be a limit to the toughening if the microcrack zone size approaches the specimen size and a 'yield-limit' to the strength if the critical microcracking stress becomes too small. Non-linear elastic behavior associated with microcracking has been observed in overaged partially-stabilized zirconia (Green, Embury and Nicholson, 1973, Green and Nicholson, 1973) and microcrack toughening has been considered to be an important toughening mechanism in some zirconia systems (Evans and Cannon, 1986).

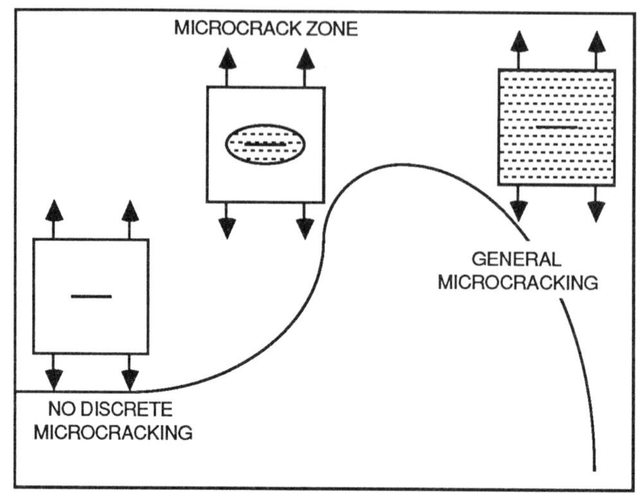

Fig. 17   Influence of particle size that gives rise to
microcrack formation on toughening. For small
particles, it is difficult to induce the
microcracks while large particles give rise
to spontaneous microcracking during fabrication.

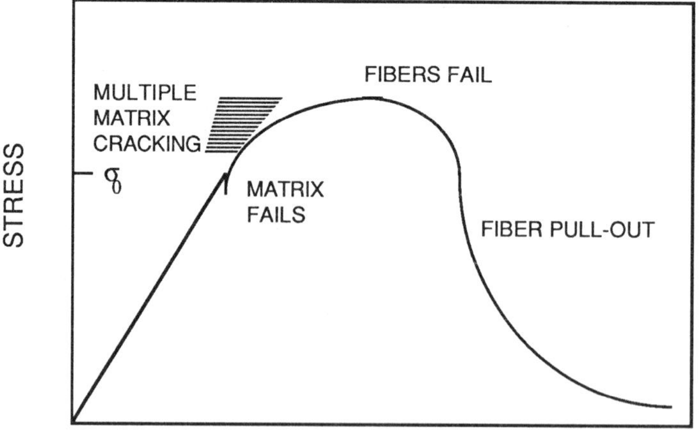

Fig. 18.   Stress-strain behavior for a ceramic fiber/
ceramic matrix unidirectional fiber composite
that undergoes matrix cracking and fiber
pull-out.

<u>C. Crack Bridging</u>.

In the discussion of crack tip interactions we indicated that a crack
may by-pass an obstacle, leaving it intact. In such cases, the obstacle
is left as a ligament behind the crack tip. In other cases, ligaments
may formed by the mechanical interlocking of grains. These ligaments
will make it more difficult to open the crack at a given applied stress
and will increase fracture toughness. This mechanism has been shown to
be important in frictionally bonded fiber composites (Aveston, Cooper
and Kelly, 1971; Marshall, Cox and Evans, 1985), in large-grained Al_2O_3
(Steinbrech, Khehans and Schaarwachter, 1983; Swanson and co-workers,
1987) and possibly whisker- and metal-reinforced ceramics. In order to
determine the effect of crack bridging on fracture toughness, it is
necessary to know the force-displacement relationships for the
ligaments. The effect of crack bridging is best understood for
frictionally bonded fiber composites and we will briefly discuss the
progress in this area.

The addition of continuous carbon or SiC fibers to glasses or ceramics
has been studied since about 1970 and it has been found, in some cases,
that unidirectional fiber composites do not undergo catastrophic failure
in tensile loading (e.g., Marshall and Evans, 1985a). This 'ductile'
type of behavior for a material composed of two brittle components is
particularly attractive for structural applications and is illustrated
in Fig. 18. In the optimum materials, the tensile loading behavior is
initially elastic until at a particular stress, a crack passes through
the matrix. This crack, however, by-passes the fibers and leaves them
available for load carrying and they completely bridge the crack.
Further loading causes the formation of regularly-spaced, bridged,
matrix cracks until at the peak load, the fibers fail. The ensuing
failure, however, is not catastrophic as the fibers continue to pull out
of the matrix.

In these materials the final failure is not the result of the
propagation of a single crack and thus, a fracture toughness value
cannot be defined. It is, however, possible to use fracture mechanics to
describe the conditions at which the first crack passes through the
matrix (Marshall, Cox and Evans, 1985). In this analysis, it was found
under some circumstances that the matrix cracking stress (σ_0) is
independent of the pre-existing flaw size and is a material property.
The analysis showed that σ_0 depends on the matrix fracture toughness,
the interfacial shear strength, the volume fraction and radii of the
fibers and the elastic constants of the two components. Techniques to
increase σ_0 can be ascertained from the analytical expression, but if
this stress approaches the fiber strength, fiber failure will accompany
matrix cracking and there will be a transition to a brittle type of
failure (Marshall and Evans, 1985b). For these cases, a bridging zone of
limited extent, is formed behind the crack and moves with the crack. The
overall effect of crack bridging in fiber composites is shown in the
normalized plot, Fig. 19. The normalization parameters take into account
the specific material properties involved in the process. For a fully
bridged crack, it was found that above a certain crack size, the
strength becomes independent of crack size, as discussed earlier. For
partially bridged cracks, the bridging increases strength for a given
crack size but the strength is no longer independent of crack size.

FRACTURE MECHANICS

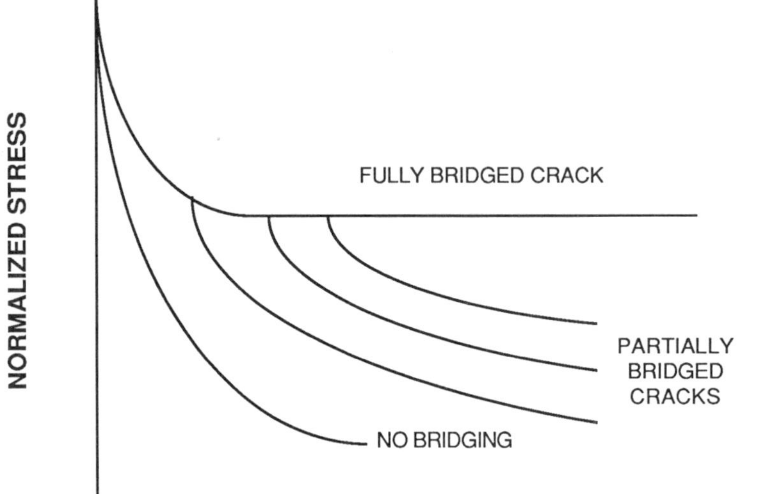

Fig. 19.   Normalized strength/crack size relationships for unidirectional ceramic fiber/ceramic matrix composites. For fully bridged cracks, the matrix cracking stress can be independent of crack size. After Marshall and Evans (1985b).

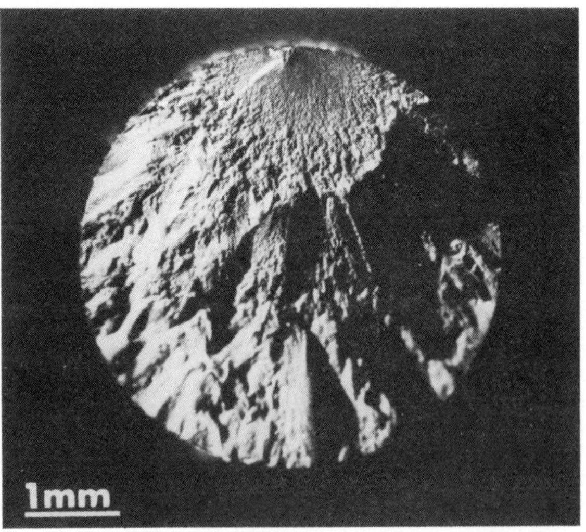

Fig. 20.   Fracture surface of silicon nitride broken in tension showing mirror region around failure origin (top) and subsequent branching region.

CRACK FORMATION

The low toughness of ceramics combined with the inability to relax high
stresses leads to relatively easy crack formation. Thus microstructural
defects when combined with high stresses are prime candidates for
precursors to crack formation. Stress concentrations will occur when
there are regions in the microstructure that have different elastic
constants than the surrounding material. For example, voids or impurity
inclusions could be the cause of such stress concentrations. The
presence of voids is particularly common and may be the result of
incomplete sintering, inhomogeneous densification or the burn-out of
organic impurities (Lange, 1986). Stress concentrations are also common
in contact events and processes such as impact, machining and wear often
give rise to surface cracks. The other common source of high stresses
are those that arise when one part of the microstructure has a volume or
shape mismatch with the surrounding material. These residual stresses
are usually associated with phase transformations, chemical reactions,
thermal expansion anisotropy or mismatch in thermal expansion
coefficients.

The use of fractography is often a key element in identifying the
defects that lead to crack formation. For example, as shown in Fig. 20,
there is a relatively smooth region at the top of the figure. This
region is known as the fracture mirror and the failure origin is located
at its focus.In addition, the crack often branches at higher velocities
and the radial ridges in the branched region 'point' back to the origin.
There are other features, such as fracture steps and river patterns that
aid in determining the direction of crack propagation.In Fig. 20, this
process showed that failure was associated with a pit at the surface,
that had lead to the formation of a surface crack. It was postulated
that some unknown contact event had caused the pit and surface crack
formation. Failure origins can often be traced to steps in the
processing. For example, Fig. 21a shows the presence of a zirconia
agglomerate in an Al_2O_3/ZrO_2 composite. The agglomerate is an indication
of poor mixing and it appears to have densified more in firing than the
surrounding material. Fig. 21b shows a void failure origin that was
probably the result of the burn-out of an organic impurity. This type of
information can be a key in identifying routes for improving the
processing, leading to substantial strength improvements (Lange, 1986).

It is important to understand the micromechanisms involved in crack
formation. For contact-induced flaws, a major step in understanding the
micromechanisms has been the use of indentation cracks in simulating the
crack formation process (Lawn, 1983). A schematic illustration of the
cracking produced by a sharp indenter is shown in Fig. 22. Associated
with the indentation impression is a residually-stressed inelastic zone
that has undergone permanent deformation. The surrounding elastic
material 'pushes' back on this zone creating a set of residual stresses.
In addition, above a threshold load, two types of cracks are formed. One
set of cracks form radially from the indentation site, while the other
set is formed approximately parallel to the surface. The radial cracks
are related to strength degradation produced by the contact, while the
lateral cracks, if they extend to the surface, are related to erosion
processes. It has been shown that the extent of cracking is related to a
set of material parameters, including fracture toughness (Lawn, 1983).
Moreover, the residual stresses play a key role in subsequent crack
extension, in that they lead to stable crack growth and influence the
failure stress (Lawn, 1983).

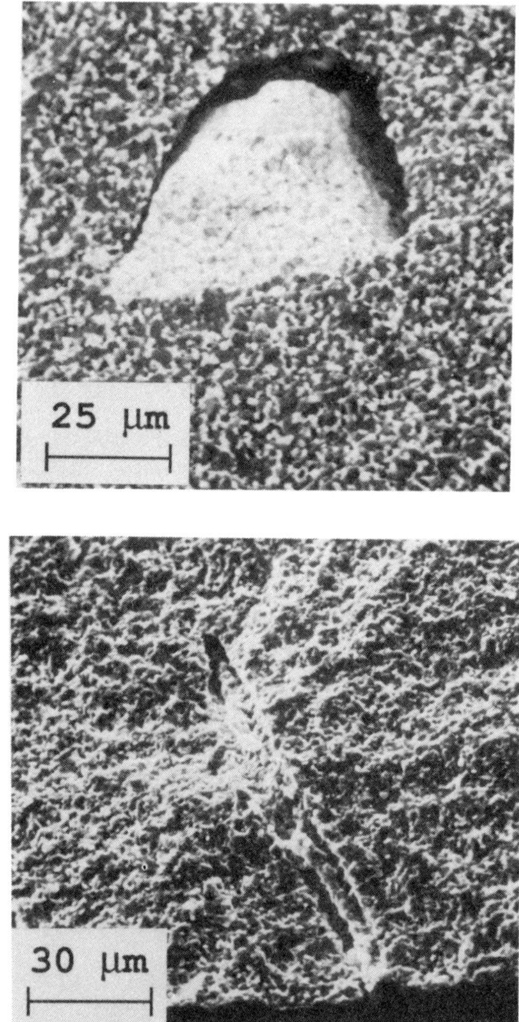

Fig 21.   Failure origins in zirconia-toughened alumina; a)
           zirconia agglomerate that led to a 'crack-like'
           void, b) lenticular void, probably produced by
           burn-out of an organic impurity. Micrograph b)
           courtesy of F. F. Lange, University of California,
           Santa Barbara.

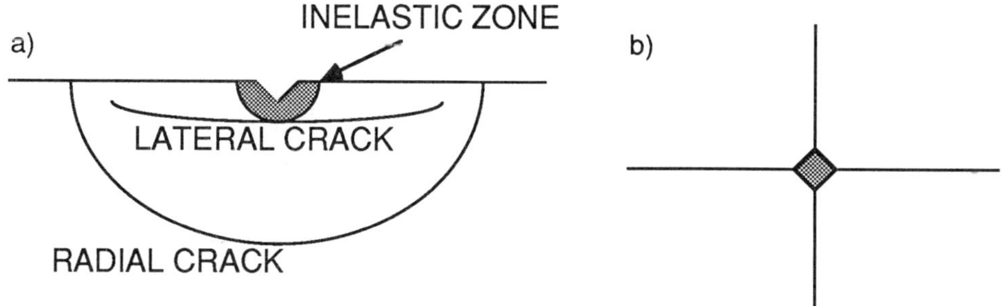

Fig. 22. Schematic of cracking associated with Vicker's indenter; a) side view, b) top view.

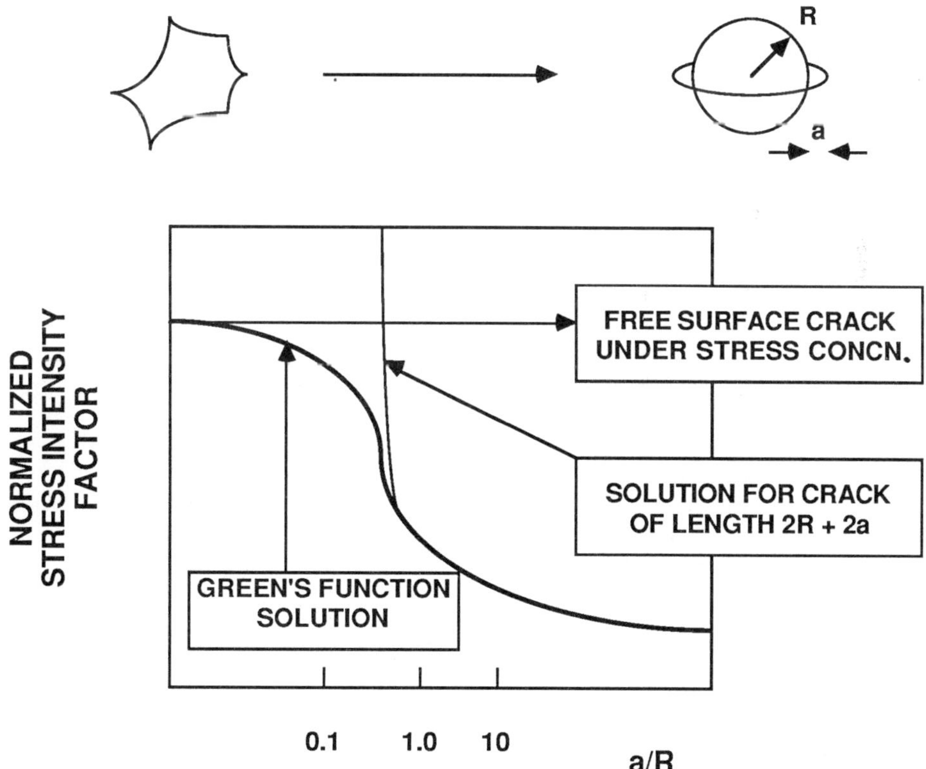

Fig. 23. Stress intensity factor solution for annular crack around a spherical void. After Green,1980.

Progress has also been made in understanding the failure process associated with inclusions or voids. For example, voids have been modelled as a spherical void with a surrounding 'saturn' crack. The saturn crack is assumed to be a 'crack-like' defect, such as a grain boundary cusp, that is present at the boundary of the void. The results of such an analysis are shown schematically in Fig. 23 (Green, 1980). For small crack sizes, it was shown that the stress intensity factor should be equivalent to that of an edge crack, which is acted on by the stress concentration. For larger cracks, however, one can simply assume that the void/crack combination is simply equivalent to a penny-shaped crack. Crack formation at inclusions under a state of residual stress has been modelled as shown in Fig. 24 (Green, 1983b).The type of crack formed depends on the sign of the residual stress in the inclusion and the relative toughness of the inclusion compared to the matrix. The results for the stress intensity factor for a radial crack are shown schematically in Fig. 25. It was found that for failure to be possible, the inclusion must be above a critical size (2 μm in figure). The actual inclusion size required to form the microcrack, however, depends on the size of the defects that exist near the inclusions. For this radial crack configuration, the failure process is analogous to the formation of radial indentation cracks, and again one expects that stable crack growth occurs in the residual field but in addition, one can often predict the stress for catastrophic failure from the inclusion size (Green, 1983b). This type of understanding about crack formation is often of importance in non-destructive evaluation, if one wishes to relate defect size to fracture stress. For example, this type of modelling was used for relating the size of the various types of failure origins observed in a commercial Si_3N_4 to the expected fracture stress (Evans, 1985).

REFERENCES

Aveston, J., G. A. Cooper, and A. Kelly (1971). Single and multiple fracture. In _Properties of Fiber Composites,_ IPC Science and Technology Press Ltd., Surrey, England, pp. 15-26.

Budiansky, B., Hutchison, J., and Lambroupolos, J. C. (1983). Continuum theory of dilatant transformation toughening in ceramics, _Int. J. Solids Struct.,_ _19,_ pp. 337-55.

Chen, I-W., and Y. H. Chiao. (1985a). Theory and experiment of martensitic nucleation in ZrO_2-containing ceramics and ferrous alloys, _Acta Metall.,_ _33,_ pp. 1827-45.

Chen, I-W., and Y. H. Chiao. (1985b). Statistics of martensitic nucleation, _Acta Metall.,_ _33,_ pp. 1847-59.

Clarke, D. R., and F. Adar. (1982). Measurement of the crystallographically transformed zone produced by fracture in ceramics containing tetragonal zirconia, _J. Am. Ceram. Soc.,_ _65,_ pp. 284-88.

Claussen, N., Microstructural design of zirconia-toughened ceramics. In N. Claussen, M. Ruhle, and A. H. Heuer (Eds.), _Science and Technology of Zirconia, II,_ _Advances in Ceramics_, Vol. 12, The American Ceramic Society, Columbus, Ohio, pp. 325-51.

Evans, A. G. (1972). The strength of brittle materials containing second-phase dispersions, _Phil. Mag.,_ 26, pp. 1327-44.

Evans, A. G., and K. T. Faber (1984). On the crack growth resistance of microcracking brittle materials, _J. Am. Ceram. Soc.,_ _67,_ pp. 255-60.

Evans, A. G. (1985).Engineering property requirements for high performance ceramics, _Mater. Sci. Eng.,_ _71,_ pp. 3-21.

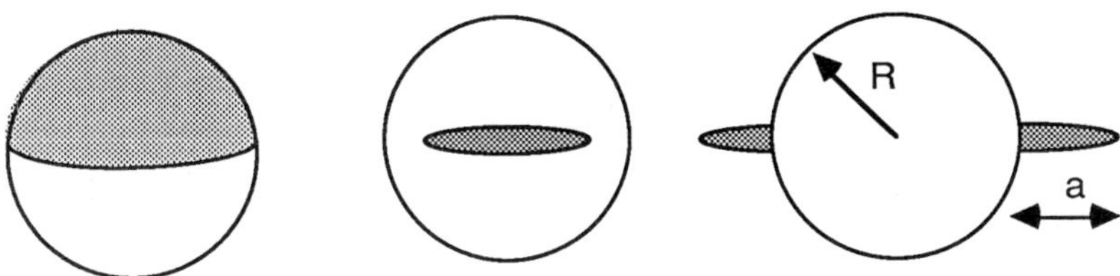

Fig. 24.   Possible failure mechanisms at a spherical misfit
            inclusion; a) decohesion (particle in tension),
            b) particle failure (particle in tension), c)
            radial cracking (particle in compression).

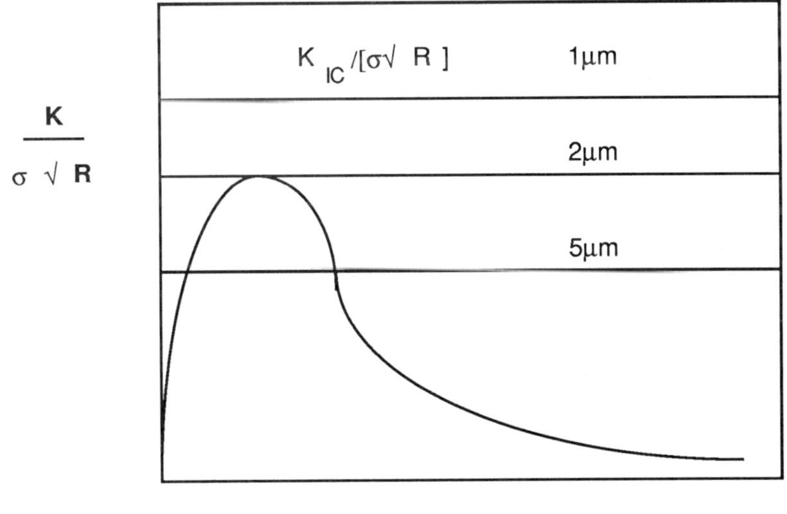

$$\frac{K}{\sigma\sqrt{R}}$$

a/R

Fig. 25.   Stress intensity factor solution for radial
            cracking at a misfit inclusion. For small
            particles, crack formation cannot occur but
            there is a critical particle size above which
            crack formation is possible.(σ is magnitude
            of residual stress in inclusion)

Evans, A. G., and R. M. Cannon. (1986). Toughening of brittle solids by martensitic transformations, _Acta Metall.,_ _34,_ pp. 761-800.

Faber, K. T., and A. G. Evans (1983a). Crack deflection processes - I. Theory, _Acta Metall.,_ _31,_ pp. 565-76.

Faber, K. T., and A. G. Evans (1983b). Crack deflection processes - II. Experiment, _Acta Metall.,_ _31,_ pp. 577-84.

Faber, K. T., and A. G. Evans (1983c). Intergranular crack deflection processes in silicon carbide, _J. Am. Ceram. Soc.,_ _66,_ pp. C94-96.

Green, D. J., J. D. Embury, and P. S. Nicholson (1973) Fracture toughness of a partially stabilized ZrO_2 in the system $CaO-ZrO_2$, _J. Am. Ceram. Soc.,_ _56,_ pp. 619-23.

Green, D. J., and P. S. Nicholson (1973). Microstructure development and fracture toughness of a calcia partially-stabilized zirconia. In, R. C. Bradt, F. F. Lange and D. P. H. Hasselman (Eds.), _Fracture Mechanics of Ceramics,_ Vol. 2, Plenum Press, New York, pp. 541-54.

Green, D. J. (1976). Crack-particle interactions in brittle composites, Ph. D. Thesis, McMaster University, 1976.

Green, D. J., (1980) Stress intensity factor estimates for annular cracks at spherical voids, _J. Am. Ceram. Soc.,_ _63,_ pp. 342-44.

Green, D. J. (1983a). Fracture toughness predictions for crack bowing in brittle particulate composites, _J. Am. Ceram. Soc.,_ _66,_ pp. C4-5.

Green, D. J. (1983b) Microcracking mechanisms in ceramics In, R. C. Bradt, A. G. Evans, F. F. Lange and D. P. H. Hasselman (Eds.), _Fracture Mechanics of Ceramics,_ Vol. 5, Plenum Press, New York, pp. 457-78.

Green, D. J., R. H. J. Hannink, and M. V. Swain (1988). Transformation Toughening of Ceramics, CRC Press, Boca Raton, Florida.

Griffith, A. A. (1920). The phenomena of rupture and flow in solids. _Phil. Trans. Roy. Soc., (London),_ _A221,_ pp. 163-98.

Hasselman, D. P. H. (1969). Unified theory of thermal shock resistance of ceramic materials, _J. Am. Ceram. Soc.,_ _52,_ pp. 600-604.

Heuer, A. H. (1987).Transformation toughening in ZrO_2-containing ceramics, _J. Am. Ceram. Soc.,_ _70,_ pp.689-98.

Heuer A. H., and M. Ruhle (1985). On the nucleation of the martensitic transformation in zirconia (ZrO_2) _Acta. Metall.,_ _33,_ pp. 2101-12.

Hutchinson, J. W. (1987) to be published, cited in Evans and Cannon (1986).

Irwin, G. R. (1958). Fracture, _Handbuch der Physik,_ Vol. 6, Springer Verlag, Berlin.

Kosmac, T., R. Wagner, and N. Claussen. (1981). X-ray determination of transformation depths in ceramics containing tetragonal zirconia, _J. Am. Ceram. Soc.,_ _64,_ pp. C72-73 .

Lange, F. F. (1970). Interaction of a crack front with a second phase dispersion, _Phil. Mag.,_ _22,_ pp. 983-92.

Lange, F. F., and D. J. Green. (1981). Effect of inclusion size on the retention of tetragonal ZrO_2: theory and experiments. In A. H. Heuer and L. W. Hobbs, (Eds.), _Science and Technology of Zirconia, Advances in Ceramics,_ Vol. 3, The American Ceramic Society, Columbus, Ohio, pp. 217-25.

Lange, F. F. (1986).Processing-related fracture origins: IV, elimination of voids produced by organic inclusions, _J. Am. Ceram. Soc.,_ _69,_ pp. 66-69.

Lawn, B. R. (1983).The indentation crack as a model surface flaw.In, R. C. Bradt, A. G. Evans, F. F. Lange and D. P. H. Hasselman (Eds.), _Fracture Mechanics of Ceramics,_ Vol. 5, Plenum Press, New York, pp. 1-26.

Marshall, D. B., M. D. Drory, and A. G. Evans. (1983). Transformation toughening in ceramics. In, R. C. Bradt, A. G. Evans, F. F. Lange and D. P. H. Hasselman (Eds.), <u>Fracture Mechanics of Ceramics,</u> Vol. 6, Plenum Press, New York, pp. 289-307.

Marshall, D. B., B. N. Cox and A. G. Evans (1985) The mechanics of matrix cracking in brittle-matrix fiber composites, <u>Acta Metall.,</u> <u>33,</u> pp. 2013-21.

Marshall, D. B., and A. G. Evans (1985a) Failure mechanisms in ceramic-fiber/ceramic matrix composites, <u>J. Am. Ceram. Soc.,</u> <u>68,</u> pp. 225-31.

Marshall, D. B., and A. G. Evans (1985b). Tensile failure of brittle-matrix fiber composites. In W. C. Harrigan, J. Strife and A. K. Dhingra (Eds.), <u>Proceedings of Fifth International Conference on Composite Materials,</u> The Metallurgical Society, Warrendale, pp. 557-68.

Marshall, D. B. (1986). Strength characteristics of transformation-toughened zirconia, <u>J. Am. Ceram. Soc.,</u> 69, pp. 173-80.

McMeeking, R. M., and A. G. Evans. (1982). Mechanics of transformation toughening in brittle materials, <u>J. Am. Ceram. Soc.,</u> <u>65,</u> pp. 242-46.

McMeeking, R. (1986). Effective transformation strain in binary elastic composites, <u>J. Am. Ceram. Soc.,</u> <u>69 [12],</u> pp. C301-02.

Porter, D. L., and Heuer, A. H. (1977) Mechanism of toughening partially stabilized zirconia (PSZ), <u>J. Am. Ceram. Soc.,</u> <u>60,</u> pp. 183-84.

Ruhle, M., and A. H. Heuer. (1984). Phase transformations in ZrO_2-containing ceramics: II, The martensitic reaction in t-ZrO_2. In N. Claussen, M. Ruhle, and A. H. Heuer (Eds.), <u>Science and Technology of Zirconia, II,</u> <u>Advances in Ceramics</u>, Vol. 12, The American Ceramic Society, Columbus, Ohio, pp. 14-22.

Ruhle, M., A. Strecker, D. Waidelich, and B. Kraus. (1984). In Situ Observations of stress-induced phase transformation in ZrO_2-containing ceramics. In, N. Claussen, M. Ruhle, A. H. Heuer, (Eds.), <u>Science and Technology of Zirconia, II,</u> <u>Advances in Ceramics,</u> Vol. 12, The American Ceramic Society, Columbus, Ohio, pp. 25674.

Schoenlein, L. H. and A. H. Heuer, Transformation zones in Mg-PSZ. In, R. C. Bradt, A. G. Evans, F. F. Lange and D. P. H. Hasselman (Eds.), <u>Fracture Mechanics of Ceramics,</u> Vol. 6, Plenum Press, New York, pp. 309-26.

Steinbrech, R., R. Khehans, and W. Schaarwachter (1983). Increase of crack resistance during slow crack growth in Al_2O_3 bend specimens, <u>J. Mater. Sci.,</u> <u>18,</u> pp. 265-70.

Swain, M. V., and L. R. F. Rose. (1986). Strength limitations of transformation-toughened zirconia alloys, <u>J. Am. Ceram. Soc.,</u> <u>69,</u> 511.

Swanson, P. L., C. J. Fairbanks, B. R. Lawn, Y-W Mai, and B. J. Hockey (1987). Crack-interface grain bridging as a fracture resistance mechanism in ceramics, <u>J. Am. Ceram. Soc.,</u> <u>70,</u> pp. 279-89.

FRACTURE CONTROL USING ELASTIC-PLASTIC FRACTURE MECHANICS

L.A. Simpson*, R.R. Hosbons**, P.H. Davies** and C.K. Chow*

*Whiteshell Nuclear Research Establishment
Pinawa, Manitoba   ROE 1L0
**Chalk River Nuclear Laboratories
Chalk River, Ontario   KOJ 1J0

ABSTRACT

Fracture control is the prevention of failure in engineering structures due to
crack propagation. It requires a thorough knowledge of the susceptibility of the
structural material to all pertinent crack growth mechanisms, an ability to
determine critical crack sizes in the structure and the ability to detect, through
inspection, flaws which develop in service. Established Engineering codes governing failure by unstable crack propagation in metallic structural components are
biased mainly towards preventing brittle cleavage crack propagation in ferritic
steels. The near future should see new procedures established for more ductile
materials based on the emerging understanding of sub-critical crack growth mechanisms and the development of elastic-plastic instability assessment procedures.
This paper describes the philosophy of fracture control by using quantitative data
on crack growth behavior combined with state of the art methodologies for
instability assessment.

KEYWORDS

Fracture, fatigue, stress corrosion cracking, critical crack length, instability
assessment, J-integral, tearing modulus, crack tip opening displacement.

INTRODUCTION

Failure of components costs the economy of North America more than $100 billion
per year, (Duga and co-workers, 1983). There are ways of reducing this cost but
often they are bypassed. Since the early 1960's fracture mechanics has been
available and since the early 1970's it has even been in design codes. In spite
of this it is still only used in the 'high-tech' industries such as aero-space and
nuclear. In this paper we will attempt to point out the principles of using
fracture mechanics for fracture control purposes in the design and operation of
load bearing components in service.

The basis of fracture mechanics is the use of a parameter which describes the
condition at a crack tip, be it displacement, energy or stress field, as a
description of the force applied to a crack or, in terms of a critical value, a
material property signifying resistance to propagation. Through an equilibrium
between this applied force and resistance, the response of the crack in terms of
sub-critical, stable or unstable growth can be described. Sub-critical crack
growth is usually described by a growth law pertaining to a specific mechanism for
instance, fatigue or stress corrosion cracking. Unstable growth occurs when the

mechanical loads on the crack exceed its material resistance (toughness) to
propagation. It can be either brittle or ductile. The main difference between
linear elastic and elastic-plastic fracture mechanics is that the latter considers
larger crack tip plasticity with possibly some stable ductile crack growth ocur-
ring prior to attaining a global state of mechanical instability.

In general, fracture mechanics can be used in the design stage to

    1) Reduce or eliminate the possibility of failures.

    2) Determine the period between in service inspections to detect cracks
       prior to growing to a critical size.

The traditional approach is to postulate that the largest flaw which can exist in
the component is actually in the region of highest stress. This flaw is usually
taken as the smallest flaw that non-destructive examination (NDE) can detect, but
this is not an absolute number and some judgment must be made. The expected
operating conditions should be considered and the growth of the crack, by any
means which might be expected in service, calculated. Growth laws for mechanisms
occurring in typical structural metals will be described later. This then gives a
crack size at the anticipated end-of-life and it must be compared to the critical
size to determine the possibility of failure of the component. If the critical
size is not significantly larger than the calculated end-of-life crack size, then
the component must be either redesigned or inspected at appropriate periodic
intervals.

The latter approach is often used in the aircraft industry. An operating period
is set so that any hypothetical flaw of a size equal to the inspection limit,
cannot grow to some pre-determined fraction of the critical size and the
components are inspected. If flaws are found, remedial action is taken which
could either be repair or a special analysis to show fitness for purpose. The
fitness assessment for the detected flaw is the same as for the hypothetical flaw.
The decision as to whether the flaw should be repaired or the operation of the
component continued for a specified period will be made on the basis of this
analysis. Such decisions can have great economic impact for instance, shutting
down a nuclear power plant for repairs will result in a loss of revenue of the
order of a half million dollars per day.

In this paper we will give the parameters which describe the forces on the crack,
the laws which describe various crack growth mechanisms, and instability criteria
for cracks. We will concentrate on emerging technologies and procedures rather
than those which have already been adopted into codes. Thus we will be describing
the underpinnings of the fracture control procedures of the future.

FITNESS FOR PURPOSE

To exercise fracture control through a demonstration of fitness-for-purpose, we
can address fracture behaviour at any or all stages in the life cycle of a crack,
which we define as follows:

         I - Initiation

        II - Stable Growth

        III - Instability

        IV - Dynamic Behaviour

If one could show from our knowledge of crack initiation behaviour that cracks positively will not initiate in a particular structural component, then we need not concern ourselves with the next three stages of the life cycle. While it may be possible to rule out specific initiation mechanisms by demonstrating non-susceptability, it is not practical to deny initiation in general or the presence of flaws which escape inspection. Hence, we must also understand the growth laws for cracks and the factors affecting the critical crack size. Safety considerations may also require that we understand the dynamic behaviour sufficiently (as a function of the metallurgical condition of the component) to demonstrate that the consequences of an unstable failure are acceptable if we are to take any sort of a risk in operating in a non-design condition.

One example of a fitness criterion used in the nuclear industry and others is 'leak-before-break'. Recently, the U.S. Nuclear Regulatory Commission (Piping Review Committee, 1985), has recommended as a first priority that loss-of-coolant accident analyses for U.S. reactors incorporate leak-before-break criteria. This can be expected to lead to changes in engineering codes covering nuclear piping to recognize leak-before-break.

Demonstrating leak-before-break entails the satisfactory description of three items:

(1)   The maximum size to which a crack can grow subcritically prior to detection through leakage.

(2)   The laws governing growth of that crack from the leakage size to a mechanically unstable size (configuration).

(3)   The unstable size (configuration) of a crack for the particular material state and loading conditions of the tube under assessment.

One then proceeds to show that the leak detection/location capability is sufficient to allow the operator to safely shut down the system prior to attainment of an unstable crack size. While this approach relies specifically on leakage to signal impending failure, the spirit of all fitness procedures is to demonstrate that growth of a defect to a critical size is not possible prior to end-of-life or the next inspection period.

Clearly the major pieces of information required to do a fitness assessment are the maximum flaw size that can escape detection (by non-destructive examination or leakage), the laws governing relevant sub-critical crack growth mechanisms and the minimum unstable crack size (critical crack length). As NDE and leak detection are outside of the expertise of the authors, the remainder of this paper will deal with growth laws and crack instability assessment.

CRACK TIP PARAMETERS

In order to describe the response of a crack to applied forces (for growth laws or instability assessment) we need a parameter which describes the intensity of the elastic or elastic-plastic deformation field in the crack-tip region. We can then measure the kinetics of stable growth in terms of this parameter and express the instability condition of the crack in terms of a critical value of it. It is often not possible to do laboratory tests on full size structural components and therefore methods which use small specimens have been developed to gather data on crack behavior in pieces of material cut from a component. The viability of such an approach depends on the development of a reliable methodology for ensuring that the fracture parameter can be calculated in both the small specimen and the postulated crack in the structure and that its critical value is not affected by the differences in geometric configuration.

Three options for crack tip parameters are shown in Fig. 1. This parameter might describe the local stress, an example being the linear elastic stress intensity factor, K. Alternatively it could describe the local strain as does the crack tip opening displacement, δ. A third possibility is a parameter like the J-integral, describing the flow of energy into the crack tip region per unit crack extension. All three are being used to some extent in various industries and each have their advantages and disadvantages depending on the condition of the material under test, and the specific nature of the application of the technology. Some of the features of those parameters are described below.

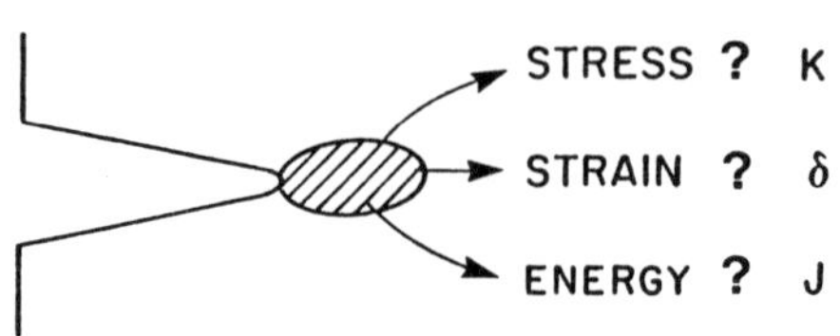

Fig. 1   Choice of crack-tip parameters for describing local deformation state.

Stress intensity factor, K.

- Proportional to the crack tip stress field.

- Based on linear elasticity, hence is not appropriate for large scale crack tip plasticity.

- Most suitable as a crack initiation parameter, not appropriate if stable crack extension (ductile fracture) occurs.

- Crack driving forces for cracks in structures are easy to calculate in terms of K.

Crack tip opening displacement, δ.

- Proportional to the local strain near the crack tip.

- Rules for geometry independence are least stringent.

- Difficult quantity to measure properly.

- Can be adapted to allow for stable crack growth.

- Computation of crack driving forces is difficult.

- Widely used for piping and off-shore structures in the oil and gas industry.

J-integral, J.

- Elastic-plastic crack tip parameter.

- Rigorous basis in solid mechanics if rules for validity are satisfied. (These rules are frequently not satisfied).

- Methodology to describe stable crack growth most advanced (J-resistance curve).

- Crack driving force may be easily calculated for cracks in a structure.

- J is easy to determine on certain small specimens.

- Nuclear industry, world-wide prefers J.

We now proceed to show how these characterizing parameters are used to describe crack growth behavior and estimate crack instability in structures.

CRACK GROWTH LAWS

In many applications, cracks can be tolerated as long as they are below a critical size. However, catastrophic failure can occur, if the initially small defects grow during operation. If the crack growth laws can be defined under the operating conditions, the time from the detection of the crack to unstable failure can be calculated. The existence of the crack growth law greatly facilitates the assessment of growth rates in a structure, because relatively simple laboratory tests can be used for a fitness assessment such as the time to detect a leak and discontinue operation of a component. In this section we shall describe some crack growth laws for several mechanisms which are common in industrial applications.

Fatigue Crack Growth

Crack growth laws for fatigue comprise the most common application of this approach and are accepted by many code procedures. A crack can grow if it is subjected to an alternating stress, even when its peak amplitudes are less than the strength of the material (Smith, 1986). The relation between peak stress S and number of cycles N of tensile and compressive stress needed for failure has the form shown in Fig. 2, in which the number of cycles increases as the peak stress decreases.

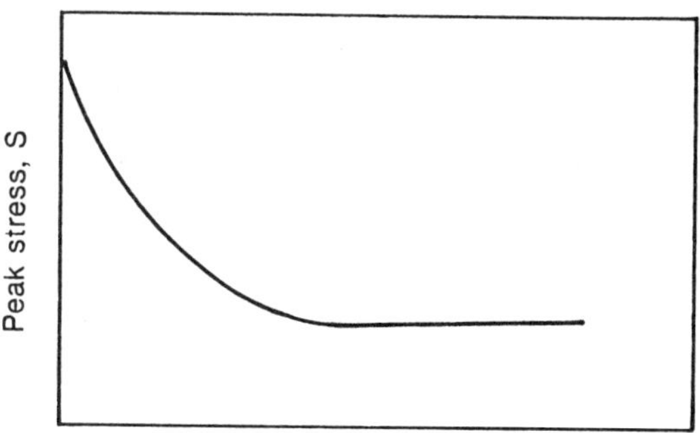

Fig. 2 Typical S/N curve used to determine fatigue life.

Conventional fatigue tests are performed on smooth specimens producing an S/N curve as shown in Fig. 2. The fatigue properties can be divided into two regimes: initiation controlled and propagation controlled. If the fatigue resistance of

                    FRACTURE MECHANICS

the material is initiation-dominated, the S/N curves are used to determine the
fatigue life. Once the fatigue crack is formed, or if flaws are presumed present
or detected in service, then the growth law becomes important. The propagation
rate will depend on the alternating stress intensity factor ΔK, which is defined
as:

$$\Delta K = K_{max} - K_{min}. \tag{1}$$

For most materials, fatigue growth begins at an apparent threshold, and rises
rapidly with increasing ΔK (Regime A in Fig. 3). As the stress intensity factor
increases, the propagation rate can be described by the Paris equation (Regime B,
Fig. 3), (Paris and Erdogan, 1963):

$$da/dN = c\ \Delta K^m \tag{2}$$

where c and m are constants.

When K_{max} approaches K_{1c}, the crack growth will increase rapidly, and fast
fracture can result.

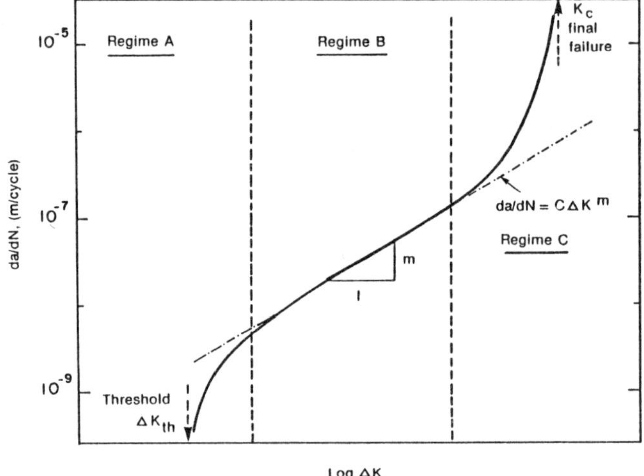

Fig. 3 Plot of fatigue crack growth rate against
cyclic stress intensity factor

Generally, the average of the exponent, m, in the Paris equation is close to 3-4
for most materials. Some high strength steels exhibit values as high as 10.
Figure 4 summarizes the exponent m with K_{1c} of some steel, titanium and aluminum
alloys (Ritchie and Knott 1973).

Stress Corrosion Cracking

The crack propagation can be influenced by the environment. A material can be
ductile in one environment, but brittle in another, and the crack velocity can be
drastically different under the same stress. This phenomenon is called stress
corrosion cracking.

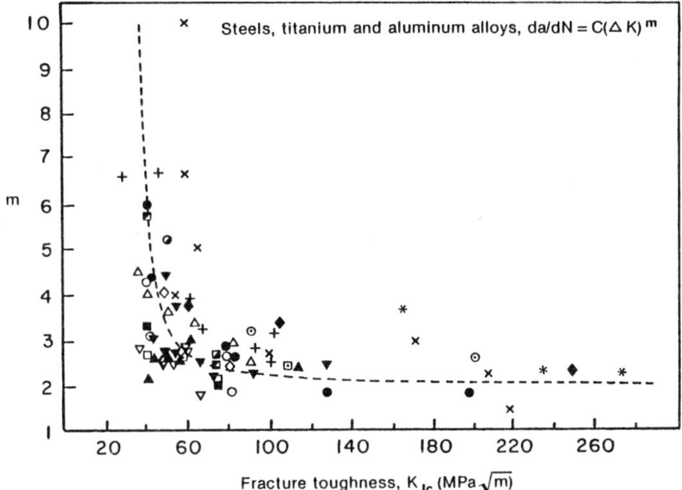

Fig. 4 Typical values for m in Eqn. 2 for selected alloys
(after Ritchie and Knott, 1973).

Stress corrosion cracking includes both anodic dissolution processes, which preferentially remove stressed material by chemical means, and cathodic processes,
such as the evolution of hydrogen which can diffuse into a material and cause
fracture ahead of a crack tip by a hydrogen embrittlement mechanism (Ford, 1984).
SCC growth rates can depend on many factors including temperature, chemistry of
the environment, microstructure and the applied force at the crack tip usually
described by the stress intensity factor, K.

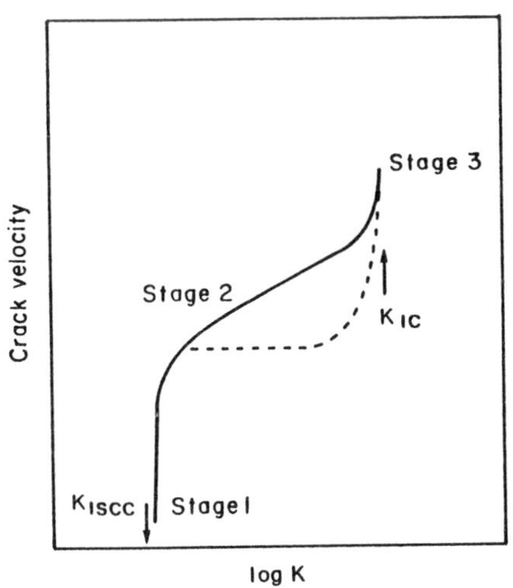

Fig. 5 Schematic plot of SCC velocity
against stress intensity factor.

We now consider how fracture mechanics can be used to characterize
stress corrosion cracking. In this
case, it is assumed that a crack has
been initiated at the beginning of
the analysis. Figure 5 shows the
crack velocity as a function of
stress intensity factor. The crack
velocity curve has a sigmoidal shape.
The crack growth begins at the
threshold stress intensity factor
K_{1scc}, below which crack growth does
not occur. K_{1scc} is a constant characteristic of the mechanism but may
depend on environment, microstructure
and composition of the material. As
the applied stress intensity factor
increases, the slope of the curve
often decreases to a constant value
(stage 2), in some cases with zero
slope indicating that the process is
controlled by transport of corroding
species to the crack tip. As the
stress intensity factor approaches K_{1c},
the crack velocity increases rapidly,
as the structure becomes mechanically
unstable.

There is no unified, single theory to explain SCC, since the crack tip process can
be very different for different materials and environments. However, the crack
growth rate in most cases can be described by a curve similar to Fig. 5. Once
this curve is determined experimentally, together with the stress at the crack tip
of the structure, the time to failure of a structure containing a crack can be
assessed. Laboratory tests must be carried out under conditions which closely
simulate those in service. So far, uncertainties in the detailed growth behavior
of SCC cracks and characteristically high growth rates have prevented these growth
laws from being adopted into formal code procedures, and fracture control is
usually ensured by selection of materials and operating environments which
guarantee non-susceptability. However, as specific mechanisms become better
characterized, we can expect that to change.

Corrosion Fatigue

The propagation of a fatigue crack can be enhanced by stress corrosion. The
increase in crack propagation rate in an aggressive environment under alternating
load is called corrosion fatigue. Figure 6 shows the crack propagation rate of
2014-T6 aluminum alloy (used in aircraft wheels) in three environments. The
service data are much higher than the laboratory results, reflecting the corrosive
effect of the operating environment which could be wet, chemically contaminated
and subject to heating from tires and brakes.

There is not yet a generally accepted growth law for corrosion fatigue. The da/dN
vs ΔK curves have the sigmoidal shape like fatigue and SCC. However, the
threshold can be lower and the crack growth rate can be much higher than when a
single mechanism is operating. Speidel (1979), suggested a simple superposition
model, in which the observed crack growth rate in corrosion fatigue is the sum of
the fatigue growth rate in an inert environment and the stress corrosion rate under
static load conditions. However, this equation was found to be inadequate (Holroyd
and Hardie, 1984). Thus, it is important to test the material in the appropriate
environments intended for service.

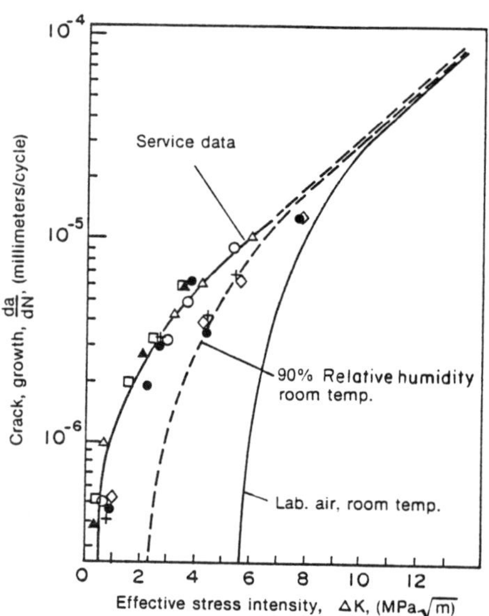

Fig. 6.   Effect of environment on
          fatigue growth rate to an
          aluminum alloy (after
          Simenz, 1984).

Creep Crack Growth

Under a uniform stress field in the creep
range, defects often develop as voids and
microcracks are formed in the material.
These defects are usually uniformly
dispersed. An analysis which considers
an average strain response to an average
applied stress is usually adequate to
describe the creep behaviour. However,
there are some cases where failure occurs
by the initiation and propagation of a
single macroscopic crack. We shall
consider these cases and investigate how
fracture mechanics can be used to describe the creep cracking behaviour.

A fracture mechanics approach to creep cracking behavior must identify a parameter
which characterizes the crack tip behaviour. The stress intensity factor has been
shown to be adequate for characterizing fatigue and SCC as the crack tip is
subjected to only small scale yielding conditions. However, in the temperature
regime where creep conditions are significant, plastic zone sizes are such that an
elastic-plastic description is required to describe the crack tip conditions. A
modification of the J-Integral called C* (essentially a time derivative of J), has
been shown to permit a good correlation of growth rate over a wide range of C* and
materials. A review of this form of time-dependent fracture can be found in
Kanninen and Popelar (1985).

Creep crack growth can also be affected by environment. It can not be emphasized
enough that the service environment should be considered when measuring the crack
growth rate in the laboratory. Since the understanding of creep crack growth phenomena is still in it's infancy, growth laws have not yet been accepted in formal
codes.

STABILITY ASSESSMENT

Having described some typical crack growth laws, we can now determine the rate at
which a crack can grow from a size at which leakage of the contained fluid occurs
to an unstable size or configuration. If we are also able to calculate that
unstable size, we can then determine if there is time to detect the leak or to
reach the next shutdown/inspection period before catastrophic fracture occurs.
Assuming our fracture parameter is geometry independent, a stability assessment
for a crack in a structural component is carried out in the following way. The
fracture toughness (FT) is determined by measuring the critical value (if there is
one) of the chosen crack tip parameter on a small specimen. The next step is to
calculate a crack driving force (CDF) curve for the crack in a component. This is
just an expression for the crack tip parameter in the structure as a function of
the crack size, a, and applied load. If the crack is unstable at the first onset
of crack extension, the instability condition is

$$\text{CDF} > \text{FT} \tag{3}$$

In general, components which fail at this condition show linear elastic behavior
and the failure can be described in terms of a critical value of the stress
intensity factor, K_{1c}. For many structural materials, which fail by ductile
fracture mechanisms, the fracture toughness can increase as the crack extends, and
the crack will extend stably under rising load for some distance prior to becoming
unstable. In this situation, Equation 1 is a necessary but not sufficient
condition for instability. Sufficiency requires also that

$$\frac{d}{da}\,(\text{CDF}) > \frac{d}{da}\,(\text{FT}) \tag{4}$$

which simply says that as the crack extends, the driving force rises faster than
the fracture toughness.

Figure 7 shows an example of an instability assessment for an axial crack in a
pipe using a crack driving force curve (calculated for an applied hoop stress of
120 MPa) and an experimentally determined J-resistance curve. The J_R curve
represents measurements of crack growth resistance (toughness) at increasing
amounts of stable crack extension as measured on a small specimen. By
superimposing the J_R curve until it is tangent to the CDF curve, we are satisfying
equations 3 and 4 and hence the intercept at the abscissa is the estimated
critical crack length. The validity of this process assumes that the J_R curve is a

geometry independent entity, that is, the J-resistance curve determined on the small specimen is identical to a J-resistance curve for the pipe geometry. Generally this is not the case beyond a limited amount of crack extension. Three currently used assessment procedures are described below.

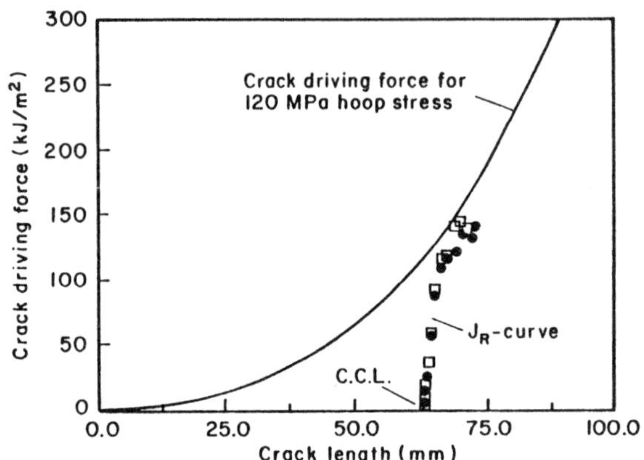

Fig. 7   Critical crack length determination using a J_R-curve and CDF for an axial crack in a pipe.

Current Elastic Plastic Assessment Procedures

At present there are three main engineering approaches to structural integrity assessment using elastic-plastic fracture mechanics (EPFM). These are the Electric Power Research Institute (EPRI) estimation schemes (Kumar and co-workers, 1981), Central Electricity Generating Board (CEGB) R-6 (Harrison and co-workers, 1976 and 1980, Milne and co-workers, 1986) and The Welding Institute (TWI) (Anderson and co-workers, 1985) methods. These methods are still at the proposal stage although the ASME Boiler and Pressure Vessel Code Section XI Flaw Evaluation Committees are in the process of developing EPFM, based on the EPRI work, for inclusion in the Code. The essential premise is that structural integrity in the presence of defects is governed by the instability relationship given previously, (Eqn's. 3 and 4).

EPRI approach. The EPRI approach allows calculation of crack driving force as a function of crack length and loading condition for different cracked geometries by combining the results of fully plastic solutions from the EPRI handbook with existing elastic solutions. The tables of results in the handbook assume material behaviour to follow a Ramberg-Osgood stress-strain power law. The compilation also includes solutions for the crack mouth opening and load point displacement. The normal procedure for stability assessment then follows that outlined in Fig. 7.

An alternative method of studying stability advocated by EPRI is by means of stability assessment diagrams or J/T plots. These are based on the work of Paris and co-workers, (1979) in which both the crack driving force and the material resistance are defined in terms of nondimensional "tearing moduli" as follows:

$$T_J = E/\sigma_y^2 \ (\partial J/\partial a)_{\Delta_T} \tag{5}$$

$$T_{JR} = E/\sigma_y^2 \ (dJ_R/da) \tag{6}$$

where Δ_T = total system displacement = $\Delta + C_m P$

    Δ = load point displacement

    C_m = system compliance.

    σ_y = yield stress

The instability criterion equivalent to Eqn. 4 can then be rewritten:

$$T_J \geq T_{JR} \tag{7}$$

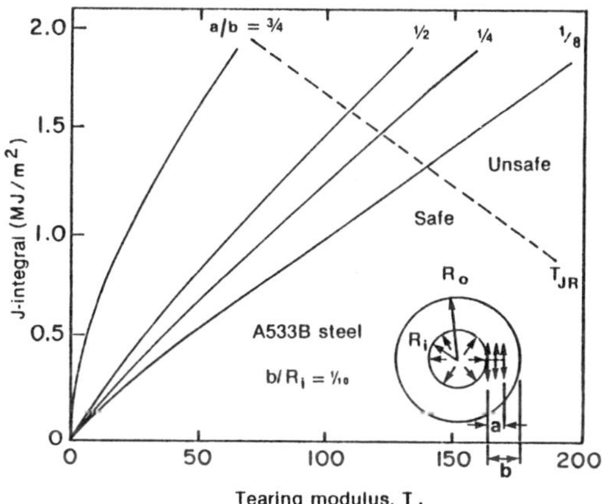

Fig. 8   J/T stability assessment diagram (after Kumar and co-workers, 1981).

Once the tearing moduli have been calculated by analytical or numerical differentiation, a stability assessment diagram can be constructed by plotting T_J and T_{JR} in J-T space. An example is shown in Fig. 8 for an internally pressurized axially cracked cylinder. The regions of stable and unstable crack growth are clearly defined since the instability condition corresponds to the intersection of the T_J and T_{JR} curves. Critical values of J are then used to determine the instability pressure from the crack driving force. Such figures are useful for representing multiple sets of data. Both applied loadings (as a function of crack size and/or loading conditions) and material data can be plotted on a single diagram in order to study the relative stability of a flawed structure.

CEGB R6 method. In this method, a failure assessment diagram (FAD) allows determination of margins of safety for a flawed structure based on an original two criteria approach (Dowling and Townley, 1975, Harrison and co-workers, 1976 and 1980, Milne and co-workers, 1986). The two-criteria approach assumed failure to occur when the applied stress reached the lower of the stress to cause fast, brittle fracture or the plastic collapse stress. The Dugdale plane stress strip yield model for a through-thickness crack in an infinite plate under remote tension was used in the transition region interpolating from the linear elastic to the fully plastic regimes. The original CEGB R6 FAD is shown in Fig. 9a where the co-ordinates are defined as:

$$K_r = K_1(a,P)/K_{1c} \tag{8}$$

$$S_r = P/P_o \tag{9}$$

where $K_1(a,P)$ = the elastic stress intensity factor

    K_{1c} = plane strain fracture toughness

    P = applied load

    P_o = plastic collapse load.

Using these co-ordinates the failure assessment curve (based on the Dugdale model) is given by:

$$K_r^2 = S_r^2 \ / \ [\ \frac{8}{\pi^2} \ (\ln \ \sec(\pi S_r/2))] \tag{10}$$

For a given flawed component the K_r and S_r co-ordinates can easily be calculated from the linear elastic and fully plastic solutions (limit analysis or nonlinear finite element analysis) for the appropriate geometry as well as the material properties. According to the R6 assessment procedure failure will not occur provided the assessment point for the flawed geometry, S_r', K_r' lies within the envelope bounded by the axes of the FAD and the failure assessment curve i.e., points within the envelope are deemed to be safe and those outside the envelope unsafe. The margin of safety is then given by the position of this point relative to the assessment line (Fig. 9a):

$$\text{Safety Factor} = OI/OA \quad (\text{crack initiation})$$

In its original form the definition of the co-ordinate, K_r in terms of the crack initiation parameter, K_{1c} did not allow for any stable crack growth before failure i.e., failure was assumed to occur at the onset of crack growth. More recently, the method has been modified to allow for stable crack growth (Milne, 1979), the parameter K_r now defined in terms of a given crack extension on the crack growth resistance curve:

$$K_r = K_1(a,P)/K_R(\Delta a) \qquad (11)$$

and the limit load, P_o based on the instantaneous crack size, $a+\Delta a$.

Using these co-ordinates a FAD is now equivalent to a crack stability diagram such as in Fig. 7, in which all instability points have been normalized and compressed to a single curve. Crack instability occurs when the failure assessment curve or crack driving force (given by the Dugdale strip yield model) is tangent to the growth locus corresponding to a maximum load (dead weight loading) as shown in Fig. 4. Load factors for a flawed geometry having assessment points $(S_r'$, $K_r')$, $(S_r''$, $K_r'')$ etc. on the growth locus for an applied load, P can then be defined by their position relative to the newly defined assessment line. For example, a load factor based on crack instability would be given by (Fig. 9b):

$$\text{Load Factor} = OU/OZ \quad (\text{crack instability})$$

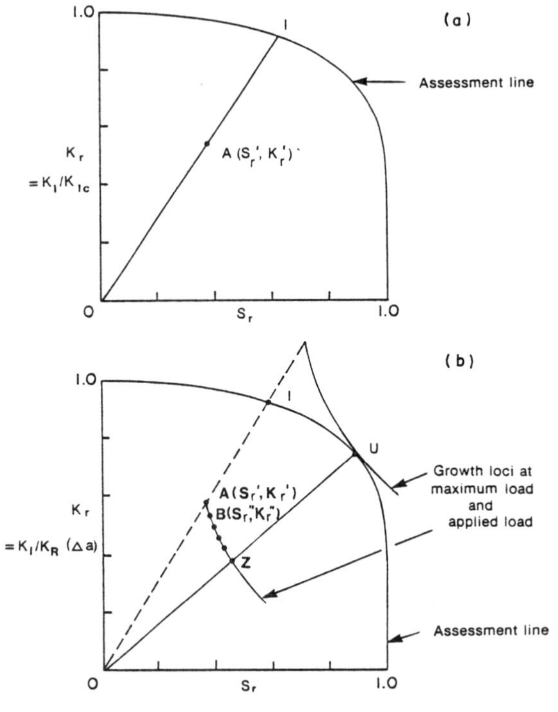

Fig. 9   CEBG failure assessment diagrams

    a.   original approach based on K_{1c}

    b.   revised version allowing for stable crack growth

These procedures are still under development and are now being extended to widen the range of material properties and conditions for which they are applicable.

<u>TWI method</u>.   The EPRI and CEGB R6 approaches should be contrasted with that of The Welding Institute in the U.K. Fracture mechanics is being used extensively for

analyzing the fatigue performance of welded joints where crack propagation is the major part of fatigue life. In particular, it is being applied to the growth of cracks from inherent discontinuities that may exist at the weld toe and weld root. Traditionally, TWI has been a strong advocate of the CTOD approach to failure. This method was developed at TWI and is currently the best documented toughness method for weldments (weld and heat affected zone) and parent steel in the ductile-to-brittle transition region.

The original method utilizes a design curve based on a non-dimensional CTOD, Φ, and is described in British Standard BSI PD6493, "Guidance on some Methods for the Derivation of Acceptance Levels for Defects in Fusion Welded Joints." The design curve is shown in Fig. 10 where

$$\Phi = \delta_c E \,/\, (2\pi\sigma_y a) = (\sigma/\sigma_y)^2 \qquad \text{for } \sigma/\sigma_y \leq 0.5 \qquad (12)$$

$$= (\sigma/\sigma_y) - 0.25 \qquad \text{for } \sigma/\sigma_y > 0.5$$

where δ_c = critical CTOD

and the other symbols have their usual meaning, already defined. Thus when the critical CTOD is specified, the curve gives the maximum allowable flaw size. However, the following limitations of the design curve approach in PD6493 have been pointed out (Anderson and co-workers, 1985).

1. It contains a large safety factor at low net section stress but may be non-conservative at high values. This makes probabilistic studies and critical crack size assessments very difficult.

2. Empirical part of the CTOD design curve has only been validated for ferritic steels. In PD6493, other materials have a more conservative assessment line based on linear elastic fracture mechanics.

3. No mechanism is available for incorporating a crack growth resistance curve.

4. Plastic collapse is not an integral part of PD6493 analysis.

In view of these limitations a more flexible three level approach has been proposed by TWI (Anderson and co-workers, 1985) which allows more accurate flaw evaluations:

Level 1   Current PD6493 approach with closer control over plastic collapse conditions. This is the simplest and most conservative approach.

Level 2   The Dugdale strip yield model is used for Φ instead of Eqn. 12. The model assumes elastic perfectly plastic behaviour and is therefore inaccurate for work hardening materials above net section yield. A comparison of this model with the CTOD design curve (PD6493) is shown in Fig. 10. Note the non-conservatism of the design curve at high net section stress (σ/σ_y).

Level 3   Reference Stress Method. This is based on option 2 in revision 3 of the CEGB R6 approach. It is applicable to materials where the material true stress-true strain curve is available and allows for evaluations above net section yield.

Generally, information is required concerning the size, shape and position of the initial flaw, geometry of the welded joint, applied loadings, residual stresses as

well as fatigue crack growth and fracture toughness data relevant to the
metallurgical condition of the welded joint where the defect is located. In the
absence of such information the most conservative assumptions must be made.
Guidelines are provided for choosing the appropriate assessment level depending
upon the material behaviour and information available.

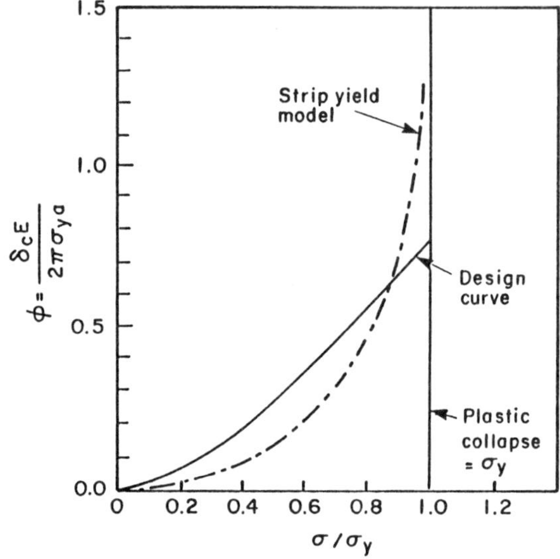

Fig. 10   CTOD design curve and level 2 strip yield model

Revisions to PD6493 are currently under discussion to include a three level
approach to flaw assessment, incorporate J-integral procedures and also to
introduce developments contained in the latest revision to R6.

SUMMARY

This paper has presented a brief summary of the essential information required to
carry out fracture control of loaded structures. While important, non destructive
examination has been omitted from this review as it is outside the expertise of
the authors. We have concentrated on the establishment of a crack-tip
characterizing parameter and its application to the determination of growth laws
and instability conditions for cracks. These are the important stages of the
crack life cycle in a fitness assessment.

Growth laws vary in sophistication and reliability and, except for fatigue, have
not yet reached the stage of inclusion in formal engineering codes for most
materials. This should not preclude one from applying such information to
fracture control in specific cases where growth mechanisms are well characterized.
The important rule is that you must "know your material" and the effects of
operating environment on the crack growth mechanisms.

Three examples were given of methodology for ensuring that crack instability does
not occur. We shall not pass judgement on which may be best as the particular
application will determine the most appropriate approach. If the material is very
crack tolerant, then a method with built-in conservatism such as the original CTOD
design curve or an R6 approach with crack initiation as the control point should
suffice. A more demanding application may require an accurate instability
analysis such as an advanced level of the CTOD or R6 methods or the EPRI approach.

The philosophy of fitness-for-purpose was described and the steps to be taken in
its demonstration. Because of the broad scope of this topic, this summary has
been necessarily cursory and the reader is referred to the references for further
instruction.

REFERENCES

Anderson, T.L., R.H. Leggatt and S.J. Garwood (1985). The use of CTOD methods in
fitness for purpose analysis. WI staff paper of presentation at the workshop on
CTOD methodology, GKSS research centre, Geesthacht, 23-25 April.

Dowling, A.R. and C.H.A. Townley (1975). The effects of defects on structural
failures: a two-criteria approach. *Int. J. of Pres. Ves. and Piping*, Vol. 3,
p. 77.

Duga, J.J., W.H. Fisher, R.W. Buxbaum, A.R. Rosenfield, A.R. Buhr, E.J. Honton and
S.C. McMillan (1983). The economic effects of fracture in the United States.
U.S. Department of Commerce, National Bureau of Standards, Report No. NBS SP
647-2, Washington D.C.

Ford, F.P. (1984). Current understanding of the mechanisms of stress corrosion
and corrosion fatigue, *Environment-Sensitive Fracture: Evaluation and Comparison of Test
Methods*, ASTM STP 821, S.W. Dean, E.N. Pugh and G.M. Ugiansky Eds., ASTM,
Philadelphia, pp. 32-51.

Harrison, R.P., K. Loosemore and I. Milne (1976). Assessment of the integrity of
structures containing defects. Central Electricity Generating Board, Report No.
R/H/R6, CEGB Research Division, London.

Harrison, R.P., K. Loosemore, I. Milne and A.R. Dowling (1980). Assessment of the
integrity of structures containing defects. Central Electricity Generating Board,
Report No. R/H/R6-Revision 2, CEGB Research Division, London.

Holroyd, N.J.H. and Hardie D. (1984). 'Corrosion fatigue of 7000 series aluminum
alloys *Environment-Sensitive Fracture: Evaluation and Comparison of Test Methods*, ASTM
STP 821, S.W. Dean, E.N. Pugh and G.M. Ugiansky Eds., ASTM, Philadelphia, pp. 534-
547.

Kanninen, M.F. and Popelar C.H. (1985). *Advanced Fracture Mechanics*, Oxford
University Press, pp. 437-497.

Kumar, V., M.D. German and C.F. Shih (1981). An engineering approach for elastic-
plastic fracture analysis. EPRI Report No. NP-1931, Research Project 1237-1,
General Electric Company, Schenectady, N.Y.

Milne, I. (1979). Failure analysis in the presence of ductile crack growth.
Central Electricity Generating Board, Report No. RD/L/N 179/78, CERL, Leatherhead,
England.

Milne, I., R.A. Ainsworth, A.R. Dowling, and A.T. Stewart (1986). Assessment of the integrity of structures containing defects. Central Electricity Generating Board, Report R/H/R6-Revision 3, CEGB Research Division, London.

Paris, P.C. and Erdogan, F (1963). "A critical analysis of crack propagation laws," *J. of Basic Engineering*, Vol 85, pp. 528-534.

Paris, P.C., H. Tada, A Zahoor, and H. Ernst (1979). The theory of instability of the tearing mode of elastic-plastic crack growth. *Symposium on Elastic-Plastic Fracture, Atlanta 1977*, in ASTM STP 668, American Society for Testing and Materials, Philadelphia, pp. 5-36.

Piping Review Committee (1985). U.S. Nuclear Regulatory Commission Report No. NUREG 1061. Washington D.C.

Ritchie, R.O. and Knott, J.F. (1973). Mechanisms of fatigue crack growth in low alloy steel, *Acta Metallurgica*. Vol 21, p. 639.

Simenz, R.F., Pengra, J.J. and Langenbeck, S.L. (1984). Application of laboratory test data to engineering design in *Environment-Sensitive Fracture: Evaluation and Comparison of Test Methods*, S.W. Dean, E.N. Pugh and G.M. Ugiasky, Eds., ASTM STP 821, American Society for Testing and Materials, Philadelphia, pp. 52-71.

Smith, R.A. Ed., (1986). *Fatigue Crack Growth - 30 Years of Progress,* Pergamon Press, Oxford.

Speidel, M.O. (1979). Stress corrosion cracking of cast aluminum alloys, *Stress Corrosion Research*, M. Arup and R.N. Parkins, eds., Sijthoff and Noordoff, Alphen and Den Rijn, the Netherlands, p. 117.

FRACTURE OF WELDMENTS

J.T. McGrath and J.E.M. Braid
Physical Metallurgy Research Laboratories, CANMET,
Energy, Mines and Resources Canada,
Ottawa, Canada

ABSTRACT

The complexity in assessing the fracture behaviour of weldments is discussed in
terms of the factors contributing to the variations in fracture toughness and the
micromechanisms of failure. The control of weldment toughness through the selection of weld process parameters, welding consumables and post weld heat treatment
is reviewed.

KEYWORDS

Fracture toughness; metallurgical heterogeneity; residual stress; strain aging;
weld process variables and consumable selection; post weld heat treatment.

INTRODUCTION

The heterogeneity of structure and properties in the vicinity of a weld can
make fracture mechanics analysis of such an area difficult compared with that
of base material. Although welding may be an ideal cost effective fabrication
process, welds usually suffer from reduced fracture toughness and/or a propensity towards the creation of defects which can act as crack initiation sites.
Defects can be present in both the weld metal and heat-affected zone (HAZ)
regions in the form of lack of penetration, lack of fusion, slag entrap-
ment, porosity and solidification and hydrogen cracking in the weld metal as
well as hydrogen and liquation cracking in the heat-affected zone.

The microstructure of weldments varies widely from an as-cast type structure
in the weld metal to coarse grain and fine grain structures in the HAZ. In
thicker weldments multiple passes are necessary to fill the joint, and the
variations in microstructure increase in complexity to include coarse grain and
fine grain reheated weld metal as well as HAZ material which has undergone
multiple thermal cycles. Hence, in evaluating the toughness of these regions
the position of the test specimen and the specimen notch or starter crack must
be carefully considered since the type of microstructure sampled can influence
toughness. Examples of the variation of toughness with notch location and
microstructure will be given for ferritic weldments. Examples of variations
in measured toughness arising from residual stresses and dynamic strain aging
in weld root regions will also be given.

Micromechanisms of failure, specific to welds, will be reviewed including the
role of weld metal inclusions in the initiation of both cleavage and ductile
failures. Finally, the control of the factors which influence weldment toughness will be discussed. Specifically, this will include the selection of weld
process parameters, welding consumables and post-weld heat treatment.

FRACTURE TOUGHNESS OF WELDMENTS

Fracture mechanics analysis of weldments can be complex because of the variation
in fracture toughness properties that can occur. Factors which contribute to the
variation in fracture toughness include metallurgical heterogeneity, residual
stress and dynamic strain ageing.

Metallurgical Heterogeneity

Weldments, which include the weld metal and HAZ regions, generally exhibit a
heterogeneity of microstructure. Macroscopic heterogeneity exists in the evaluation of fracture toughness of a welded joint when the crack front of the test
specimen resides within the weld metal, HAZ and base metal. Another example of
heterogeneity can occur when the crack front, residing totally within the weld
metal, of a multi-pass joint, can contain as-deposited or reheated microstructure
or combinations of both. The interpretation of fracture toughness of welded
joints from test specimens containing macroscopic heterogeneity requires an understanding of what features of heterogeneity contribute to the overall fracture
toughness of the weldment. In this regard the work of Satoh and Toyoda will be
examined. Through controlled experiments they have sought to define the effects
of the size, location and toughness of local embrittled regions on the overall
fracture toughness of welded joints. Examples will also be given of the effects
of microstructural heterogeneity on the toughness of welded joints in Cr/Mo steel.

A quantitative analysis of the effect of macroscopic heterogeneity on fracture toughness evaluation of weldments

The experiments of Satoh and Toyoda [1,2,3] have attempted to quantitatively
define the factors which control the fracture toughness of welds containing macroscopic heterogeneity. They prepared a series of specimens with controlled heterogeneity along the crack front, as shown in Fig. 1. In the Series I specimens (Fig.
1a), the effect of the size of an embrittled region on the fracture toughness was investigated. The embrittled region in the center of the weld was produced by depositing a few layers of high strength alloy steel (DF2A-300-B) in a
lower strength (D6216) C/Mn weld metal. In the second series the size of the embrittled region was kept constant and the specimen thickness was varied (Fig. 1b).
The local brittle region in Series II was an electron beam weld in high strength
(784 MPa) HT80 steel. The third series (Fig. 1c) of experiments were conducted
to investigate the fracture toughness in a cross-bond-type specimen in which the
notch was located in weld metal, HAZ and base metal. The amount of embrittled
weld metal along the notch length was varied. In addition to measuring the fracture toughness of the specimens containing macroscopic heterogeneity, toughness
measurements were made in macroscopically homogeneous material. As shown in Fig.
1 this allowed an assessment of the fracture toughness of all the individual
regions contained in the specimens with heterogeneity.

In a COD test of a specimen containing heterogeneity along the crack front, the
load-deflection curve will contain a pop-in associated with a cleavage fracture
initiation in the local embrittled region (HT80) crack front. The detailed

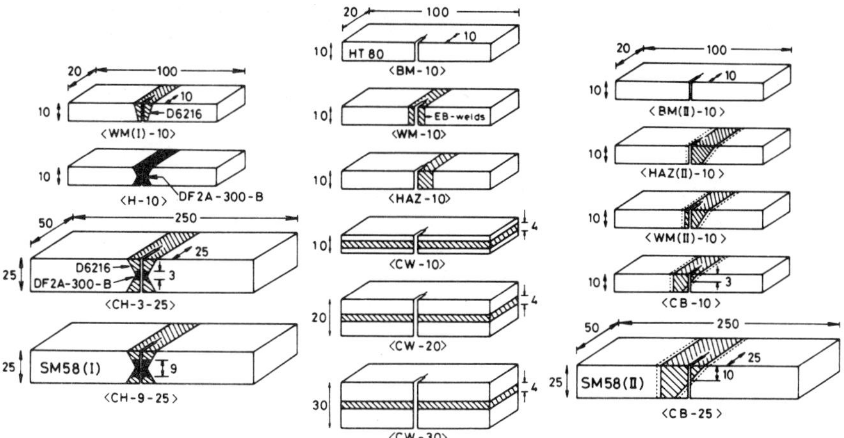

(a) Specimens extracted from shielded metal-arc welded joints of SM58 steel (Series I)

(b) Specimens extracted from electron beam welded joints of HT80 steel (Series II)

(c) Specimens extracted from submerged-arc welded joints of SM58 steel (Series III)

Fig. 1. COD specimens for evaluation of fracture toughness of welds with heterogeneity [1]

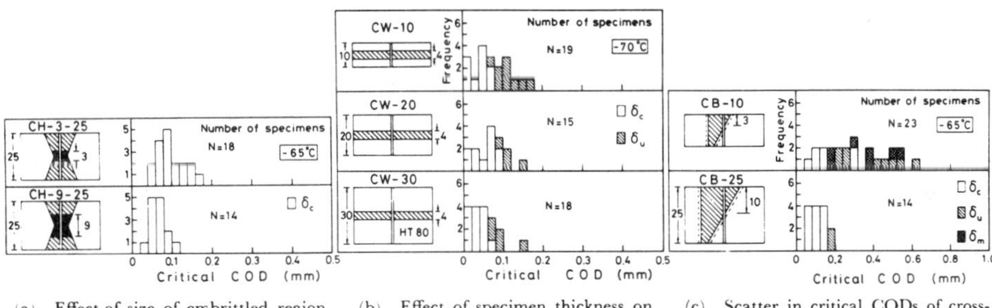

(a) Effect of size of embrittled region along crack front on scatter in critical CODs under constant specimen thickness (Series I)

(b) Effect of specimen thickness on scatter in critical CODs under constant size of embrittled region along crack front (Series II)

(c) Scatter in critical CODs of cross-bond-type notched specimens (Series III)

Fig. 2. Effect of macroscopic heterogeneity on fracture toughness of welded specimens. [1]

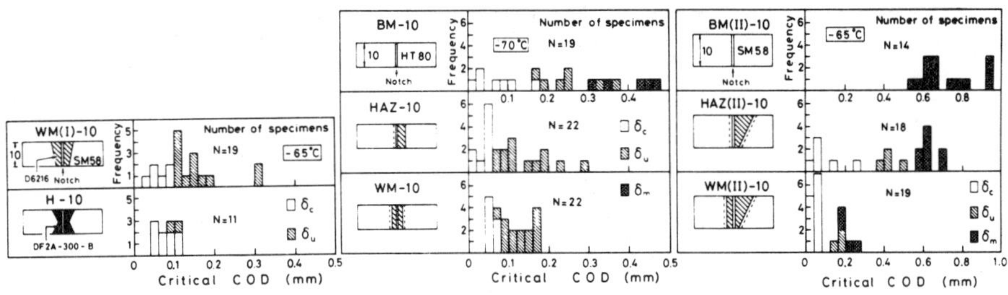

(a) Shielded metal-arc welds of SM58 steel (Series I)

(b) Electron beam welds of HT80 steel (Series II)

(c) Submerged-arc welds of SM58 steel (Series III)

Fig. 3. Fracture toughness results for macroscopically homogeneous material. [1]

results of COD tests on the specimens with heterogeneity are shown in Fig. 2.
The COD value at fracture initiation was defined as δ_c (critical COD for unstable fracture with no prior stable crack growth) or δ_u (critical COD for
unstable fracture with prior stable crack growth). The effect of the size of a
locally embrittled region on fracture toughness of specimens with constant thickness is shown in Fig. 2a. The scatter in critical COD was much greater in specimens containing the narrower (3 mm) embrittled region. It was in the specimen
containing a 9 mm thickness of embrittled material (30% the crack front) that the
average value of critical COD coincided with the critical COD of the individual
embrittled zone (compare with Fig. 3a). In the Series II experiments the COD
results (Fig. 2b) showed that the average value of critical COD for specimens
with a constant width of embrittled region compared to that measured for the embrittled weld metal (Fig. 3b). There was little change in scatter of COD with
change in specimen thickness. Finally, in the cross-bond type specimens it was
found that (Fig. 2c) the specimen containing the larger (10 mm or 40% of the
crack front) amount of embrittled weld metal had a significantly lower average
critical COD and smaller scatter. Once again the fracture toughness of the
specimen containing the larger amount of embrittled weld metal was similar (Fig.
3c) to that of the all-weld metal test specimen. In summary the results of Satoh
and Toyoda have shown that in specimens which have a locally embrittled region
along the crack front, the overall fracture toughness, and scatter of toughness
are mainly controlled by the toughness of the individual embrittled region and
by its size along the crack front.

Let us now consider briefly the application of the Charpy impact test in the
evaluation of toughness of weldments containing heterogeneity. In the evalualion
of fracture toughness of macroscopically homogeneous materials there can be a
reasonable correlation between the Charpy absorbed energy and the COD value.
Whether such a correlation exists in weldments containing macroscopic heterogeneity has also been examined by Satoh et al. [2]. It has already been shown
that fracture initiation in welds containing heterogeneity is controlled by the
fracture toughness of a local embrittled region in the vicinity of the crack tip.
Thus the COD test was appropriate to evaluate fracture toughness in this situation
because it is sensitive to local deformation at the crack tip. On the other hand,
the standard Charpy test measures the energy associated with crack initiation and
propagation through the net section of the specimen. A comparison of results
between the Charpy and the COD tests of specimens with heterogeneity along the
crack front is shown in Fig. 4. Specimens were prepared in which the crack front
for both Charpy and COD test specimens contained an embrittled, low toughness
region (designated by HT80 steel) and a lower strength, tougher region (SUS 304
stainless steel). The results of the Charpy test showed (Fig. 4a) that the
heterogeneous specimen had a notch toughness approximately mid-way between the
toughness of the embrittled HT80 and the more ductile SUS 304 steel. On the other
hand, the COD results in Fig. 4b, indicated that the critical COD for a specimen
in which the fraction of embrittled HT80 was more than 30% of the thickness is
nearly equal to that of the HT80 steel. Thus, although the Charpy test is sensitive to heterogeneity, a good correlation between Charpy absorbed energy and COD
for weldments containing macroscopic heterogeneity may not be possible.

<u>Examples of microstructural heterogeneity in weldments</u>

The vast majority of welded joints in engineering structures contain multi-de-
posits as for example, the schematic representation of a narrow-gap submerged-arc
weld in a thick section pressure vessel shown in Fig. 5. As the joint is filled
each weld bead will have a reheating effect on the previous bead thus resulting
in a heterogeneity of microstructures which can range from a coarse-grained structure at the highest reheating temperature to a fine-grained reheated structure and

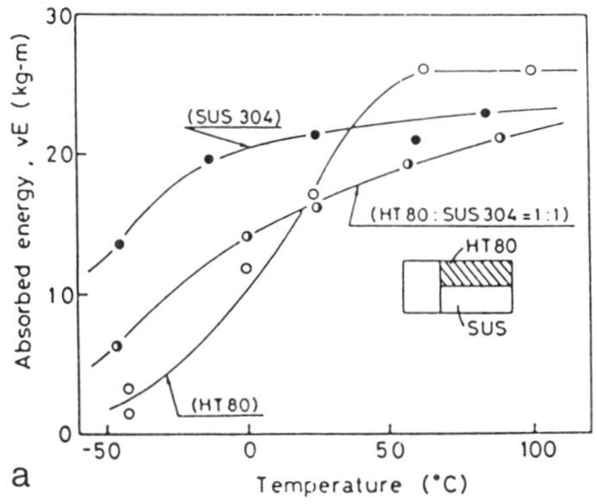

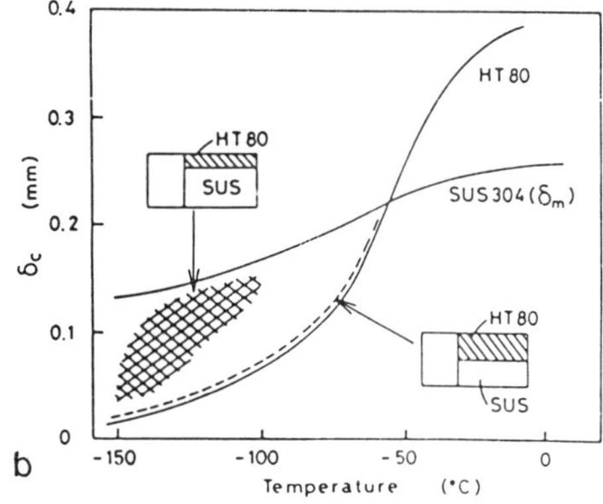

Fig. 4. Comparison of results between Charpy and COD tests for specimens with heterogeneity along the crack front.
(a) Charpy absorbed energy   (b) Critical COD [2]

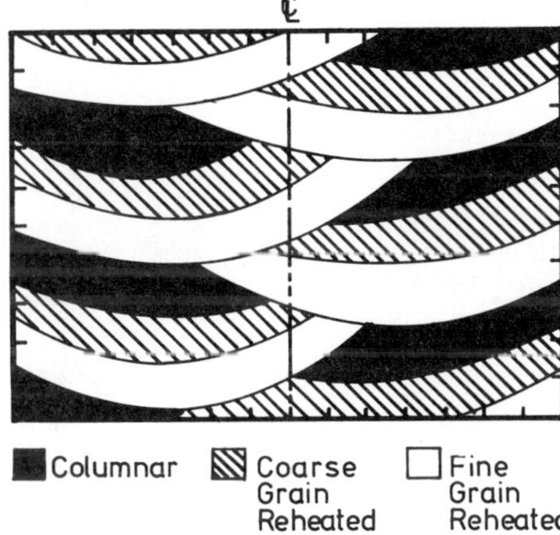

Fig. 5. Schematic representation of microstructure in a two-pass per layer narrow gap weld.

a relatively undisturbed as-deposited structure. In overlap regions, particularly at the weld centreline a microstructure containing mainly reheated weld metal can result.

An example of the variation in microstructure in multi-pass narrow-gap submerged-arc weld metals can be shown in two-pass per layer welds in 2.25 Cr-1 Mo steel. The 2.25 Cr-1 Mo steel is used primarily in a high temperature, high pressure hydrogen environment in heavy-wall hydro-treatment reactor vessels in the Petro Chemical industry. The microstructure produced in both base metal and weld metal upon cooling from the austenite is bainite. The CCT diagram for 2.25 Cr-1 Mo steel indicates that bainite can form over a wide range of cooling rates [4]. It has been shown [5,6] that the bainite transformation temperature, which controls the grain size of the bainite's microstructure, is greatly influenced by carbon content. For a carbon level of 0.10 the microstructure of 2.25 Cr-1 Mo weld metal in the center of individual weld beads (1/4 w position as per Fig. 5) is composed of a mainly as-deposited structure of fine bainite (Fig. 6a). In the overlap region at the 1/2 w position, a reheated structure of fine bainite is also found. If Charpy impact tests are performed with notches located at the 1/4 w and

1/2 w positions no significant difference in notch toughness is observed as indicated in Fig. 7. When the carbon level was reduced to 0.075 the bainite transformation temperature was increased [5] and the microstructure at the 1/4 w position is now a much coarser bainite as shown in Fig. 6b. At the 1/2 w position the reheated microstructure is much finer by comparison. When Charpy tests were performed with specimens notched at the 1/4 w and 1/2 w positions a much higher toughness at low temperatures was observed at the 1/2 w position. The results of these studies indicate that in order to achieve a more uniform toughness within a multi-pass weld an effort should be made to reduce the differences in microstructure i.e. grain size, between reheated regions.

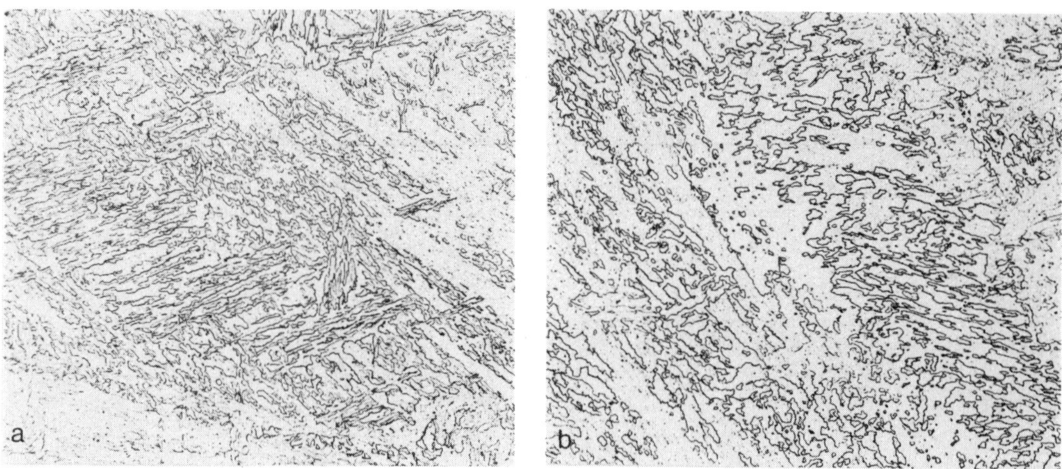

Fig. 6. Bainitic structures in 2.25Cr-1Mo narrow gap welds at 1/4 W position
  (a) Weld W1 C = 0.1%   (b) Weld W2 C = 0.075%

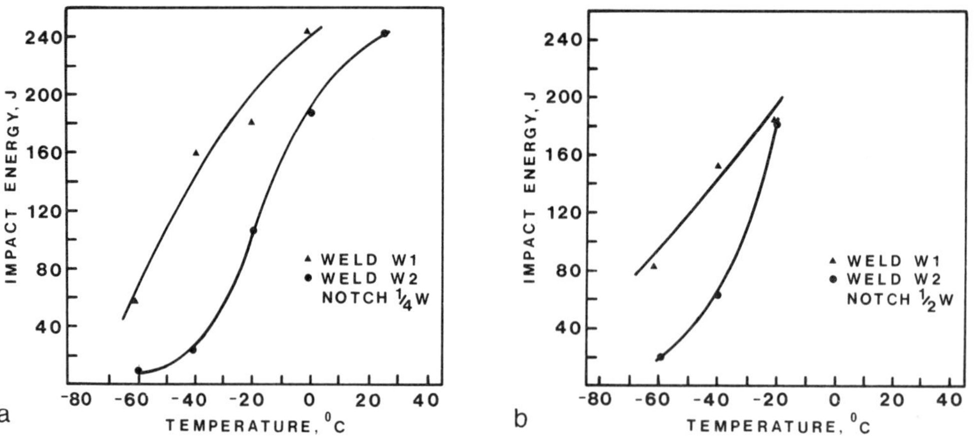

Fig. 7. Charpy transition curves for 2.25Cr-1Mo narrow gap welds W1 and W2.
   (a) notch located at 1/4 W position
   (b) notch located at 1/2 W position.

Residual Stresses

Welding results in residual stresses which can be described by a superposition of
three basic sources [7] shown in Fig. 8 as: (a) thermal and uniform yielding,
(b) non-uniform yielding, and (c) transformation. As a weld pool solidifies and
cools the weld metal contracts. This is resisted by the cooler surrounding material and gives rise to an elastic stress given by $E\alpha\Delta T$. Because of the reduced
yield strength at elevated temperatures, the thermally induced stresses exceed the
yield strength of the weld metal and uniform yielding of the weld metal occurs.
As the temperature decreases, weld metal yield strength increases, thus increasing
the balancing stresses in the surrounding cooler plate. The resulting tensile
residual stresses in the weld can be as high as the yield stress and of a multiaxial nature. The temperature across a section of a weld bead will not be constant owing to the transfer of heat to the surrounding cooler material. Steep
temperature gradients within the weld metal can form depending on the original
rate of heat input and coefficient of thermal conduction. This can result in
local thermal contractions which lead to non-uniform yielding in the weld metal.
Also, on cooling a phase transformation may occur. In steels this will be from
austenite to ferrite, bainite or martensite depending on the cooling rate and
composition of the steel. This transformation results in an expansion which is
hindered by the cooler surrounding material. Thus the transformed material is
subjected to a net compressive residual stress. Generally the residual stresses
from uniform yielding increase with increasing yield strength, but in steels, an
increase in yield strength is often achieved by alloying which changes the transformation temperature of the material. Hence, in alloy steels, transformation
stresses can become more significant resulting in complex residual stress distributions [7,8].

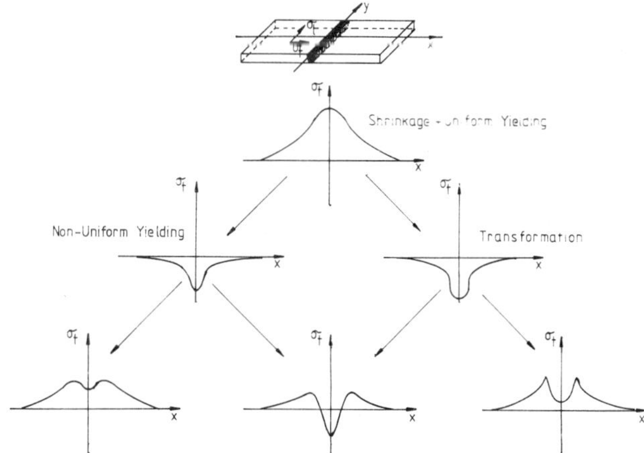

Fig. 8. Superposition of three basic sources of residual stress in welds.[7]

The three sources are not independent but by using the principle of superposition
the three can be added together to indicate regions of maximum tensile residual
stresses. The above discussion is based primarily on single-pass welds. In
multipass welds the build-up of residual stresses become somewhat more complex.
Heat from the deposition of a weld bead causes a partial relaxation of the residual stresses formed by previous passes. Studies have shown that residual stresses in multipass welds are built-up slowly with each pass and can vary considerably depending on joint preparation, heat input, welding process, pass sequence,
preheat and interpass temperatures.

The large number of variables affecting residual stress formation and build-up make their prediction difficult. The most accurate predictions have been made by transient finite element analysis. Currently this is expensive and time consuming, although advances are underway which may make this type of analysis a practical tool for the engineer. Hence, for a fracture mechanics analysis, simplified distributions and conservative levels of residual stresses are used. Where more accuracy is required, such as for nuclear pressure vessels, residual stresses may be measured (also expensive and time consuming) and the measured distribution used. In applications where linear elastic fracture mechanics can be used there are several methods available for calculating the stress intensity factor due to residual stresses, K_r. Superposition is then used for the overall assessment of the defect [9]. For elastic plastic fracture mechanics applications the residual stresses or strains are combined with those due to applied loading and the resultant membrane and bending stresses or strains are derived by a linearization process. Using methods such as found in the British Standards PD6493 the critical defect size or crack tip opening displacement can be evaluated [10]. The various methods mentioned above are themselves a large topic and the reader is referred to the literature for further information.

In evaluating the toughness of weldments the effects of residual stresses are seen only in large scale tests such as the wide plate test or tests of full scale components. Residual stresses are substantially relaxed in small-scale specimens for K_{Ic}, CTOD or Charpy tests during removal/machining of the test piece, from the larger weld. The remaining residual stresses have a negligible effect on measured toughness but may affect the shape of the fatigue crack front in K_{Ic} and CTOD specimens. To maintain the fatigue crack shape within the limits of the various standards a local compression treatment is often used [11]. Here, the ligament below the notch is compressed on each side (typically 0.5% per side) resulting in a redistribution/relaxation of residual stresses which has a negligible effect on measured toughness.

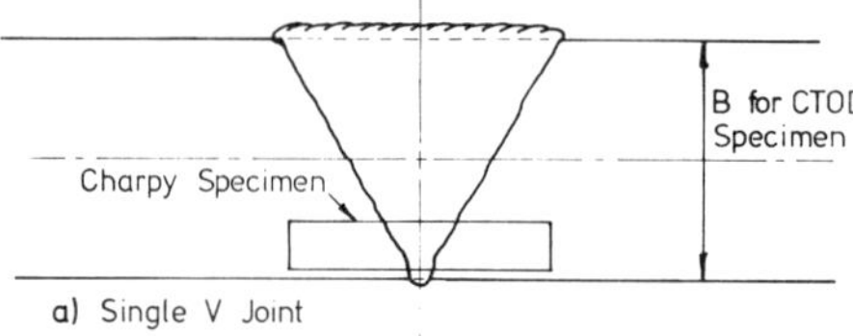

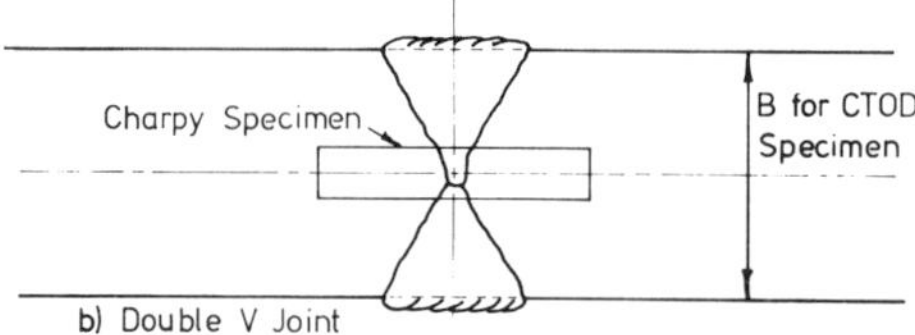

Fig. 9   Location of root region Charpy specimen and full thickness CTOD specimen for (a) single V joint column and (b) double V joint.

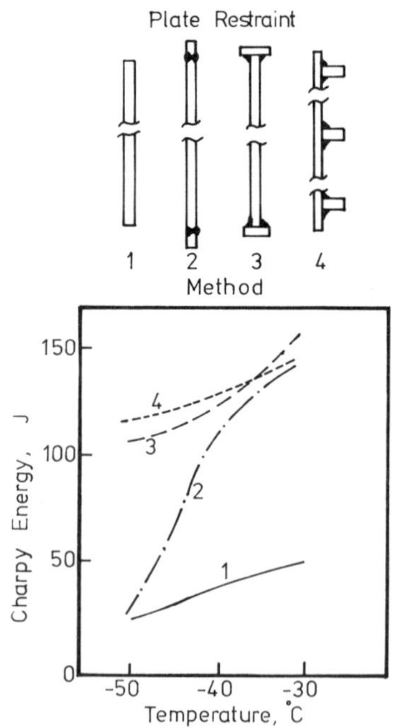

Fig. 10   Effect of plate restraint on impact toughness [18]

Dynamic Strain Aging

Strain aging arises from the segregation of solute atoms to dislocations producing
pinning of the dislocations [12]. As the segregation of solute atoms to dislocations becomes more extensive, frictional resistance to the passage of unpinned
dislocations increases and a higher strain-hardening rate may also occur [13].
In iron and steel, strain aging is a result of the presence of interstitial
solutes carbon and nitrogen [12] after aging at 200-250°C [14,15]. The overall
result is an increase in yield strength and/or flow stress on aging after
("static") or during ("dynamic") straining beyond yield. Dynamic strain aging
involves diffusive jumps of solute atoms during deformation which makes the process both temperature and strain-rate sensitive. Dislocations are rapidly pinned
during dynamic strain aging causing others to be generated to allow deformation
to continue. This results in a larger dislocation density for a given strain
compared with static strain aging with a greater resultant loss in ductility and
toughness [13].

In the previous section, the formation and build-up of residual stresses was described. Yielding occurs (at elevated temperatures) owing to contraction during
cooling of the weld metal, which is resisted by the cooler surrounding material.
As the weld cools dynamic strain aging can occur particularly for welds in steels
cooling from 300°C to 200°C with uncombined nitrogen in solution. Cochrane et al.
[14], Berkhaut and van den Brink [15] and McRobie and Knott [16] have shown that
the dynamic strain aging phenomenon is responsible for the loss of toughness in
the root region where straining in the 200-250°C temperature range occurs.
Through-thickness CTOD tests are affected by root properties particularly for
double "V" joints [17] where the root region is located in the centre of the CTOD
test piece (Fig. 9), the area of highest constraint. Dawson and Judson [18] point
out the dilemma that this can create in comparing CTOD values for single "V" and
double "V" joints. Test results showed that single "V" joints had higher CTOD
values than double "V" joints, but root region CVN values for single "V" welds
were lower than those for the double "V" welds. Single "V" CTOD values were
higher because the root region was located near the specimen surface (Fig. 9), a
low constraint (plane stress) region. The single "V" joint often results in
larger distortions and hence higher root region strain than double "V" joints for
plates of the same thickness.

Temperature, plastic strain and chemical composition can affect the intensity of
dynamic strain aging in steel welds. During multipass welding, the root region
weld beads are reheated by subsequent passes. Also, the use of preheat and high
maximum interpass temperatures, for the avoidance of cold cracking, contributes
to the total time that the root region is held within the critical dynamic strain
aging temperature range. Hence, minimizing the time spent in this critical temperature range should aid in minimizing dynamic strain aging. This could be
achieved by reducing heat input and using low preheat and interpass temperatures.
The benefits of lower heat input shown by Cochrane et al. [14] may, however, also
be attributed to increased proportions of fine grain reheated weld metal with its
consequent improvement in toughness. Reduction of preheat and interpass temperatures will increase the susceptibility to hydrogen cracking and reduce the proportion of reheated weld structure, lowering weld metal toughness.

The contractions during cooling which lead to plastic straining have three basic
components: longitudinal, lateral and angular (rotational). Resistance to
these contractions are, respectively, the length of the plate (joint), transverse
restraint and bending restraint. Of the three, angular (rotational) distortion
has been shown to have the largest influence on dynamic strain aging damage in
weld metal [18,19]. This form of distortion arises as a consequence of asymmetry

of the joint detail or fill procedure resulting in greater shrinkage on one side
of the plate, or joint, compared to the other side. Dawson and Judson [18] have
investigated the effects of restraint on root region toughness. Several methods
of providing angular restraint were used and toughness measured using both CTOD
and root region CVN. Strong backs of method 4 type illustrated in Fig.10 were
found to be the most effective method of restraint for resisting angular distortion and improving toughness. Reduction of the width of weld preparations and
smaller preparation angles as well as using balanced welding techniques also
reduces angular distortion.

In structural steels nitrogen is more effective than carbon in dynamic strain
aging owing to its higher solubility in iron [13]. In welds the main source of
nitrogen is from the surrounding atmosphere. Flux formulations, efficiency of gas
shielding and, in particular, operator skill affect the nitrogen content of the
weld metal. Additions of Ti, V or Al could be used to reduce free nitrogen
levels. However, too large a quantity could lead to a deterioration of toughness
due to precipitation hardening and/or production of unsuitable microstructures
[14].

MICROMECHANISMS OF FAILURE IN WELDS

As in homogeneous metals static failure in welded joints can be ductile, cleavage
or intergranular in nature. It is the heterogeneity of microstructure and chemistry combined with local stress concentrations and the increased propensity of
defects in the vicinity of welds that increases the probability of failure in the
weld region compared with the base material. Although the basic micromechanisms
of ductile, cleavage or intergranular failure are similar to those of base metals
there are some aspects that are particular to weldments.

Segregation of impurity elements such as S, P and Sn can occur at the grain boundaries in the coarse grain HAZ. In high strength structural steels such as HY-80
or 2-1/4 Cr-1 Mo intergranular cracking can occur. McGrath et al. [20] have found
instances of sulfide liquation cracking in the HAZ of welds in HY-80 steel.
Hippsley and co-workers [21] have studied reheat cracking in 2-1/4 Cr-1 Mo and
A533B steels. They found intergranular cracking to occur during the heating phase
of stress relief heat treatment as a result of P and Sn segregating to the grain
boundaries in the CG HAZ aided by the tensile residual stresses that the heat
treatment was designed to relieve. These microcracks can lead to complete fracture of the joint under external loading.

Welds made with processes such as SMAW, FCAW, GMAW or SAW result in weld metals
with high levels of oxy-sulphide inclusions compared with the base metal. The
inclusions arise from the interaction of oxygen with oxide forming elements such
as Ti, Al, Mn and Si either in the molten weld metal or from slag/metal reactions.
The composition, number and size distribution of inclusions affect the composition
and microstructure of the weld metal which can lead to either improvements in
toughness through microstructure refinement or deterioration of toughness through
the generation of coarse microstructures. In C-Mn structural steels the presence
of the Ti-bearing inclusions between 0.5 µm and 0.75 µm dia. has been shown to
aid the production of favourable acicular ferrite microstructures [22]. Inclusions, however, are the prime initiation sites for ductile failure in welds and
recent evidence has shown that they also act as cleavage crack initiation sites
in ferritic weld metals [23]. The effects of non-metallic inclusions on ductile
fracture are well known (see, for example, [24]). This can be seen in the effect
of volume fraction of inclusions on the upper shelf Charpy energies. The upper
shelf Charpy energies increase for decreasing O+S level. The lower the O+S content the lower the total number of inclusions which act as void nucleation sites.

Recent work in ferritic weld metals has shown that non-metallic inclusions act as sites for cleavage crack initiation [23]. Examination of fractured surfaces revealed that larger non-metallic inclusions in the 1-2 μm dia. range had fractured and that river lines radiated from them. Tweed and Knott [23] have postulated that plastic deformation in coarse grain boundary proeutectoid ferrite produced dislocation pileups and high stresses at inclusions. If the stresses were high enough, the inclusions cracked. At a critical crack tip stress brittle fracture would initiate from the broken inclusions.

Work in support of the Tweed and Knott model of cleavage crack initiation at large inclusions in weld metal has been reported by Judson and McKeown [25]. They sought to modify the size distribution of weld metal inclusions by deoxidant additions to the SMAW electrode coverings. The deoxidant materials were not identified because their compositions were of a proprietary nature. As shown in Fig. 11a, deoxidant material G, when added to the electrode covering, produced a much finer distribution of weld metal inclusions (max. particle dia. 0.5 μm) than the reference electrode covering. The volume fraction of inclusions in each weld was similar. The Charpy notch toughness results in Fig. 11b indicated significantly less energy to fracture at temperatures < −40°C in the weld with

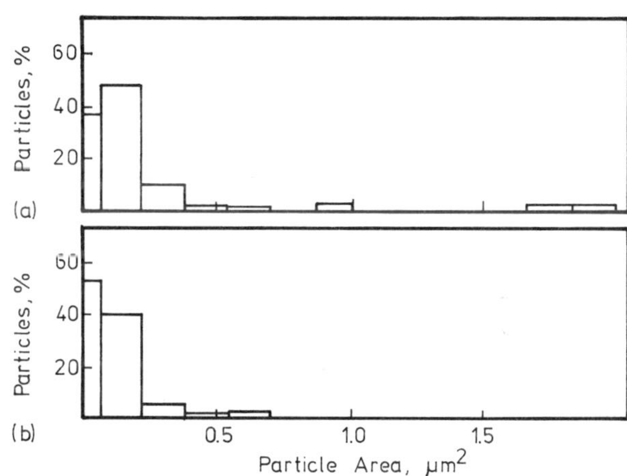

Fig. 11(a)

Non-metallic inclusion size distribution from reference electrode and optimized electrode [25]

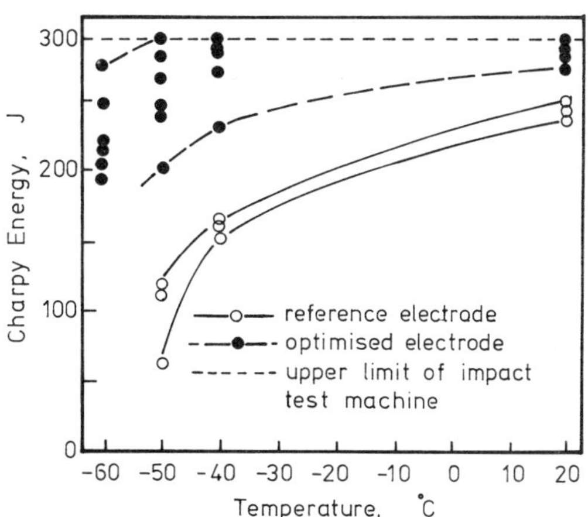

Fig. 11(b)

Charpy energy transition curves welds prepared by the reference electrode and one optimized with respect to inclusion size distribution. [25]

the coarse inclusion particles thus supporting the idea that large weld metal
inclusions can initiate cleavage failure. Further work is required to confirm the
role of inclusion size distribution on the fracture toughness of weld metals.

CONTROL OF WELD PROCESS VARIABLES FOR OPTIMUM FRACTURE PROPERTIES

Weld Process Parameters

As an example of how weld process parameters can be used to control weld fracture
properties the work of Evans on multi pass C/Mn weldments can be cited. In
shielded metal-arc welds Evans has shown the effects of interpass temperature [26]
electrode diameter [27] heat input [28] and welding position [29] on the relative
proportions of as-deposited and reheated regions and the subsequent effect on
Charpy notch toughness properties. These studies were done on weld deposits containing 0.05% C, 0.30% Si and Mn in the range of 0.6 - 1.8%.

Increasing the interpass temperature from 20 to 300°C resulted in an extension of
the heat-affected zone in the underlying weld metal, thus increasing the amount
of reheated structure and decreasing the proportion of the as-deposited region
(Fig. 12a). This resulted in an improvement in notch toughness, represented by
the decrease in Charpy-V test temperature to give a level of 100 J, as shown in
Fig. 12b. While the improvement of Charpy impact properties with increasing interpass temperature is the likely consequence of the progressive elimination of
the as-deposited region, it was pointed out by Evans [26] that a reversal in notch
toughness properties would probably occur at higher interpass temperatures if the
grain size of the reheated region continued to increase.

Increasing the electrode diameter from 3.25 to 6.0 mm decreased the amount of
reheated weld metal and reduced the notch toughness properties (Fig. 13).

Raising the heat input from 0.6 to 4.3 kJ/mm generally reduced the proportion of
as-deposited structure in a weldment. For the same plate thickness and joint
preparation the number of beads per layer decreased from four to one as the heat
input was increased to the maximum value. The effect on notch toughness is shown
in Fig. 14. There was an improvement in toughness up to 2 kJ/mm and a significant
drop thereafter. Although the amount of as-deposited structure was reduced with
increasing heat input, there was an increase in the coarseness of the transformation products within the reheated region which accounted for the reversal in
toughness properties at heat inputs above 2 kJ/mm.

Welding in the vertical and flat positions affected the proportions of as-deposited and reheated structure. Welds were made with single (full weave) and two
passes (limited weave) per layer. The amount of reheated structure was much
greater in the flat position weld compared with the vertical position in the two
pass per layer welds. The notch toughness data indicated higher toughness for the
flat position compared with the vertical for the two pass per layer welds. There
was no difference in toughness between flat and vertical position welds in the
higher heat input (4 kJ/mm) single pass per layer welds. Harasawa et al. [30]
also found no effect of weld position on toughness in welds prepared at 5 kJ/mm
heat input.

Consumable Selection

The selection of welding consumables i.e., wire/flux for the SAW process, wire
and flux coating for the SMAW process or wire and gas for the GMAW process, is
essential in the control of weld metal strength and toughness properties. The
weld metal microstructure, including the type of transformation products and the

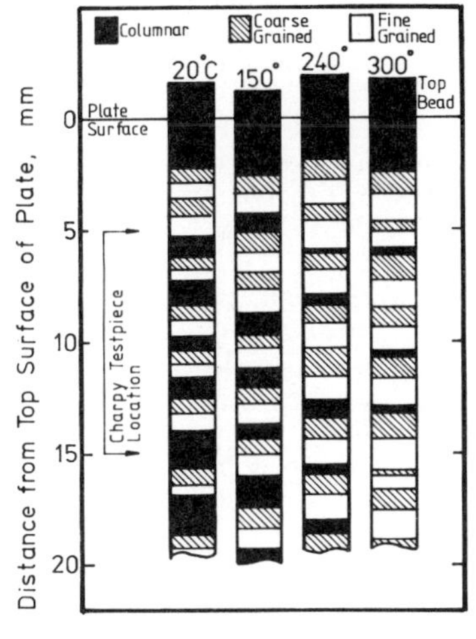

Fig. 12(a)   Proportions of columnar (as deposited) and reheated microstructuress as a function of interpass temperature [26]

Fig. 12(b)   Effect of interpass **temperature on the** 100 J Charpy temperature. Electrodes designed to have the following Mn contents A-0.6%, B-1.0%, C-1.4%, D-1.8%. [26]

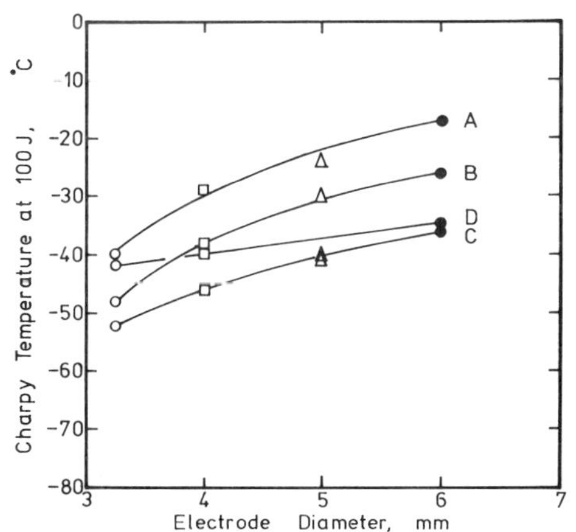

Fig. 13   Effect of electrode diameter on the 100 J Charpy temperature see Fig. 12(b) for electrode designation. [27]

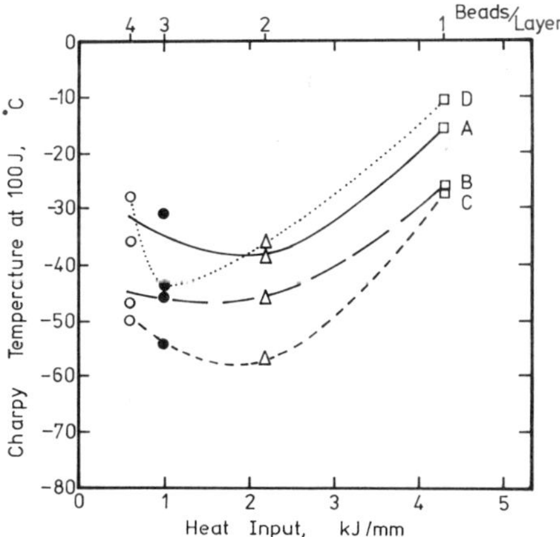

Fig. 14   Effect of heat input on the 100 J Charpy temperature see Fig. 12(b) indicates electrode designation.[28]

volume fraction and size distribution of inclusions, depends upon the composition of the weld. The alloying elements in the weld are supplied primarily from the welding wire with dilution from the base metal also acting as a source. The slag/metal reactions taking place in the SAW and SMAW process and the oxidation potential of the shielding gases in the GMAW process control the formation of inclusions in the weld metal.

In the welding of C/Mn and HSLA steels, it has been found that the ideal weld
metal microstructue for low temperature fracture resistance should contain a high
proportion of acicular ferrite in the as-deposited region and fine, equiaxed polygonal ferrite in the reheated region. The transformation products that can form
during continuous cooling of a weld metal of a given composition are expressed by
the CCT diagram shown in Fig. 15. The effect of such elements as C, Mn, Mo and
Ni on the relative positions of the boundaries for the transformation of austenite
to various constituents will dictate how much of the desired acicular ferrite will
form for a given cooling rate. For example, Evans [31,32] has shown that increasing C (optimum level 0.1%) and Mn (1.5%) in weld metal resulted in an increase
in acicular ferrite and a reduction in pro-eutectoid ferrite and ferrite with
aligned second phase. Mo will promote the formation of acicular ferrite in additions up to about 0.1%. Higher levels of Mo result in the formation of coarse
bainite or ferrite with aligned second phase which are deleterious to fracture
toughness [33]. The addition of Ti-B to fluxes and/or wires primarily in SAW
consumables but also in SMAW consumables, has led to increased weld metal toughness [34-37]. The mechanism for improvement has been attributed to increased
amounts of acicular ferrite at the expense of pro-eutectoid ferrite. It is felt
that the precipitation of TiN allows solute B to segregate to prior austenite
grain boundaries reducing the grain boundary energy and hence suppressing the
nucleation of pro-eutectoid ferrite [38]. This allows various small (0.5 - 0.75
μm) Ti-containing oxides to become acicular ferrite nucleation sites [39].

Finally, alloying elements such as Nb and V, which enter the weld pool primarily
by dilution from the base metal, can precipitate as carbides at temperatures below
1000°C. At the high cooling rates normally associated with welding there may not
be time for carbides to form. However in the reheated regions associated with
multi-pass welding precipitation may occur which can result in matrix hardening
and a decrease in toughness.

The control of inclusion content in welds deposited by the SMAW and SAW processes
comes from the basicity of the fluxes. The flux basicity (B) is calculated in
terms of the ratio between basic oxides and acidic oxides as follows [39]:

$$B = \frac{CaO + 1.4\ MgO + 0.60\ K_2O + 0.90\ Na_2O + 0.45\ ZrO_2(+\ 0.55\ Al_2O_3)}{SiO_2\ (+\ 0.59\ Al_2O_3)}$$

The values for B can range from 0.5 to 3.0. The use of more highly basic fluxes
results in lower inclusion levels in the weld metal.

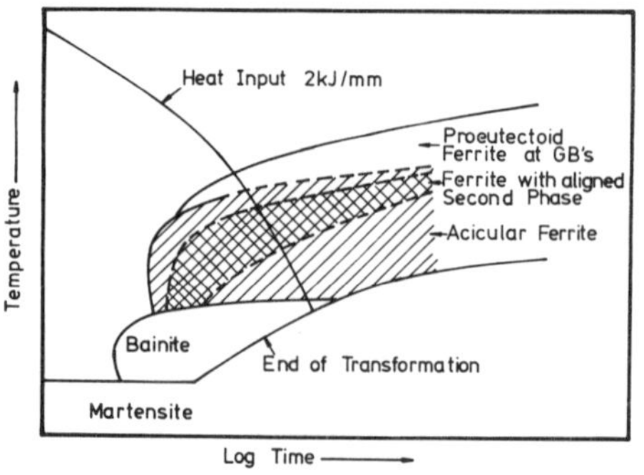

Fig. 15   Schematic of continuous cooling transformation (CCT) diagram for a C-Mn
          weld metal.

<u>Post Weld Heat Treatment</u>

Post weld heat treatment (PWHT) is primarily used to reduce the level of residual stresses present in a welded joint. The reduction of residual stresses will lead to a reduction of brittle fracture susceptibility and improvements in fatigue resistance, stress corrosion cracking and dimensional stability. Such treatments are carried out at temperatures below those where phase transformations take place. For ferritic steels this is below Ac_1 and is normally between 550°C and 625°C. By elevating the temperature the yield stress is lowered allowing stresses to redistribute via creep deformation. Where residual stresses are essentially uniaxial, PWHT is effective in relieving them. However, for thick section welds where residual stresses are multiaxial the situation is not clear. Residual stresses, measured after PWHT, in thick welds indicate that relaxation may not be complete, with peak tensile residual stresses up to 40% of yield remaining [40]. Longer holding times may be required, however, this could result in other, detrimental changes to the properties of the weldment.

Apart from the problem of reheat cracking exhibited by some alloys, as discussed earlier, other problems associated with PWHT can include loss of yield strength, and loss of toughness by embrittlement from precipitation of Nb, V, Ti (C-N) in microalloyed steel weldments or carbide formation and coarsening in weldments from the decomposition of martensite-austenite phase [41]. In general, PWHT results in improvement in weld properties apart from the reduction of residual stresses. Toughness is improved by tempering of martensitic structures in alloy steel welds and by a softening of ferrite in ferritic welds. In some Al alloys PWHT may be a solution treatment followed by a precipitation treatment to restore full strength to precipitation strengthened alloys. In other Al alloys, this is unnecessary as the precipitation strengthening occurs at ambient temperature.

The subject of post weld heat treatment and stress relaxation of welds is extensive and is an active area of research in two commissions of the International Institute of Welding (IIW). The reader is referred to the literature for further information on the topic and to the review presented by Salkin [42] at the 1987 Annual Assembly of IIW.

REFERENCES

1. K. Satoh, M. Toyoda, F. Minami. <u>Trans. Japan Welding Soc.</u>, <u>16</u>, (1), 70, (1985).
2. K. Satoh, M. Toyoda, F. Minami, M. Nakanishi, K. Arimochi and S. Satoh. <u>Trans. Japan Welding Soc.</u>, <u>16</u>, (2), 74, (1985).
3. K. Satoh and M. Toyoda. <u>Int. Inst. of Welding</u>, Doc. X-1113-86, (1986).
4. T. Wada and G.T. Eldis, <u>ASTM STP 755</u>, 343 (1982).
5. R.S. Chandel, R.F. Orr, J.A. Gianetto, J.T. McGrath and B. Wright. <u>Proc. Int. Conf. on Welding for Challenging Environments</u>, Pergamon Press, Toronto, 97, (1985).
6. M. Tokuhisa, Y. Hirai, N. Nishiyama, I. Yamashita, K. Nishio and K. Nakatsuyi. <u>Kawasaki Steel Tech.</u>, Rep. No. 15., (1986)
7. E. Macherauch and H. Wohlfahrt. <u>Proc. Int. Conf. on Residual Stresses in Welded Construction and their Effects</u>. Welding Institute, Abington, Cambridge, (1978).
8. K. Satoh. <u>Trans. Japan Weld. Soc.</u>, <u>1</u>, (1972).
9. A.P. Parker. <u>The mechanics of fracture and fatigue: An introduction</u>, E.&F.N. Spon, London, (1981).
10. R.H. Leggatt. <u>Proc. 6th Int. Conf. Fracture</u>, New Delhi, Pergamon Press, Oxford, (1984).

11. O.L. Towers and M.G. Dawas. Proc. ASTM Symposium on User's Experience with Elastic-Plastic Fracture Toughness Test Methods, Louisville, Kentucky, (1983).

12. A.H. Cottrell and B.A. Bilby. Proc. Phys. Soc. Land., 62A, 49, (1949).

13. J.D. Baird. Metall. Rev. 149, 1, (1971).

14. R.C. Cochrane, P. Terry and J.G. Garland. Weld Metal. Fabric, 44 (4), 316 and 44 (6), 439, (1976).

15. C.F. Berkhaut and S.H. Van den Brink. Weld. Metal. Fabric, 46 (4), 316 and 44 (6), 439, (1976).

16. D.E. McRobie and J.F. Knott. Metal. Sci. Technol., 1, 357, (1985).

17. R.E. Dolby. Metal. Constr., 13, (1), 43, (1981).

18. G.W. Dawson and P. Judson. Proc. 2nd Int. Conf. Offshore Welded Structures, Paper 2. Welding Institute, Abington, Cambridge, (1982).

19. J.L. Robinson. Weld. Inst. Report 41/1977/M, (1977).

20. J.T. McGrath, J.A. Gianetto, R.F. Orr and M.W. Letts. Can. Met. Quarterly, 25 (4), 349, (1986).

21. C.A. Hippsley, J.F. Knott and B.C. Edwards. Acta Metall., 28, 869, (1980).

22. R.A. Ricks, P.R. Howell and G.S. Barritte. J. Mater Sci., 17, 732, (1982).

23. J.H. Tweed and J.F. Knott. Metal. Sci., 17, 45, (1983).

24. J.F. Knott. Fundamentals of Fracture Mechanics, Butterworths, London, (1973).

25. P. Judson and D. McKeown. Proc. 2nd. Int. Conf. Offshore Welded Structures, Paper 3. The Welding Institute, London, (1982).

26. G.M. Evans. Weld. Res. Abroad, 29 (1), 24, (1983).

27. G.M. Evans. Weld. Res. Abroad, 29 (1), 35, (1983).

28. G.M. Evans. Weld. Res. Abroad, 29 (1), 46, (1983).

29. G.M. Evans. Weld. Res. Abroad, 29 (1), 58, (1983).

30. M.E. Harasawa, M. Hano and W.E. Bremmar. Proc. 2nd Int. Conf. Offshore Welded Structures, Paper 20, The Welding Institute, London, (1982).

31. G.M. Evans. Weld. Res. Abroad, 29 (1), 13, (1983).

32. G.M. Evans. Weld. Res. Abroad, 29 (1), 2, (1983).

33. Y. Ito and M. Nakanishi. The Sumitono Search, May (15), 42, (1976).

34. J.R. Still and J.H. Rogerson. Metall. Constr., 10 (7), 339, (1978).

35. K. Shinmyo, S. Suitô, N. Mori, T. Takami and R. Kôno. Doc. No. XII-A-138-77 (1978).

36. R. Kihno, T. Takami, N. Mori and K. Nagano. Weld. J., 61 (12), 373s, (1982).

37. J. Tsuboi and J. Terashima. Weld. in the World, 12 (11/12), 304, (1983).

38. N. Mori, H. Homma and S. Okita. Int. Inst. of Welding, Doc. No. IX-1196-81, (1981).

39. K. Easterling. Introduction to the Physical Metallurgy of Welding, Butterworths, Micrographics in Metals, 6, (1983).

40. R.H. Leggett. Proc. IIW Int. Conf. on Stress Relieving Heat Treatments of Welded Steel Constructions, International Institute of Welding, Pergamon Press, Oxford, (1987).

41. J.E.M. Braid and J.A. Gianetto. Proc. Int. Conf. on Welding for Challenging Environments, Ed. by Welding Institute of Canada, 245, Pergamon Press, Oxford, (1985).

42. R.V. Salkin. Proc. IIW Int. Conf. on Stress Relieving Heat Treatments of Welded Steel Construcitons, International Institute of Welding, Pergamon Press, Oxford, (1987).

MECHANISMS AND MECHANICS OF FATIGUE CRACK INITIATION AND GROWTH

T.H. Topper, M.T. Yu and D.L. DuQuesnay

Department of Civil Engineering, University of Waterloo.
Waterloo, Ontario. Canada. N2l 3G1

ABSTRACT

This paper reviews the mechanisms by which a fatigue crack grows and the metallurgical and loading variables that influence these mechanisms.

Fatigue crack growth regimes for constant amplitude fatigue are characterized as a metallurgically short crack regime, a mechanically short crack regime and a long crack regime. In the metallurgically short crack regime the crack is short compared to metallurgical variables, such as the grain size, and growth is strongly affected by microstructure. In the second regime the crack is long enough that the resistance to growth by microstructural barriers is averaged out, but it is not long enough for crack closure due to the plastic wake of the crack to have reached a stable level. During this stage the crack grows faster than a long crack having a similar linear elastic stress intensity. In the final long crack regime crack closure has reached a steady state level and the crack growth rate can be characterized by the range of stress intensity and the stress ratio independent of crack length.

The effect of stress or stress intensity level, mean stress or stress ratio, microstructure and crack closure on crack growth in each of these regimes is examined. The manner in which crack growth resistance to short cracks develops is shown to lead to short cracks in sharp notches which stop propagating at low stress levels, and to reduced fatigue concentration factors for small notches and defects.

Work illustrating some of the ways in which variable amplitude loading changes crack closure, thresholds, and other characteristics of fatigue behaviour is reviewed. Particular attention is paid to closure changes that cause current fatigue analysis procedures to give unconservative life predictions.

KEY WORDS

Fatigue; crack growth; crack closure; short cracks; notches; variable amplitude loading.

INTRODUCTION

Current fatigue analysis, excluding applications involving elevated temperatures and corrosion, can be divided into three different approaches, each of which is used for solving a distinct type of fatigue problem. In the first approach, aimed at the prevention of fatigue crack initiation, critical areas of a component are chemically, thermally or mechanically treated to increase the local hardness and to introduce compressive residual stresses. The success of these treatments relies on the fact that the fatigue limit for metals increases with hardness and compressive mean stress. The design criterion is that peak stresses which may occur millions of times during a component's lifetime should not exceed the fatigue limit so that fatigue cracks do not initiate.

The second approach deals with components where, for economic reasons, peak stresses at machined notches or other stress raisers are allowed to exceed the fatigue limit at occasional high load levels. In this approach the local stresses and strains at the root of the stress raiser are first determined for an estimate of the service load history to be experienced by the component. Adopting the conservative assumptions that the strain range at the notch root prevails as the crack advances into the component, and that most of the fatigue life is expended in the early stages of crack growth, fatigue lives are determined by equating the notched component fatigue life to the total fatigue life of a small smooth laboratory specimen subjected to the same stress-strain history. Typically, this equivalent life is determined using strain-life curves for smooth specimens and a linear cumulative damage hypothesis that ignores load level interaction effects. The design criterion is that the fatigue life of the component determined by the analysis should exceed the desired service life by a selected factor of safety. This approach, often called the "critical location approach", has proved to be satisfactory for machined components where notches are not too sharp, although load interaction effects due to certain load histories give unconservative results.

The third approach, based on linear elastic fracture mechanics (LEFM) of crack growth, is applied to crack growth problems and to sharp notches and crack-like flaws. Material crack growth data from laboratory tests in the form of crack growth rate versus range of stress intensity, ΔK, or more recently the effective stress intensity (the maximum stress intensity, K_{max}, minus the crack opening stress intensity, K_{op}), is used to determine the crack growth for a given computed ΔK history in a component under service loading. This approach has proved to be powerful and versatile, although K values for some crack front shapes require considerable computation and methods of determining crack closure stress and crack opening values of stress intensity, K_{op}, are not yet adequate for service load histories. The design criterion used with this approach is again an adequate factor of safety on service life.

Work in the past decade has led to a better understanding of the early stages of the fatigue process that bridge the gap between crack initiation and fracture mechanics studies. It is now generally agreed that the processes by which a crack initiates and grows through the first few grains are fundamentally different from those of the long crack described by fracture mechanics. In addition to the first regime, termed the "microstructurally short crack regime", in which the average stress required for continued crack propagation is the fatigue limit and microstructure plays a dominant role, there is a second regime often termed the "physically short crack regime" in which crack growth can occur at stress levels below the fatigue limit and below the threshold stress intensity for long cracks before cracks behave as "long" fracture mechanics governed cracks. This work, coupled with recent results concerning crack closure and the threshold stress intensity, will be reviewed in this paper.

In order to fit the large body of new work into a simplified framework useful to designers, the constant amplitude fatigue behaviour of long cracks, short cracks, and notches are discussed in that order. A final section indicates some of the phenomena caused by load interaction effects in variable amplitude loading and suggests some simplified design approaches.

LONG CRACKS

Figure 1 gives an example of the plot of fatigue crack growth rate versus stress intensity range on logarithmic scales typically used to describe low and intermediate rates of fatigue crack growth. A log-linear Paris Law (Paris, Gomez and Anderson, 1961) portion of the curve describes crack growth at levels of ΔK above the "threshold regime" but below the static fracture regime (not shown in the figure) in which crack growth accelerates as the fracture stress intensity is approached:

$$da/dN = A \, \Delta K^m \tag{1}$$

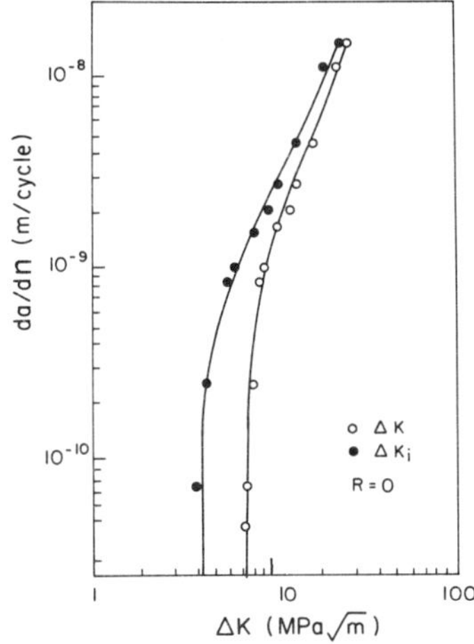

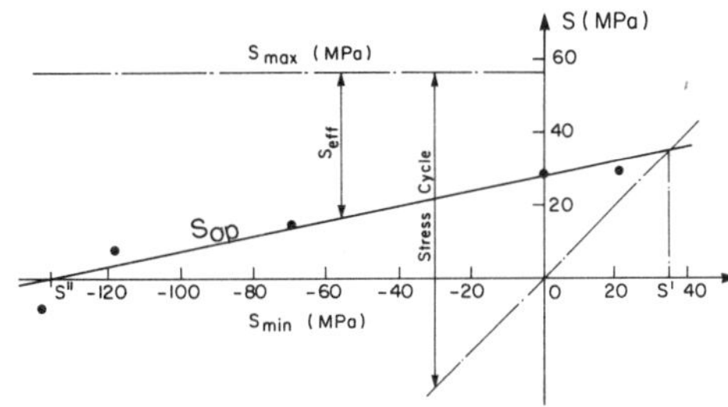

Fig. 1. Crack Propagation Rate vs. Stress Intensity Factor for a SAE1010 Steel

Fig. 2. Variation of Crack Opening Stress with Minimum Stress for a SAE1010 Steel

Where A and m are constants in equation (1). At low levels of ΔK in the threshold regime the crack growth rate curve falls below the rates given by the Paris Law and approaches zero at the threshold stress intensity, ΔK_{th}. The ΔK curve in Fig. 1 for a ratio of minimum to maximum stress, R, equal to zero gives the minimum crack growth rate in terms of the positive portion of ΔK. Higher or lower R values give curves having higher crack growth rates until at high and very low R values an upper bound on crack growth rates (given in Fig. 1 as the ΔK_i "intrinsic" curve) is reached.

If environmental influences on near threshold crack growth in certain metals are excluded, most of the variation in crack growth rate with R value can be attributed to changes in crack closure. When an effective stress intensity, ΔK_{eff}, is calculated using only the part of the K cycle during which the crack is open at the tip, and ΔK_{eff} is used to correlate crack growth rates, data for all values of R fall into a band along the intrinsic curve of Fig. 1.

Crack Closure

Elber (1970) first measured crack closure at positive stress levels above the minimum stress due to the plastic wake left behind the tip of an advancing crack. He suggested that only the range of stress above the crack opening stress, S_{op}, was effective in promoting crack growth and he introduced the corresponding effective stress intensity range, ΔK_{eff}. Subsequently, other mechanisms that cause closure including surface roughness caused by crack growth along crystallographic planes at low stresss intensities and a build up of corrosion products in the crack have been identified, and crack closure and the effective stress intensity have proved to be useful in explaining most of the phenomena observed in propagation of long cracks. Schijve (1986) gives an excellent review of crack growth mechanisms and the significance of crack closure.

For the restricted condition of constant amplitude loading considered in this section, the crack opening stress increases with increasing minimum stress as shown in Fig. 2. When the minimum stress in the stress cycle is equal to S″ the crack will open at zero stress. The crack opening stress increases continuously as the minimum stress is increased reaching about one half of the maximum stress when the minimum stress is zero and falling below the minimum stress at S′ typically at an R value of about 0.6 . Solutions for crack opening stresses (eg. Newman, 1981; Ibrahim, Thompson and Topper, 1986) due to a plastic wake show trends similar to those of Fig. 2, as do experimental results by the authors for various metals (Yu and others, 1986). In these experiments, crack opening loads even below zero stress and crack growth under fully compressive load cycling (maximum stress near zero) were observed in an aluminum alloy at high compressive values of the minimum stress.

Figure 2 shows that the fraction of the positive portion of the stress cycle (which by ASTM definition is used in calculating ΔK) that is effective is a minimum when the minimum stress and R are zero and increases with higher and lower R values until at values of $S_{min} > S'$ the full stress cycle is effective, and at $S_{min} < S''$ (if such a high compressive stress can be reached in a material before gross yielding) part of the compressive portion of the stress cycle may also be effective. When closure measurements have been made and effective stress intensities obtained, crack growth data for all R values tend to fall on the intrinsic curve of Fig. 1. This curve can, therefore, be used to predict constant amplitude crack growth rates for all stress ratios if measured or calculated values of S_{op} are available so that ΔK_{eff} can be obtained. It also provides an upper bound on crack growth rates in variable amplitude loading in which closure levels may vary on a cycle by cycle basis.

Threshold Stress Intensity

The threshold stress intensity, ΔK_{th}, which has as a lower bound the threshold, ΔK_i, of the intrinsic curve of Fig. 1, is the stress intensity below which a fatigue crack will not grow. Figure 3, which defines the various K values (equivalent definitions also apply to S values) indicates that the threshold stress intensity is made up of a closure component, ΔK_{cl}, and an intrinsic component, ΔK_i. The effective stress intensity associated with the intrinsic curve is equal to the difference between K_{max} and the greater of K_{cl} or K_{min}.

Figure 4 gives the form of the relationship between ΔS_{th} and S_{min} typical of many metals. The authors have obtained experimental results showing this trend of ΔS_{th} and the corresponding trend in K_{th} for a variety of steels with various microstructures, and commercial aluminum alloys (Yu, Topper and Au (1984), Yu and Topper (1985)). The intrinsic component of the threshold stress intensity

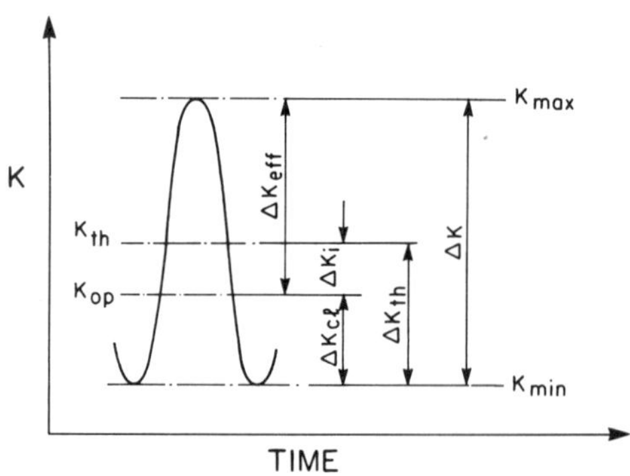

Fig. 3. Definition of K Values

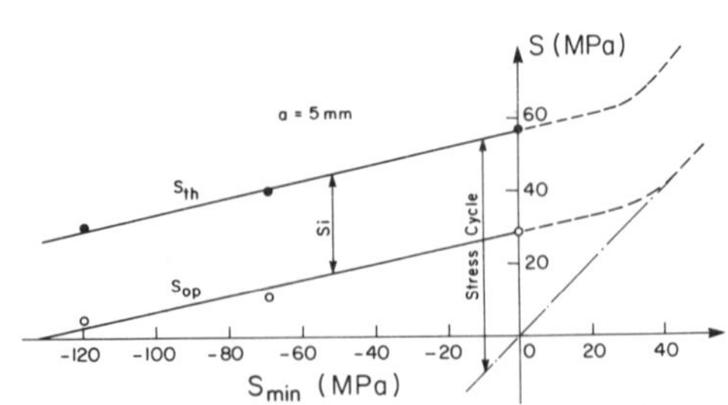

Fig. 4. Variation of Threshold Stress and Opening Stress with Minimum Stress for SAE1010 Steel

remains sensibly constant for all stress ratios. The closure component varies as indicated in Fig. 2. Again, as might be expected from Fig. 2, the curves of Fig. 4 show that at high stress ratios the closure stress falls below the minimum stress and the threshold stress intensity is equal to ΔK_i.

Although the trend in ΔK_{th} with stress ratio seems to be general, some experimental evidence is available that as the load is shed in a K-decreasing determination of ΔK_{th}, the opening stress and K_{op} can increase with a tendency for K_{op} to approach K_{max} at the threshold (Minakawa and McKevily, 1982). This observation was attributed to a change in growth mode from Mode I to Mode II at low values of ΔK. In cases where there is such a change to microstructure sensitive crystallographic crack propagation, roughness due to faceted surfaces may increase crack closure. There will still be a minimum value of ΔK_{th}, but the closure contribution may be higher than suggested by Fig. 4.

As mentioned earlier, the intrinsic curve of Fig. 1 appears to represent an upper bound on crack growth rate which will be decreased by the closure component of ΔK. In design, equating calculated values of ΔK and ΔK_{eff} in making crack growth calculations will give conservative fatigue life predictions. This greatly simplifies design, since it eliminates the extensive analytical calculations or experimental measurements required to estimate closure stresses for variable amplitude loading.

MICROSTRUCTURALLY SHORT CRACKS

Figure 5, which plots the threshold stress for fully reversed constant amplitude loading crack growth on logarithmic coordinates of threshold stress versus crack length, as first suggested by Kitagawa and Takahashi (1976), shows three regimes of crack growth. In the regime of long cracks (zone 2B) the threshold stress is governed by the fracture mechanics ΔK_{th}. However, as the crack length becomes shorter (zone 2A) the threshold stress increases but falls further and further below the value associated with ΔK_{th} until, at the end of this zone of "physically small" cracks, it reaches the fatigue limit for the material. At shorter lengths, (zone 1), "microstructurally short cracks" may grow temporarily at stress levels below the fatigue limit but will stop at barriers.

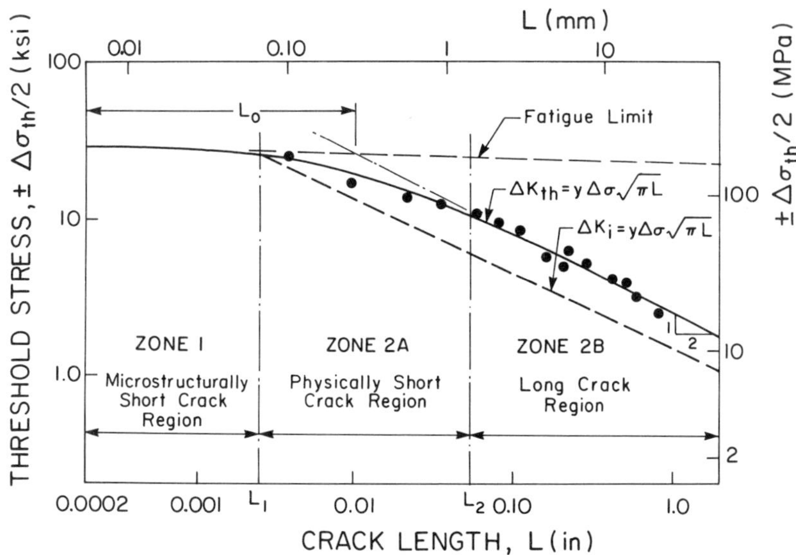

Fig. 5. Variation of Threshold Stress with Crack Length for CSA G40.21 Steel

Crack initiation in smooth specimens has been found to occur at intermetallic particles and other material defects and at grain boundaries, as well as by the classic mechanism of extension along a slip band. Initial growth is usually crystallographic and associated with local plastic cyclic strains. As the stress level increases initiation of microcracks takes place at more locations as more grains experience reversed plastic deformation.

The microstructural conditions associated with crack initiation are a defect, a slip band or a grain boundary that acts as a stress raiser located in a grain which, by virtue of its lattice orientation and shape, has a high shear stress. Observations have shown that, once initiated, the crack grows at an initially high rate but slows down as the crack tip approaches a grain boundary or other metallurgical barrier. In some cases (Lankford, 1982) the decrease in crack growth rate is repeated at the first few grain boundaries while in the case of a pearlite lath in steel only the first barrier gave a pronounced decrease in crack growth rate, (Hobson, Brown and de los Rios, 1986). These various kinds of barriers to microcrack growth establish the fatigue limit for a metal and give rise to a length, shown as L_1 in Fig. 5, over which continued crack growth occurs only at stress levels at or above the fatigue limit stress.

Extent of the Microstructurally Short Crack Regime

The length, L_1, which defines the end of zone 1 in Fig. 5 and the metallurgically short crack regime, seems to have a magnitude of at most a few spacings of the characteristic metallurgical barrier to crack growth. This seems reasonable because once a crack has traversed a few barriers the crack front will extend over many microstructural units and the resistance provided by barriers should become an average one leading to a nearly uniform average crack growth rate. From a macroscopic point of view it appears that, as shown in Fig. 5, a good estimate of L_1 is provided by the intersection of the fatigue limit stress (σ_e) with the threshold stress given by the long crack intrinsic threshold stress intensity ΔK_i. Kendall, James and Knott (1986), showed that this macroscopic measure fitted the extent of zone 1 well for various steels. Blom and co-workers (1986) reported similar results for high strength aluminum alloys. There seems to be an inconsistency between the use of a linear elastic fracture mechanics long crack value of ΔK_i to determine threshold stresses and the high stress levels at the fatigue limit which for the steel data of Kendall, James and Knott (1986) would be of the order of the macroscopic cyclic yield stress. However, as these authors pointed out, the suggestion of Miller (1984) that at stresses above an amplitude of a third of σ_{cy} large increases in plastic zone size would invalidate LEFM did not seem to be true for their results. They found that when compared using elastically calculated values of ΔK_{eff} long and short cracks grew at the same rate for stresses less than the fatigue limit and concluded that the differences between long and short cracks in zones 2A and 2B were primarily due to crack closure.

Fraction of the Fatigue Life Spent in the Microstructurally Short Crack Regime

Recent observations have confirmed earlier suppositions that the fraction of the fatigue life spent in crack initiation represented most of the fatigue life at stress levels near the fatigue limit but decreases rapidly to a small fraction as stress levels are increased. Hobson, Brown and de los Rios (1986), who modelled crack growth in zone 1 and subsequent crack growth separately, predicted and verified experimentally for a medium carbon steel that zone 1 crack propagation occupied a rapidly decreasing life fraction with increasing stress and concluded that for their tests at lives up to 70,000 cycles the life fraction in zone 1 could be ignored. Other investigators, such as Blom and co-workers (1986), found that at stresses near the fatigue limit millions of cycles and most of the fatigue life could be expended in propagating a crack past the first few grain boundaries. Models (Radhakrishnan and Mutoh, 1986), and microscopic observations (Lankford, 1982), relate crack advance through metallurgical barriers to cyclic plastic yielding caused by a cyclic shear stress sufficient to reinitiate a crack in a grain past a boundary or other barrier. It seems then that the conditions necessary for crack advance through zone 1 are closely related to those for local cyclic plasticity and that stresses above the fatigue limit will greatly increase cyclic plasticity in grains ahead of the crack tip and the rate of crack advance.

Figure 6, which gives a strain life curve for smooth specimens together with curves giving the number of cycles spent in zone 1 (N_i) and in post zone 1 crack propagation, illustrates the trends mentioned above. At high strain levels the fraction of the fatigue life spent in zone 1 is negligible, but zone 1 propagation dominates the fatigue life as the strain amplitude approaches the fatigue limit.

Figure 7, which is plotted on a strain versus life basis, illustrates the effect of a few large prestrain cycles, (10 cycles at ±1%) on subsequent constant amplitude fatigue. The fatigue limit of the constant amplitude curve, and presumably zone 1 crack propagation, is removed by the prestrain cycles leaving a fatigue curve determined by post zone 1 propagation and a new fatigue limit characteristic of the material's resistance to propagation of the crack initiated during prestraining.

Dependence of the Fatigue Limit on Mean Stress

Although, as the previous section has shown, cyclic plastic straining due to high stress levels can rapidly drive a crack through zone 1 making constant amplitude fatigue limits of limited importance to finite life fatigue design, they are of great interest to designers concerned with preventing crack initiation. Figure 8 shows the changes of the maximum stress at the fatigue limit with minimum stress as well as the fatigue limit stress cycle given by the range between the minimum stress and the maximum stress for a high strength 2024-T351 aluminum alloy and for an SAE1045 steel. The fatigue limit stress range shows the expected increase as the minimum stress becomes more compressive. For moderate tensile mean stresses the trends generally follow the mean stress rules suggested by Morrow (1968) or Smith, Watson and Topper (1970), also shown in the figure. For compressive mean stresses the steel, for which the fatigue limit stress is approximately equal to the cyclic yield stress, tends to follow a curve with a constant stress range equal to the cyclic yield stress presumably because gross cyclic yielding rapidly drives cracks through the initial barriers. The aluminum alloy, for which the macroscopic yield stress is almost three times the fatigue limit, shows a much more gradual decrease of the maximum fatigue limit stress and a greatly increased fatigue limit stress range as the minimum stress is decreased.

Of particular importance to designers who intend to improve the fatigue resistance of critical locations in components via hardening and the introduction of compressive residual stresses is the observation that any cyclic plasticity will rapidly initiate cracks. Thus material conditions which lead to cyclic softening under repeated stressing should be avoided.

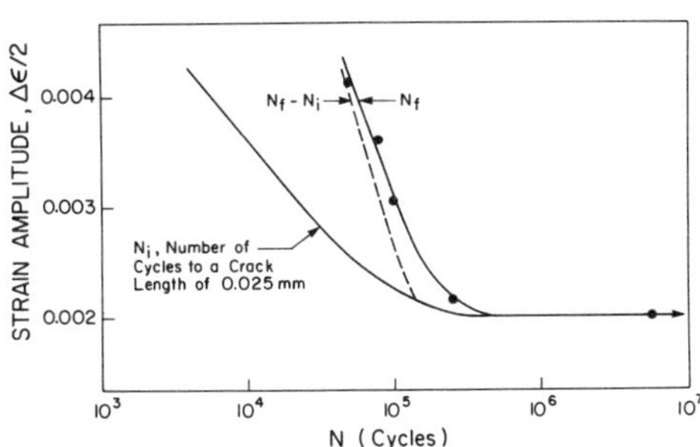

Fig. 6. Typical Strain-Life Diagram For a Low Carbon Steel (Tokaji, Ogawa and Harada, 1986)

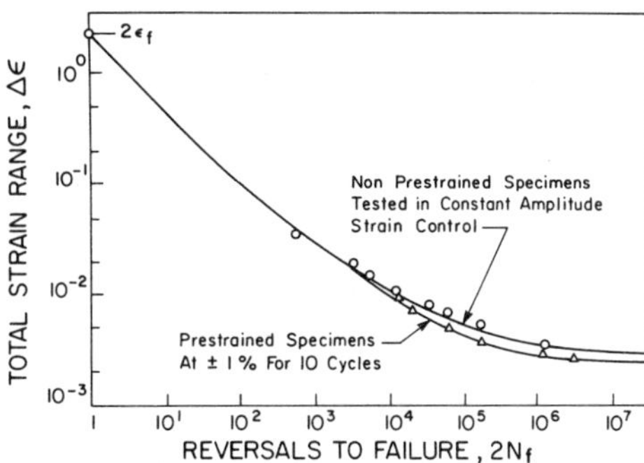

Fig. 7. Effect of Prestrain on the Strain-Life Behaviour of CSA G40.21 Steel

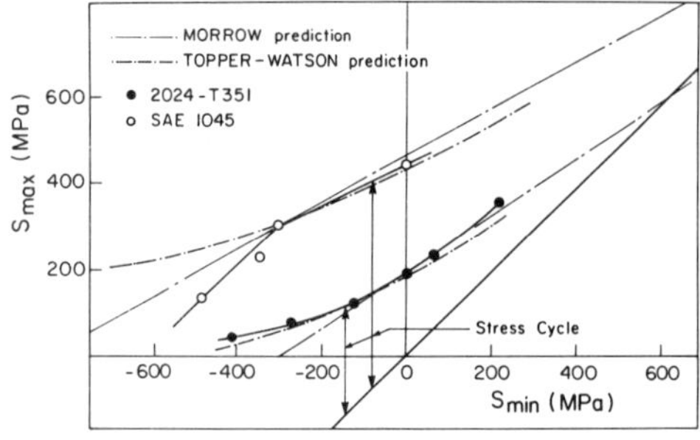

Fig. 8. Effect of Mean Stress on the Fatigue Limit

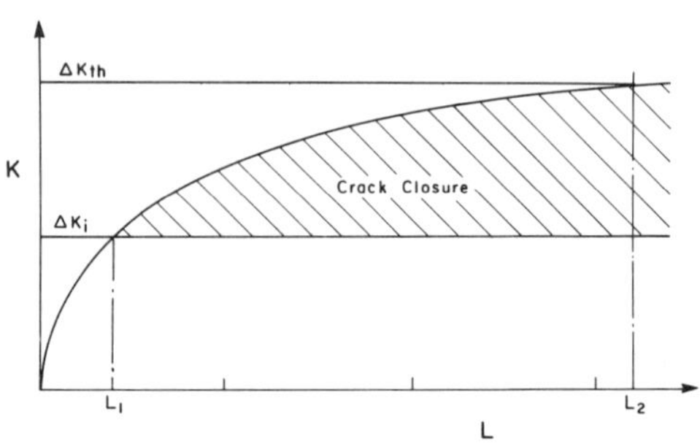

Fig. 9. Variation of Threshold Stress Intensity Factor with Crack Length

PHYSICALLY SHORT CRACKS

Cracks in zone 2A of Fig. 5 are termed "physically short" and propagate at fatigue threshold stresses below both the fatigue limit and the long crack threshold stress given by K_{th}. As noted earlier, Miller (1984) suggested that an upper limit, L_2, be placed on this region at a stress range of one third of the yield stress because higher stresses would give significant errors in LEFM solutions used for long cracks. Taylor (1986), also using criteria for the validity of LEFM, suggested that L_2 be defined by the greater of 10 times the metallurgical barrier spacing or 10 times the grain size. On the other hand, Kendall, James and Knott (1986) and Blom and and co-wokers (1986) found that for steels and high strength aluminum alloys respectively, experimental evidence indicated that LEFM analysis was adequate if closure was taken into account. Perhaps the emphasis on LEFM applicability is not very important since a strain based stress intensity (Topper and Haddad, 1979) which becomes identical to the LEFM K for elastic strains and identical to J at inelastic strain levels (Haddad and others, 1980) adequately correlated crack growth in zone 2A with that of long cracks.

Some crack closure has been noted even in zone 1 (Nisitani and Takeo, 1981; Lankford, Davidson and Chan, 1984), but due to differences in growth mode it is not clear whether its effect is comparable with that of closure in the Mode II growth relevant to longer cracks. Blom and co-workers (1986); Kendall, James and Knott (1986), and Soniak and Remy (1986) all found that measureable closure began at about L_1 and showed an initially steep buildup that tapered off to a steady state value for long cracks at L_2. When these closure measurements were used to deduce effective stress intensities for the short cracks, data for long and short cracks fell together on the intrinsic curve for long cracks. These authors all suggested that once closure was accounted for the long crack regime could be considered to begin at L_1. It appears then that zone 2A is part of the long crack regime (zone 2B) and that closure might be considered an artifact of constant amplitude long crack growth. Topper and Haddad (1979) used the length parameter L_0, shown in Fig. 5, to approximate the increase in threshold stress intensity with crack length, as shown in Fig. 9 and given by:

$$\Delta K_{tht} = \Delta K_{th}\sqrt{\frac{L}{L_0 + L}} \qquad (2)$$

where ΔK_{tht} is the threshold stress intensity at a given length, and ΔK_{th} is the long crack threshold stress intensity. This empirical expression gave a good fit to measured ΔK_{th} values for short cracks. In the form:

$$\Delta\sigma_{th} = \frac{\Delta K_{th}}{Y\sqrt{\pi(L+L_0)}} \tag{3}$$

where Y is the LEFM shape factor for a short crack and $\Delta\sigma_{th}$ is the threshold stress, equation 3 gave a good fit to measured threshold values for short cracks.

Stress Ratio
===

Since crack growth in zone 2A seems to differ from that of long cracks mainly in that closure is still building up to its steady state value in this regime, it is expected that the influence of stress ratio on closure for long cracks noted in Fig. 2 would be apparent here. This seems to be the case. Various authors including Fathulla, Weiss and Stickler (1986), and Blom and co-workers (1986), have noted that the limit for L_1 occured at the intersection of the threshold stress given by the effective stress intensity for long cracks and the maximum stress at the fatigue limit and thereafter crack growth was controlled by the effective stress intensity for long cracks. It was noted by these investigators that L_2 values, which are governed by the amount by which closure increases ΔK_{th} above ΔK_i, varied with stress ratio as might be expected.

Overview
===

In spite of the various mechanisms governing the initiation and propagation of metallurgically short cracks in zone 1 and the various forms of closure that may affect the behaviour of physically short cracks in zone 2A it appears that macroscopic simplifications are possible. The first region in which the crack grows at the fatigue limit stress probably represents a very small fraction of the fatigue life in finite life fatigue. The level of the fatigue limit as a function of mean stress seems to roughly follow conventional rules up to the point of cyclic yielding. The remainder of the fatigue life can be dealt with by conventional fracture mechanics techniques if closure can be calculated for the load history of interest. If the designer is uncertain about closure variations, Kendall, James and Knott (1986) suggest that taking L_1 as an initial crack length and calculating crack growth using the intrinsic growth curve of Fig. 1 will give conservative fatigue life predictions.

NOTCHES

The geometry constant, K', which gives the stress concentration of a crack in a notch, shown plotted against crack length divided by notch depth in Fig. 10, is equal to the theoretical stress concentration factor, K_t, at the notch root surface and becomes equal to the K' of a crack of a length equal to the combined length of the crack and that of the notch depth (also shown in the figure) at a small fraction of the notch root radius. Propagation of a crack through the length, L_1, of Fig. 5 requires that the stress in a constant amplitude test be equal to the fatigue limit up to that length. If a notch is blunt, so that K' at a length L_1 is little different than K_t, a fatigue limit stress in the notch approximately equal to the material fatigue limit divided by K_t will be adequate. If, on the other hand, the notch depth, C, becomes very small then a much higher stress will be required since the stress will decrease significantly over the distance L_1, otherwise crack propagation will stop when the crack tip stress reaches the fatigue limit. A crack may also arrest in zone 2A in which the threshold stress intensity is increasing with crack closure, if at some point the applied ΔK becomes less than ΔK_{tht}. The conditions under which this may occur are shown in Fig. 11.

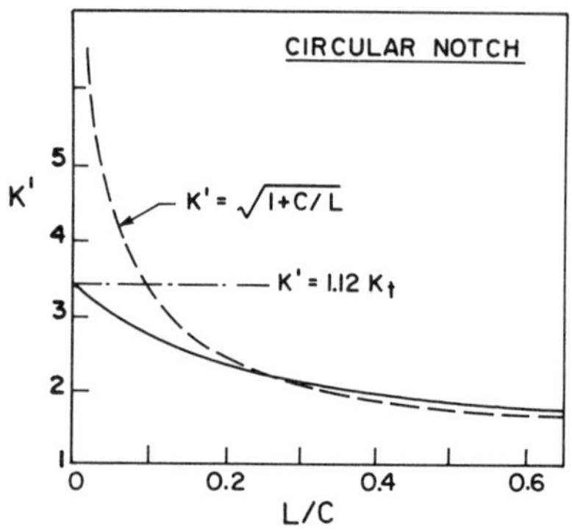

Fig. 10. Stress Concentration (K′) for a
Crack in a Notch

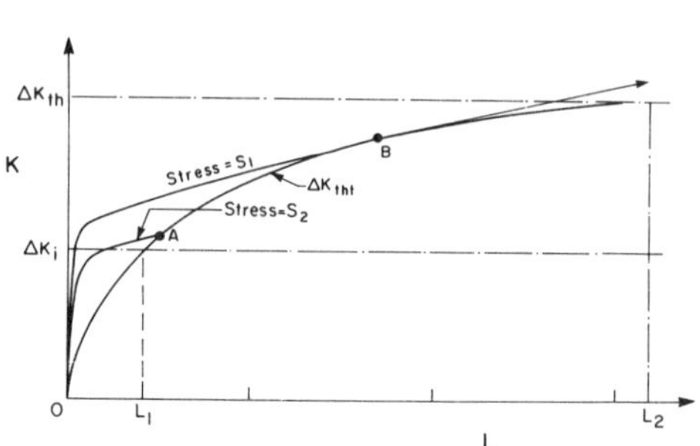

Fig. 11. Variation of Stress Intensity Factor
with Crack Length for a Crack in a
Notch

Figure 11 shows the lower boundary for crack growth defined by the fatigue limit stress up to a crack length L_1 and thereafter by the threshold stress intensity which increases from ΔK_i at L_1 to ΔK_{th} at L_2 in accordance with Fig. 5. Also shown are ΔK vs. L curves for two stress levels, ΔS_1 and ΔS_2, determined from a K′ solution such as that of Fig. 10.

$$\Delta K = K'\Delta S \sqrt{\pi L} \tag{4}$$

At ΔS_2 the stress level is high enough to drive a crack through zone 1, but the crack is arrested as increasing crack closure in zone 2A increases the threshold stress intensity above the applied ΔK. The minimum stress for continuous crack propagation, S_1 (ΔS_p), gives a ΔK vs. L curve that becomes tangent to, but does not fall below, the threshold curve. Solutions for crack propagation in notches using an estimation of the threshold stress intensity given by equation (3) have shown that notches can be divided into "blunt" notches in which the stress necessary to initiate a crack is sufficient to propagate it to failure, and "sharp" notches in which a higher stress, ΔS_p, is required for continuous propagation. The boundary between sharp and blunt notches occurs at a notch root radius equal to about $5L_0$ (DuQuesnay, Topper and Yu, 1986) for fully reversed elastic constant amplitude loading.

The same approach indicates that the fatigue notch factor, K_f, defined as the ratio of the fatigue limits of smooth and notched specimens, decreases from equality with the theoretical stress concentration factor, K_t, for blunt notches to a value of unity for small flaws as shown by equation (5).

$$K_f = 1 + \sqrt{C/L_0} \tag{5}$$

<u>Notch Fatigue Analysis for Finite Life</u>

The fatigue behaviour described above is of interest to designers wishing to prevent crack initiation where constant amplitude fatigue limits for notches are important. However, in finite life fatigue it is assumed that the design will allow a crack to initiate and be propagated at least by the high load excursions in a service history through all the zones of Fig. 5. The design criterion is an adequate excursions in a service history through all the zones of Fig. 5. The design criterion is an adequate service life for a component. Two of the approaches outlined in the introduction, namely the "critical location approach" and fracture mechanics crack growth calculations, have been applied to notch fatigue analysis with the former being more frequently applied to blunt notches. The latter has been the preferred approach for sharp notches and flaws. For both analyses load level interaction effects associated with closure changes in variable amplitude loading, which will be discussed in the next section, present complexities beyond present modelling capabilities (Schijve, 1986), for other than simple crack front shapes. However, an overview of current procedures is useful as a background to the more complex variable amplitude problems.

Critical location approach. Based on the assumptions that the notch is blunt and that the K' value of Fig. 10 decreases slowly enough with crack length that the value of $K_t = K'$, shown in the figure, applies over a length equal to the length of a crack in a laboratory smooth specimen at failure, the fatigue life of the notched component is calculated as if the uncracked notch root strains were applied to the smooth specimen. Cyclic plastic deformation models for notch plasticity and metal stress-strain response have been developed (Conle, Oxland and Topper, 1987) which allow notch root stress-strain histories to be calculated. Individual cycles in the local history are then resolved and the fraction of the life expended in each cycle is calculated by comparison with the fatigue life of a smooth specimen with the same strain range and mean stress. A summation of fractions of fatigue life expended, or damage, equal to unity indicates fatigue failure of the component. An extension of this approach to crack growth predictions for simple notches, given in the same article, proceeds in a similar manner except that notch plasticity solutions are updated as the crack advances and an elastic-plastic strain intensity is calculated as the crack driving force. Crack growth for closed cycles of strain intensity range is then calculated using crack growth data.

Fracture mechanics analysis. If notches or flaws are sharp enough they may be treated as cracks since the length over which K' in Fig. 10 differs from the long crack approximation is negligible. In design, stresses are usually low enough that LEFM considerations apply and, given a K solution for the geometry of interest, crack growth is calculated using data such as that of Fig. 1 for the K history corresponding to service loads. The designer may also apply LEFM to blunt notches using a starting crack length of L_1 from Fig. 5 if stresses are elastic. However, since high service stress levels usually cause local plasticity in blunt notches, inelastic K solutions and elastic-plastic fracture mechanics approaches, such as the one discussed above, are necessary in the notch plastic zone.

VARIABLE AMPLITUDE LOADING AND FATIGUE DESIGN

The preceding sections have shown that the effective stress intensity appears to be the parameter controlling fatigue crack growth for cracks of length greater than the length at which the microstructurally short crack regime ends (L_1 of Fig. 5). Methods of determining crack opening stresses necessary for calculating ΔK_{eff} for a given cycle in a variable amplitude load history are, however, still in a developmental stage. Schijve (1986) in his review of crack closure points out the complexities of crack closure for three dimensional crack fronts with varying degrees of constraint along the crack front. Further complications he asseses are contributions to closure from roughness due to faceted surfaces in some metals and corrosion products in others, as well as changes in crack growth mechanisms that sometimes occur at low stress intensities. Analytical models such as that of Newman (1981) have been developed to the point of predicting crack closure stresses due to a plastic wake, even for short cracks (Newman, Swain and Philips, 1986), but require calibration for specific metals. A combination of experimental closure data and calculations has been useful in other cases for given load histories and metals (ASTM STP748, 1982).

The following sections are aimed at designers without access to the experimental facilities to perform variable amplitude tests and who are lacking already validated analytical models suitable for their problems. The way in which closure changes due to high load levels accelerate and retard crack growth will be reviewed, and some conservative approximations will be suggested for use in design against fatigue.

<u>Crack Closure and Crack Growth Rate Changes Following Overloads.</u>

A tensile overload in a constant amplitude fatigue test increases the plastic zone size and the tensile stretch in front of the crack tip. The crack opening stress is initially reduced, because the plastically deformed material ahead of the crack tip tends to keep the crack open, but the opening stress increases above the steady state level as the crack grows into the deformed material and it appears behind the crack and increases the size of the wake. The increased closure has been measured and correlated with the delay in crack growth and with the arrest that follows tensile overloads when the overload is much higher than the constant amplitude load. This delay, which may increase fatigue life significantly, has been taken into account in design cases where regular high tensile overloads are characteristic of the structure's duty cycle.

Referring to Fig. 3, we see that opening stress decreases more or less linearly with decreasing minimum stress to zero at a value of S″ which it was noted was of the order of the yield stress. As a consequence, early investigations using compressive stresses that, like the usual tensile stresses in crack growth tests were a small fraction of the yield stress, indicated that the compressive part of the stress cycle had little effect on crack growth. In components that experience primarily compressive stresses and in cracks growing out of notches, however, the compressive stresses can be a high fraction of the yield stress (often above it in notches). In these cases, compressive stresses accompanying high load levels may decrease the height of the plastic wake behind the crack tip thereby reducing closure and S_{op}. A period of greatly increased crack growth rate follows until closure again builds up. Repeated compressive overloads can maintain low average crack closure levels and drastically shorten component fatigue life.

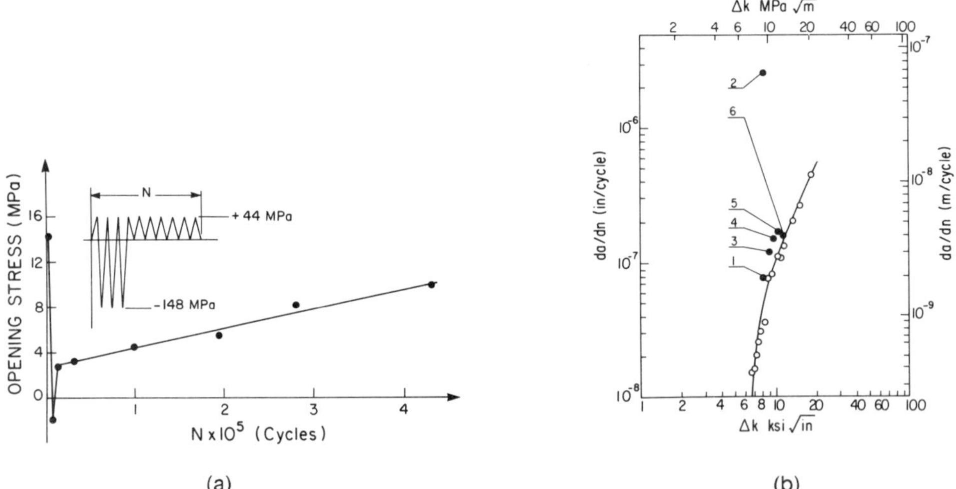

Fig. 12. Effect of Compressive Underloads on (a) Crack Opening Stress and (b) Crack Growth Rate

Figure 12 shows the reduction in crack opening stress and its recovery following compressive underloading and the accompanying acceleration in crack growth in a constant amplitude crack growth test on an SAE1010 steel (Topper and Yu, 1985). Figure 13 gives crack growth data for a test series in which a high compressive stress was applied after each η small R=0 stress cycles, with η being varied from test to test. As the frequency of application of the underload was increased (η decreasing) the crack growth curves shift towards the intrinsic curve, given by $\eta = 1$, in which every cycle included the high compressive peak stress. Closure measurements indicated, as expected, a continuous decrease in average values of S_{op} as the frequency of underload application is increased.

Similar results were obtained for steels of various microstructures and for high strength aluminum alloys. A similar phenomenon of accelerated crack growth when intermittent high compressive underloads are applied during constant amplitude fatigue testing of smooth specimens results in decreased fatigue lives and lowered fatigue limits for the fully reversed cycles even after subtracting the life fraction consumed by the compressive underload cycle. Data for an SAE1045 steel, given in Fig. 14, illustrate this trend (DuQuesnay and others, 1987). Similar results were obtained for a 2024-T351 high strength aluminum alloy. It is interesting that, like the crack growth curves used for crack growth calculations, smooth specimen fatigue life curves used in the critical location approach to fatigue analysis tend toward a limiting value as the compressive underloads become more frequent.

Design Considerations

We have seen that the base curves used in fatigue calculations in both the fracture mechanics approach and the critical location approach can be shifted to minimum levels given by the intrinsic curve of Fig. 1 and the lower bound curve of Fig. 14, respectively, by frequent compressive underloads causing compressive stresses near the yield point. Service load histories will seldom be this severe, but as shown by Conle and Topper (1980), appropriate strain-life curves, at least for service load histories, may be much lower than those usually used. It seems reasonable that designers of structural components for which the consequences of fatigue failure would be catastrophic and for which prototype testing is impractical, will choose to be conservative in their selection of the base curve. On the other hand, previous experience with prototype testing for similar load histories, access to laboratory test data, the existance of well developed closure analyses, or the design of components that are not critical may lead an engineer to less conservative assumptions. Another useful result of recent short crack studies is that the length, L_1, of Fig. 5 seems to have emerged as the appropriate

minimum starting crack length for fracture mechanics crack growth calculations in the absence of defects. Many of the problems associated with the shape of crack front development in varying stress fields and their influence on crack closure and load history interaction effects have not yet been accurately modelled. However, a fairly clear overview of the fatigue process from the initiation of a crack through the various stages of growth, and of the methods of calculating lower bounds on fatigue life have already emerged from the extensive work in recent years on fatigue mechanisms and mechanics.

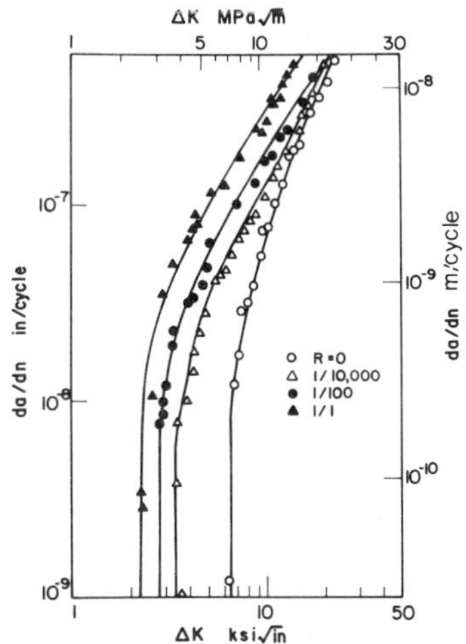

Fig. 13. Effect of Intermittent Compressive
Underloads on Crack Growth Rate

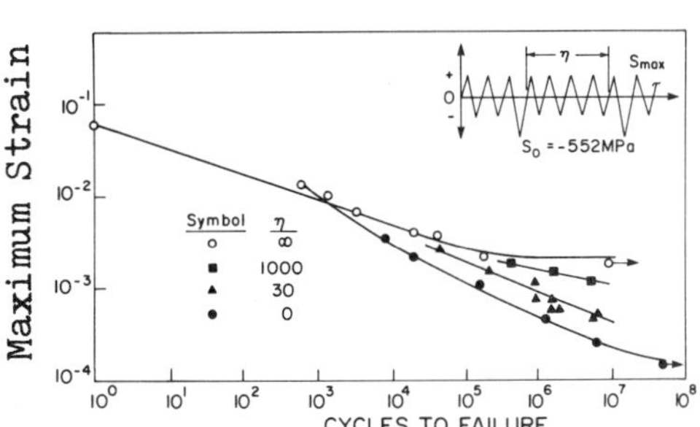

Fig. 14. Effect of Intermittent Compressive
Underloads on Fatigue Life

REFERENCES

Blom, A. F., A. Headland, W. Zhao, A. Fathulla, B. Weiss and R. Stickler (1986). In K. J. Miller and E. R. de los Rios (Eds.) The Behaviour of Short Fatigue Cracks, E.G.F. 1, Mechanical Engineering Publications, London. 37-66.

Conle, A. and T. H. Topper (1980). Int. J. Fatigue. July. 130-135

Conle, A., T. R. Oxland and T. H. Topper (1987). In Low Cycle Fatigue, ASTM STP942. American Society for Testing and Materials.

DuQuesnay, D. L., T. H. Topper and M. T. Yu (1986). In K. J. Miller and E. R. de los Rios (Eds.) The Behaviour of Short Fatigue Cracks, E.G.F. 1, Mechanical Engineering Publications, London. 323-336.

DuQuesnay, D. L., M. A. Pompetzki, T. H. Topper and M. T. Yu (1987). Low Cycle Fatigue, ASTM STP942. American Society for Testing and Materials.

Elber, W. (1970). Engineering Fracture Mechanics, Vol. 2. 37-45.

Fathulla, A., B. Weiss and R. Stickler (1986). In K. J. Miller and E. R. de los Rios (Eds.) The Behaviour of Short Fatigue Cracks, E.G.F. 1, Mechanical Engineering Publications, London. 115-132.

Hobson, P. D., M. W. Brown and E. R. de los Rios (1986). In K. J. Miller and E. R. de los Rios (Eds.) The Behaviour of Short Fatigue Cracks, E.G.F. 1, Mechanical Engineering Publications, London. 441-460.

Ibrahim, F. K., J. C. Thompson and T. H. Topper (1986). Int. J. Fatigue, Vol. 8, No. 3. 135-142.

Kendall, J. M., M. N. James and J. F. Knott (1986). In K. J. Miller and E. R. de los Rios (Eds.) The Behaviour of Short Fatigue Cracks, E.G.F. 1, Mechanical Engineering Publications, London. 241-258.

Kitagawa, H. and S. Takahashi (1976) In Applicability of Fracture Mechanics to Very Small Cracks or Cracks in the Early Stages. Proc. ICM-2, ASM. 627-631.

Lankford, J. (1982). Fatigue of Engineering Materials and Structures, Vol. 5. 223-248.

Lankford, J., D. L. Davidson and K. S. Chan (1984). Metallurgical Transactions, Vol. 15a. 1579-1587.

Miller, K. J. (1984). Proc. Eshelby Memorial Symposium. Sheffield. 477-500.

Minakawa, K. and A. J. McKevily (1982). In Fatigue Thresholds, Fundamentals and Engineering Applications, Vol. 1. EMAS. 373-390.

Morrow, J. D. (1968). In Fatigue Properties of Metals - Fatigue Design Handbook. Society of Automotive Engineers, Section 3.2.

Newman, J. C. (1981). In Methods and Models for Predicting Fatigue Crack Growth under Random Loading. ASTM STP748, American Society for Testing and Materials. 53-84.

Nisitani, H. and K. I. Takeo (1981). Engineering Fracture Mechanics, Vol. 15. 445-456.

Paris, P. C., M. P. Gomez and W. E. Anderson (1961). The Trend in Engineering, Vol. 13. 9-14.

Radhakrishnan, V. M. and Y. Mutoh (1986). In K. J. Miller and E. R. de los Rios (Eds.) The Behaviour of Short Fatigue Cracks, E.G.F. 1, Mechanical Engineering Publications, London. 87-100.

Schijve, J. (1986). In Proc. Int. Symposium on Fatigue Crack Closure. May 1986, Charleston.

Soniak, F. and L. Remi (1986). In K. J. Miller and E. R. de los Rios (Eds.) The Behaviour of Short Fatigue Cracks, E.G.F. 1, Mechanical Engineering Publications, London. 133-142.

Taylor, D. (1986) In K. J. Miller and E. R. de los Rios (Eds.) The Behaviour of Short Fatigue Cracks, E.G.F. 1, Mechanical Engineering Publications, London. 479-490.

Tokaji, K., T. Ogawa and Y. Harada (1986). Fatigue Fract. Engrg. Mater. Struct. Vol. 9, No. 3. 205-217.

Topper, T.H. and M. H. El Haddad (1979). The Canadian Metallurgical Quarterly, Vol. 18. 207-213.

Topper, T. H. and M. T. Yu (1985). Int. J. Fatigue, Vol .7, No. 3. 159-164.

Yu, M. T., T. H. Topper, D. L. DuQuesnay and M. S. Levin (1986). Int. J. Fatigue, Vol. 8, No. 1. 9-15.

CHAIRMAN: J. MASOUNAVE
Institut de génie des matériaux,
Boucherville

WIDE PLATE FRACTURE TOUGHNESS EVALUATION OF THE WELD HAZ
OF LOW CARBON MICRO-ALLOYED STRUCTURAL STEEL WELDMENTS.

Present status of the art illustrated with a worked example.

Dr. Ir. Rudi M. DENYS
St. Pietersnieuwstraat, 41, B-9000 GENT
Rijksuniversiteit Gent - Belgium.

ABSTRACT

In order to demonstrate that the low HAZ "local fracture toughness" properties as measured in CTOD testing does not always reflect the potential structural performance, extensive use is being made of (surface) notched wide plate tests to assess the significance of such low toughness regions within the HAZ. These regions are normally labeled as local brittle zones (LBZ) and are related with the CGHAZ microstructure.

If any comparison between the results of the CTOD test and the wide plate test is to be valid, it is an essential requirement that the test notch in the wide plate specimen should be fatigue precracked and that utmost care should be exercised to ensure that the fatigue crack tip is located in the coarse grained region of the HAZ. Additionally, a detailed metallographic and fractographic examination is needed if the wide plate test result is to be compared with the CTOD data base. Such post wide plate test metallographic examination of the regions of microstructures sampled by the fatigue crack tip embodies a major deviation from the wide plate testing procedures used previously.

The purpose of this paper is to review the specific notching procedures and the method of post test examinations used currently in wide plate testing. Furthermore, these procedures are extensively discussed and illustrated by a series of fatigue crack tip photomicrographs taken from a 50 mm thick weld HAZ notched wide plate test specimen.

KEYWORDS

Wide plate testing, CTOD, Local brittle zones (LBZ), surface defect.

WIDE PLATE TESTING

As such, the wide plate test is in some way more suited to provide the experimental evidence that occasional low CTOD values do not necessarily imply a critical situation. The rationale behind this observation is related to numerous factors. Essentially, in the wide plate specimen due account can be taken of the differing stress-strain characteristics of the weld metal, the HAZ and the base metal (effect of mutual interaction), whereas it is important to emphasise that the CTOD test is to be considered as a reference test in order to simplify the testing preparations and that (a) this test provides a very much lower bound value of fracture toughness (modelling of the worst possible situation), (b) the test was principally designed for the situation of brittle material behaviour (contained yielding deformation mode at fracture) and (c) the crack depth as well as the crack orientation in the CTOD and wide plate specimens are different [1].

Throughout its development the wide plate test has always been considered as an empirical means of verifying the relevance of the result of any small scale type test. In the early '60s, the wide plate test was extensively used to examine the significance of strain ageing embrittlement in the subcritical HAZ. This wide plate test design, known as the Wells test, contains a longitudinal weld which is mechanically notched prior to welding [2]. Because of the improved steel making practices the problem of strain ageing embrittlement can be eluded, whereas the use of micro-alloyed steels shifted the suspect regions of low toughness towards the transformed HAZ [2-5]. In recognition of this problem, it is then necessary to consider a wide plate specimen design in which the orientation of both the defect and the weld is transverse to loading direction. Two other important implications are that the wide specimen is to be notched after welding, where defect designs can be used which are more representative of the types of defect that are seen in a real structure. In this connection it is to be noted that in more recent times, the fatigue pre-cracked surface notched transversely loaded wide plate test is often used to examine the importance of local brittle zones (LBZ's) when low CTOD values are obtained from procedural tests [6].

The effect of a defect located parallel to the fusion boundary in the HAZ on weld joint performance can be evaluated with a flat wide plate specimen in which the weld is arranged transversely to the applied load. The testing conditions can be made more structure specific when it is of interest to examine the effect of residual stresses (e.g. by means of a cross welded specimen), the effect of geometric discontinuities (e.g. by means of a specimen in which the effect of an angular distortion, misalignment or transverse stifferners are directly integrated), etc .

When it is the aim to produce representative wide plate test data and especially, when it is attempted to establish a correlation between the results of any type of small scale tests and the results of a wide plate test, it is needed to apply an expedient testing procedure. Together with the development of the HAZ notched CTOD testing procedures, the American Petroleum Institute (API), the Welding Institute and several other organisations and oil companies put a lot of effort into the development of specific metallographic post test examinations [7,8]. Similar developments with regard to wide plate testing have evolved in a few European laboratories and, the main aspects of this expertise was summarised recently in an IIW document [9].

Clearly the recommendations which cover the post test metallographic examination requirements [7-9] are essential and should be more widely applied if the results obtained from CTOD and wide plate tests are to be compared. Moreover, the establishment of a uniform and commonly accepted wide plate testing procedure will offer the possiblity to appreciate fully the wide plate test results produced by the various testing laboratories.

LOCATION OF LBZ AND WIDE PLATE TESTING ASPECTS

A truthful comparison between CTOD and wide plate test results, implies that a substantial part of the fatigue crack front in a wide plate test specimen samples the same microstructural features as in a CTOD specimen. As for the CTOD test specimen, the wide plate test welds should have the same weld preparation. Although this is a realistic requirement, it should be realized that the procedural steps to attain that objective are associated with a number of difficulties. The main aspects of these problems are discussed below.

Selection of weld preparation.

The near fusion line or transformed HAZ in C-Mn steels presents a special problem in evaluating the notch toughness because it contains a variety of microstructures each having its own properties [10]. The coarse grained HAZ (CGHAZ) close to the fusion line has potentially, due to its large (coarse) grain size and the nature of its transformation products, the lowest toughness. In a multipass weld, part of the CGHAZ produced by the preceeding weld bead is substantially modified in lower temperature reheated CGHAZ regions which are identified as LBZ's and consist of (a) the intercritically reheated CGHAZ (ICCGHAZ) and (b) the subcritical reheated CGHAZ (SCCGHAZ). The entire extent of any LBZ in a multipass weld i.e. its width as well as its height is dependent on the detailed chemistry of the steel, the heat input, the bead size and the weld bead geometry. Between the (double) coarse grained regions produced by succeeding weld beads one will also recognise regions of grain refined CGHAZ which have good notch toughness properties.

The pre-occupation of the user is that the LBZ's cause a further reduction in notch toughness of the original CGHAZ. Because of both the complex shape and the position of the LBZ regions, and because the width of the grain coarsened HAZ is less than 0,5 mm, there is no doubt that exact placement of a test notch in the weld LBZ's will be difficult. To facilitate an easy placement of the fatigue crack tip of a surface notch in the LBZ regions of a wide plate test panel, it is advised to use a straight and perpendicular fusion boundary (i.e. a single V or a K weld preparation) and to use a strictly controlled welding procedure [10]. Although the use of a "special" weld geometry/ procedure increases the likelihood of sampling larger LBZ portions, it is to be emphasized that such artificially generated LBZ regions create a potential problem in that the test results might not be relevant at all when no similar conditions (i.e. weld joint preparation, bead placement and heat input) occur in real welds [6,11].

Fusion boundary profile.

The impossibility to deposit a weld bead with a consistent cross section (shape) over larger lengths implies that variations in the position of the LBZ regions will occur in three different directions i.e. along the length of the weld, in the through thickness and in the transverse (perpendicular to the weld) direction. Deviations of more than 1 mm are not exceptional [10]. As a result of this, a correct placement of a test notch in a wide plate specimen (as well as in a CTOD specimen) will be difficult and, it is quite possible that an area of grain coarsening aimed for on the basis of the outer (end) macrospecimens (see Fig. 1) may not exist in the area of the central portion of the test specimen. This implies already that when the waviness of the weld bead/fusion boundary contour is taken into account, it is not possible for a straight fatigue crack to occur consistently in the CGHAZ. Only small LBZ portions will be sampled by a fatigue crack tip, unless the coarse grained regions are both very long and very wide.

Fatigue crack propagation

In a wide plate test specimen it is not always possible to maximize the depth of the machined starter notch so as to prevent fatigue crack deviation from the appropriate LBZ regions. The experimental information reported in [11], has illustrated that (a) the plane of fatigue crack growth can give rise to a pronounced deflection at the junction between the weld bead and the HAZ, (b) the plane of fatigue crack extension in HAZ regions can take up any direction with respect to the fusion boundary and (c) there is the tendency for the occurrence of crack growth towards the parent material side of the weld.

On the basis of this experience it can be concluded that, when the combined effects of crack path wandering and the variation of the weld bead contour (in the direction perpendicular to the fusion boundary, along the length of the weld as well as in the through thickness direction) are taken into account, the use of a

simple straight notch in wide plate testing implies that the LBZ regions can be
completely missed, and special measures are needed to allow for a proper sampling
of the LBZ regions.

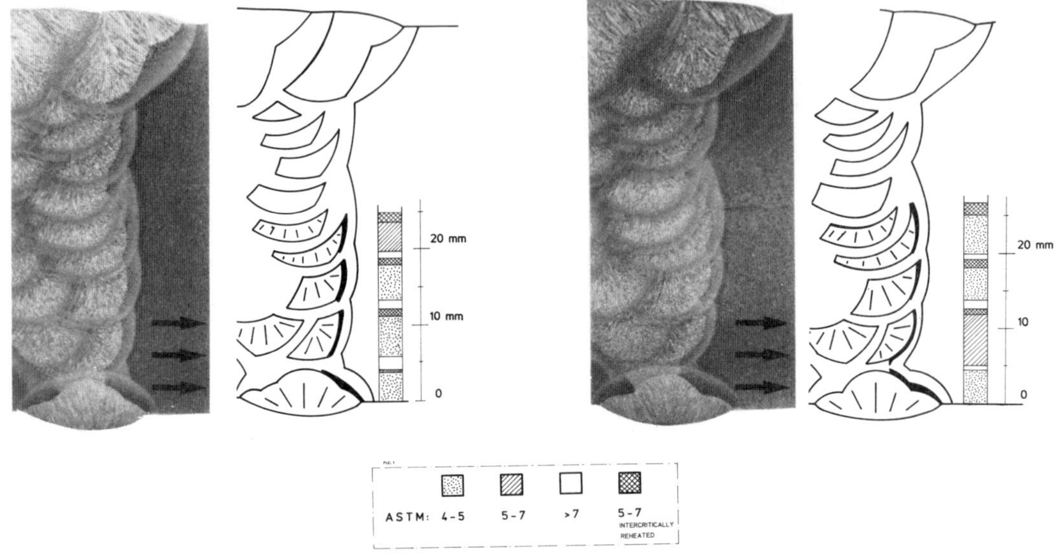

Fig. 1. End macrographs of the weld root area with corresponding bar chart
        of HAZ micostructures along the fusion boundary.

DEFECT DESIGN

Whilst testing of a through thickness defect is (a) not so structurally relevant,
and (b) it is extremely difficult, if not impossible to sample a significant
amount of LBZ portions in a wide plate test specimen, it is justifiable to confine
the following considerations to the orientation of the crack likely to occur in a
welded structure. In this connection, the use of a surface breaking fatigue sharp-
ened defect is considered to be the most appropriate defect design to simulate the
effect of natural cracks on weld joint structural performance.

In order to ensure the leading edge of a surface defect samples representative
amounts of LBZ's (this region is not exclusive with regard to the proposed
notching procedure), it is needed to make a proper design of the mechanical
starter notch configuration. A selection can be made between either a multiple
step or a saw tooth shaped crack. The multiple step notch consists of a series of
individual notches so that the distance between their longest axes is varied in
proportion to the width of the coarse grained region (Staggered or echelon notch)
– Fig. 2a. The saw tooth shaped notch consists of a series of short shallow
notches, placed in such a way that their longest axes make a slight angle to the
assumed plane (AB, see Fig. 4) of the notch (Zig-zag notch) – Fig. 2b.

The step or zig-zag starter notch geometries are preferred over a straight notch
because they increase the probability of sampling the desired LBZ's in the through
thickness or crack depth direction. Moreover, when the depth of the starter notches is also slightly varied, an irregular fatigue crack front can be produced. The
combined measures, i.e. staggering or zig-zagging and crack depth variation will
then produce a highly irregular crack front because of the overlapping of each
individual starter notch during fatigue pre-cracking. On the other hand, it is
felt that the step or zig-zag crack may model real fatigue cracks as well.

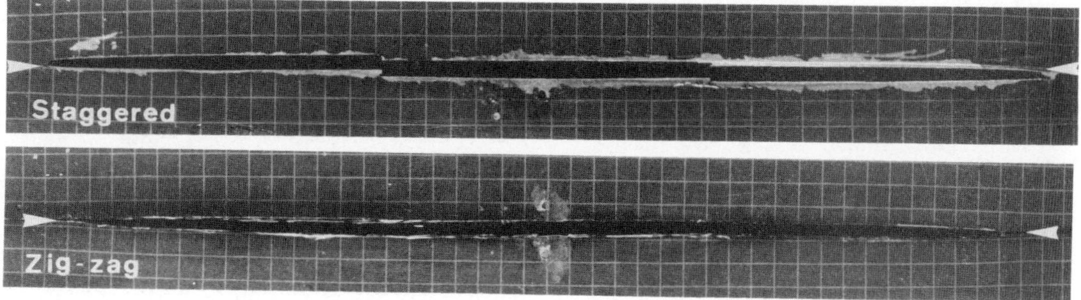

Fig. 2. Extent of crack zig-zagging/staggering at the plate surface (distance between grid lines : 5 mm). Photographs were taken after testing.

Examples of the alternative notch designs are shown in Fig. 3. It can be appreciated from Fig. 3 that the extent of zig-zagging/staggering is less than 1,5 mm. The photographs further illustrate the non-uniform shape of the fatigue crack front, whereas it can further be observed that by adding a shallow starter notch to the crack ends, it is quite easy to sample the CGHAZ regions of the cap layer.

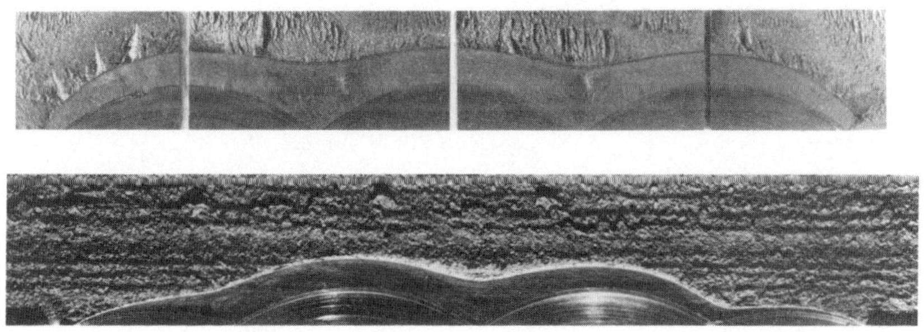

Fig. 3. Examples of fatigue crack profiles (crack depth approx. 15 mm)

NOTCH POSITIONING IN A WIDE PLATE SPECIMEN.

The experimental problems which arise in placing the fatigue crack tip in the CGHAZ and whose length and depth would be fairly close to the predetermined target values, may be overcome by following a careful experimental procedure.

Since the regions of low toughness in the through thickness direction are not known in advance, a detailed metallographic examination will be needed before the test specimen can be notched. For that purpose, two cut-off macro specimens (polished to a 1 µm finish and etched in 2 % nital) are to be taken to identify the position of the LBZ regions along the straight edge of the weld on each macrograph (Fig.1). It should be noted, however, that the end macro's are extracted several hundreds millimeters away from the future notch position in the centre of the wide plate specimen.

The information extracted from these examinations is usually presented in the form of a bar chart describing the fusion boundary microstructure. Whenever possible, a sketch of the weld bead contours and the adjacent regions of CGHAZ would be made in order to facilitate the selection of the final notch tip location. The macro specimens and the schematic presentation of the LBZ regions are further used for reference. As indicated, Fig.1, the examinations are generally directed towards the root side of the weld because the most prominent regions of grain coarsening are associated with the first deposited weld beads adjacent to the straight sided weld edge. Additionally, the largest LBZ regions are located within the outer 15 to 20 % of the specimen thickness. Conversely, the areas of grain coarsening along the straight edge and towards the cap have normally smaller ligament lengths and are more located towards the centre of the plate thickness.

The position of the final notch tip at both ends of the wide plate specimen is then marked on the fusion line (Fig. 4 - reference marks A' and B'). The relevant distances from the weld metal root (or cap pass) edge to the selected LBZ region are subsequently defined. This action is to be taken at both transverse ends of the wide plate specimen. The position of points A' and B' are referenced to the plate surface and a reference line AB is drawn on the plate surface which joins the two transferred position marks. Having checked that this line is truly parallel to the weld bead, the position of the assumed plane of the notch is scribed on the plate surface in preparation for machining.

Finally, machining of the mechanical starter notch should be performed by means of a sharp cutting wheel (width of 0,15 mm) in order to facilitate fatigue crack initiation in the selected LBZ region. The mechanical starter notches are then fatigue pre-cracked in three- or four point bending.

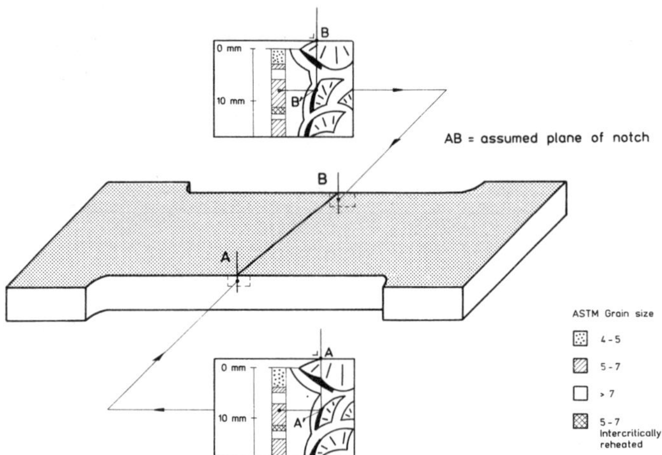

Fig. 4. Recommended procedure for making out the assumed
plane of the notch in a wide plate specimen.

WIDE PLATE TESTING

The wide plate panel is principally to be loaded up to fracture, however, in some instances testing is to be interrupted when the applied elongation reaches the maximum stroke of the testing machine. This corresponds normally to an overall strain in excess of some 2%. Since strains beyond this level have no direct engineering significance, it is normal practice to interrupt the test and not to apply a second straining/loading cycle.

Upon completion of the test, the wide plate test specimen is then sectioned to establish the success rate of the fatigue crack tip sampling position(s). The aspects of the sectioning technique are presented in detail in the next section.

WIDE PLATE SPECIMEN SECTIONING.

As it is virtually impossible to draw any valid conclusion on the significance of the wide plate test performance without demonstrating that the tip of the fatigue crack intercepted the selected LBZ regions, the welded HAZ surface pre-cracked wide plate specimen must be supplemented with additional macrographical and micrographical examinations (post test examination). The extent of examinations is dependent upon the behaviour of the test specimen at the end of test. Distinction is made between the examination of unfractured and fractured specimens.

Unfractured specimen.

When a complete separation of the test specimen has not been achieved, the procedure involves the access to the fatigue crack profile and the subsquent metallographic examinations.

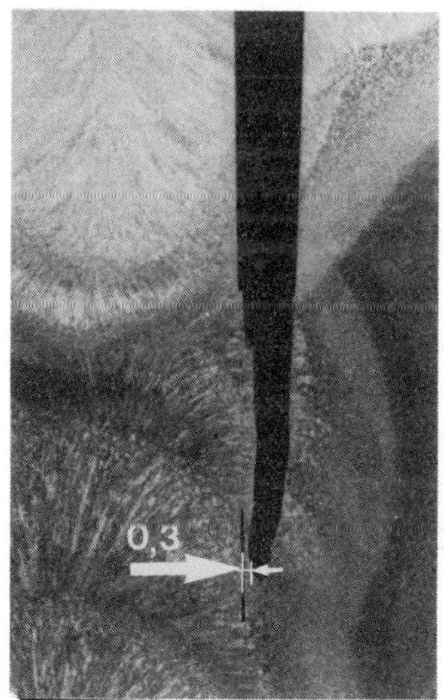
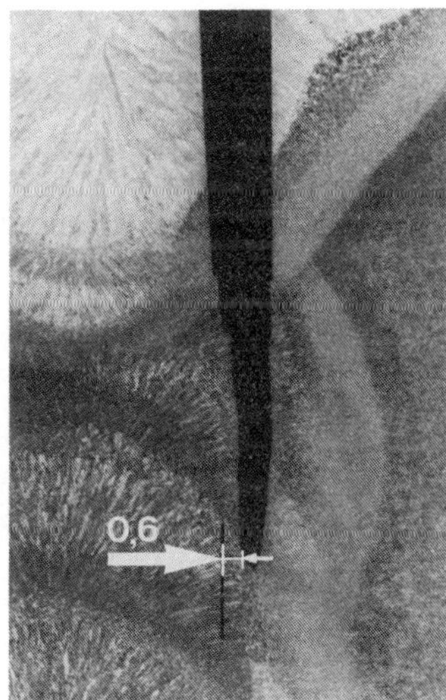
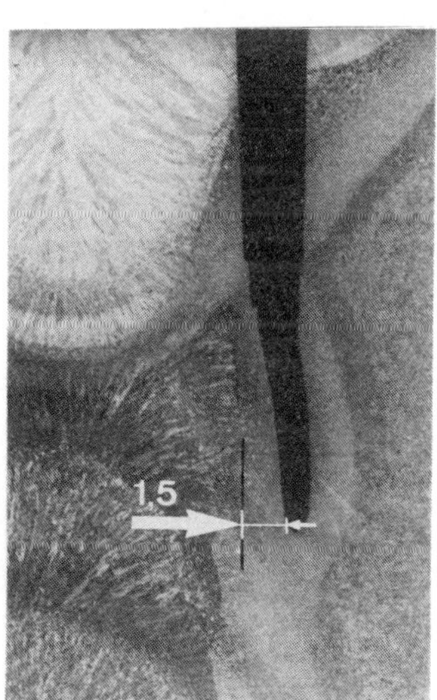

Fig. 5 . Photomacrographs showing transverse sections of an unfractured
wide plate specimen. (Magn. 4x, then reduced to 80% by printing).

For the purpose of this investigation, a coupon encompassing the whole crack is extracted from the test specimen by saw cutting. From both ends of the crack, a full thickness macrosection is prepared to provide a local record of the weld shape and to reveal the position of the coarse grained region which was intended to be sampled. As before the macro specimens are analysed in terms of fusion line microstructures, whereas the results of this examination are compared with those obtained from the end macros used for marking out the fatigue crack position. The coupon containing the test crack is then further sectioned at mid-length of each starter notch (i.e. at the deepest point of each notch). The sections thus obtained are prepared for macro-etching in order to photograph and reveal the position of the crack tip with reference to both fusion line and coarse grained HAZ

at the end of test. A typical example of this is shown in Fig.5. The macro sections were extracted at the deepest points of the mechanical starter notches. The
distance between these sections was 40 mm.

Furthermore, the photomacrographs are used to assist in the selection of the
subsequent microsections to be taken. In this connection, account shall be taken
of both crack path deviation and the original position of the crack starter
notches. Finally, access to the crack profile is achieved by breaking the reduced
sections at liquid nitrogen temperature. At that stage, enlarged photographs of
the fatigue crack profile are taken and detailed dimensional measurements of the
crack front contour are performed. That part of the various sections containing
the weld is subsequently sliced perpendicular to the fracture face at several
locations to trace coarse grained HAZ material. Further details of this procedure,
which is very similar as for the case of a fractured specimen, are given in the
next section.

<u>Fractured specimen.</u>

In the event of fracture (or when access to the fatigue crack profile has been
achieved for unfractured specimens), a photograph of the fracture face (Fig. 6) is
taken on which the fracture initiation point, when identifiable, is indicated. The
identification of this point may involve an examination by SEM.

Fig. 6. Fatigue crack profile after testing.

The weld metal side of the fracture face is then sectioned through the initiation
point and at several other locations on a plane perpendicular to both the fracture
face and the original plate surface. An example of sectioning applied to a fractured specimen is shown in Fig. 7.

All sections selected for the macro- and micro examinations are polished to 1 µm
finish and etched in 2 % nital. During the investigations, emphasis is to be laid
on a quantitative determination of the microstructures present along the length of
the fatigue crack, the grain size at the very tip of the fatigue crack and the
linear extent to which the position of the fatigue crack tip differed from the
fusion line. The importance of this information is illustrated in Figs. 8. The
microphotographs in Fig. 8 are obtained from sampling positions 13, 14 and 16
which are extracted over a length of 5,2 mm (see also Fig.7). These photographs
illustrate clearly the variation of the weld bead profile and the amount of fatigue crack deviation towards the base material. This example emphasizes also the
need to perform a rather detailed sectioning to obtain a complete picture the
microstructures sampled.

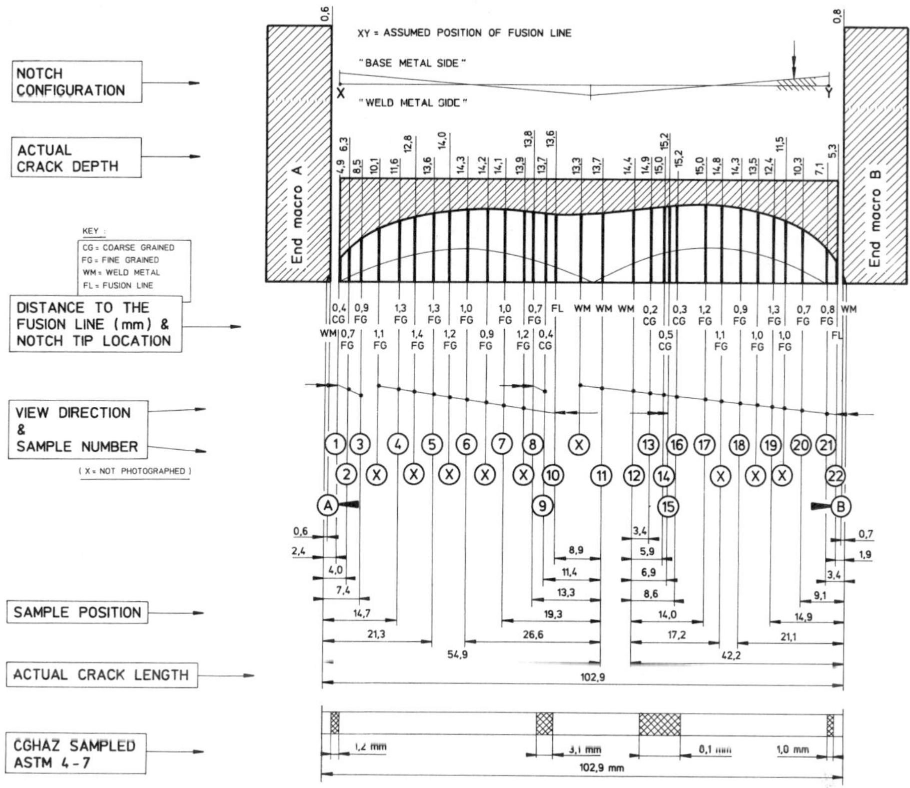

Fig. 7. Worked example of a typical metallographic examination of a fractured specimen with indication of the crack depths and sampling positions supplemented with description of notch tip position.

Furthermore, at every step during the said examinations, photographs at ten or twenty times magnification of typical sampling positions are taken to document the sectioning results. In that respect, such photomicrographs are very instructive in that they illustrate that (a) the fatigue crack tip may sample the plate material as well as the weld metal and (b) even a special notching technique does not warrant a fatigue crack to propagate into the coarse grained HAZ (Fig. 9).

When the grain coarsened HAZ is sampled, it is essential to take a picture of the crack tip area at higher magnification (e.g. 200 times) in order to determine the precise microstructural constituents sampled and to illustrate the amount of crack tip blunting/ductile tearing which occurred prior to fracture initiation. A similar action is to be taken at the point of fracture initiation. The example in Figure 10, shows the amount of plastic deformation and illustrates that the crack tip tore towards the base material side.

Finally, the relative proportions of each LBZ and their locations along the fatigue crack tip are reported. This is conveniently done in the form of a bar chart on which the intercepted CGHAZ regions only are indicated (see bottom of Fig. 7).

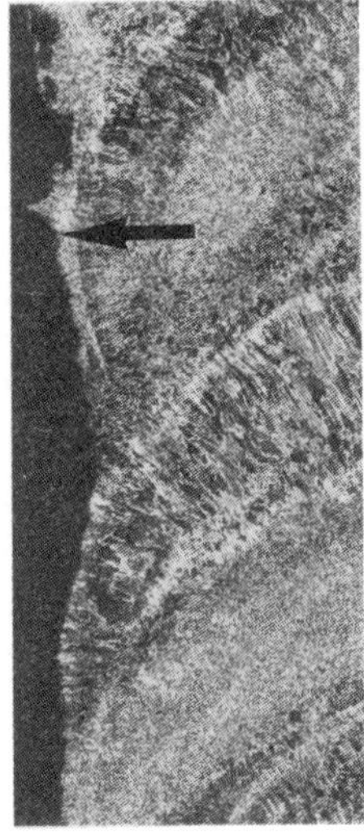 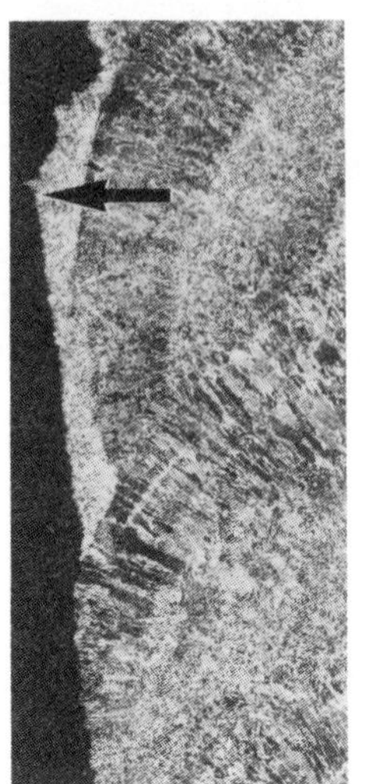 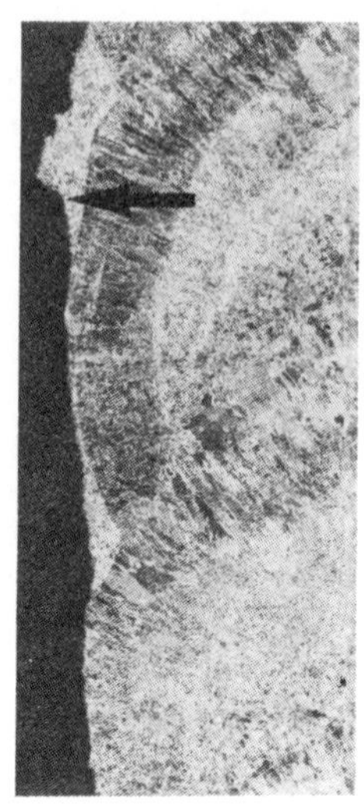

				CGHAZ	CGHAZ	
– Loc.	:	CGHAZ	CGHAZ	CGHAZ		
– Mis.	:	AC+FN	AC+FN	AC+FN		
– CD	:	14,9	15,2	15,0	[mm]	
– DFL	:	0,2	0,3	0,5	[mm]	
– GS	:	5	5–6	6–7	[ASTM]	
– HV	:	238	260	267	[HVO,5]	

Fig. 8. Optical photomicrographs illustrating the fatigue notch position, see arrow, with reference to the position of the CGHAZ region. (Magn. 8x, Loc. = location, CD = crack depth, Mis. = Microstructural constituents, DFL = distance from the fusion line, GS = ASTM grain size)

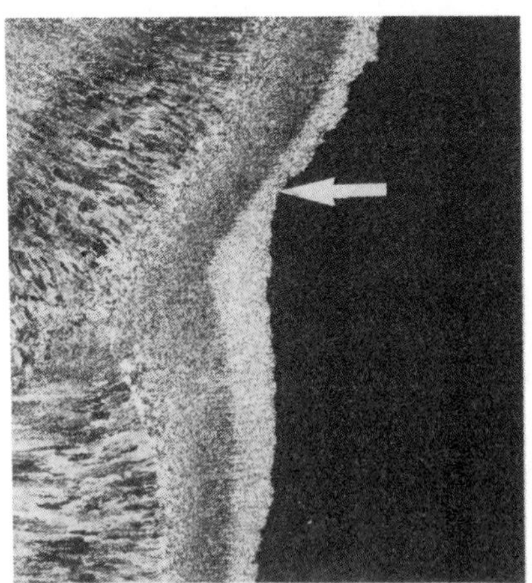 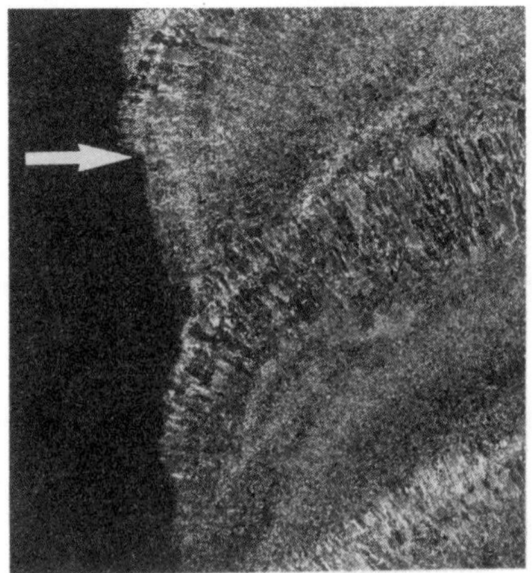

		Base metal	Weld metal	
– Loc.	:	Base metal	Weld metal	
– CD	:	14,4	15,9	[mm]
– DL	:	2,0	1,2	[mm]
– GS	:	11	–	[ASTM]
– HV	:	279	207	[HVO,5]

Fig. 9. Optical photomicrographs showing the position of the fatigue crack in the base material and the weld metal. (Key is given in Figure 8).

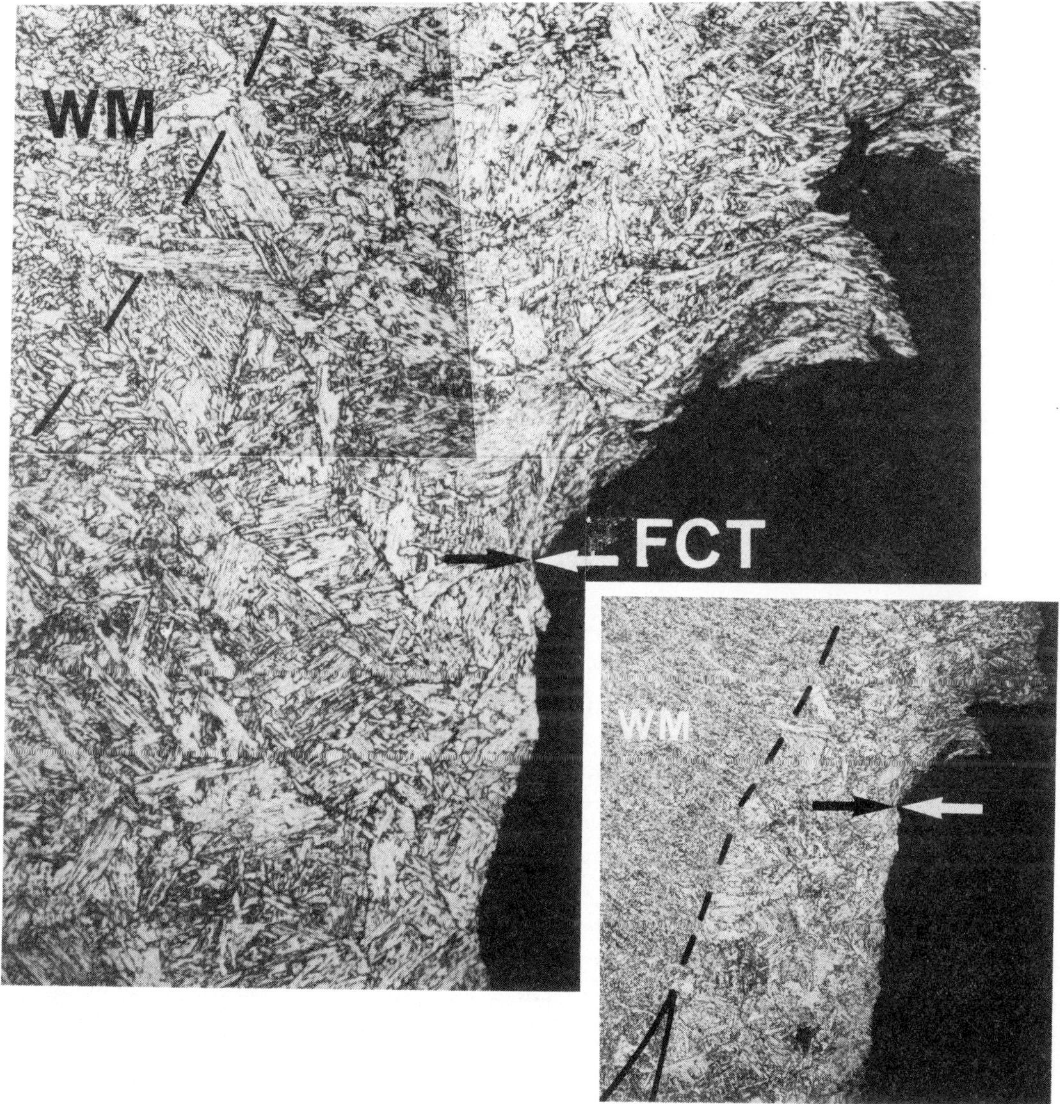

Fig. 10. Photomicrograph illustrating the plastic deformation at the fati-
         gue crack tip (Magn. 50 & 200x, then reduced to 80% by printing).
         (FCT = position of the fatigue crack tip, WM = weld metal)

         CONCLUDING REMARKS.

The results of this study, which has been principally concerned with notching and
sectioning of wide plate test specimens, lead to the following conclusions:

     1. Since the weld bead contour and consequently the positions of the coarse
grained HAZ regions vary considerably along the length of the weld, a careful
notching procedure is needed to locate the test notch in the regions of suspect
low notch toughness.

2. Since the likelihood of sampling a representative amount of LBZ regions by using a straight crack starter notch is small, it is believed that the rate of success of sampling these regions can be enhanced by using a multiple starter notch. The zig-zag notch is to be preferred to the straight notch because it is possible with the former to sample a reasonable amount of LBZ regions in a wide plate test specimen, whereas because of its special shape the measured toughness characteristics can be directly related to the practical situation of real fatigue cracks.

3. When it is the aim to produce believable wide plate test data, it has been shown that because of the apparent variations of the weld bead/fusion boundary profile(s) the crack tip location of a HAZ crack in a wide plate test specimen should be identified upon testing.

ACKNOWLEDGEMENT

The author wishes to thank A. Vinckier, professor and director of the Laboratory Soete for Strength of Materials and Welding Technology, Gent University - Belgium, for permission to publish this paper. The financial support of IWONL and NFWO is also acknowledged.

REFERENCES.

1. Denys, R., (1987). The effect of defect size on wide plate test performance of multipass welds with local brittle zones. Proceedings of the TMS conference - Denver, Febr. 1987. To be published.
2. Pisarski, H.G. and Walker, E.F. (1986). Wide plate testing as a back up to the CTOD approach. - TWI Report for the Department of Energy 3915/4/86.
3.   Royer, C (1987). A user's perspective on heat affected zone toughness. Proceedings of the TMS conference - Denver, Febr. 1987. To be published.
4. Walker, F.E. (1987). Steel quality, weldability and toughness. Steel in Marine Structures, Elsevier Science Publ. B.V. Amsterdam, 1987 (ISBN 0-444-42805)
5. Dolby, R.E. (1979). HAZ toughness of structural and pressure vessel steels - Improvements and prediction. Weld Res. Supp. A.W.S. aug. 1979, pp. 225.
6. Walker, F.E. (1986). Fracture toughness testing. Present status of Charpy V notch impact and CTOD testing. Conf. Proc. The state of the art in materials testing. Gent, Nov. 1986
7. API Specification for Preproduction Qualification for Steel Plates for Offshore Structures. API Spec. 2Z, May 1987.
8. Steel specification for Fixed Offshore Structures (1987). EEMUA Publication No 150. Engineering Equipment and Materials Users Association, London.
9. Denys, R.M. (1987). Tentative guidance notes for wide plate testing. IIW document X- 1138 - 87 (Sofia)
10. Pisarski, H.J. (1983). The CTOD approach to the specification of fracture resistant welded steel structures for arctic service - Fracture Toughness Evaluation of Steels for Arctic Marine Use. Ottawa, Oct. 12 & 13, 1983.
11. Denys, R.M., Dhooge, A. and Lefevre, A.A. (1986). HAZ fatigue precracking of welded plate specimens. Conf. Proc. The state of the art in materials testing. Gent, Nov. 1986.

PEAK AMPLITUDE DISTRIBUTION OF ACOUSTIC EMISSION EVENTS DURING
DELAYED HYDRIDE CRACKING OF Zr-2.5% Nb ALLOY

by K.F. Amouzouvi and L.J. Clegg

Materials Science Branch
Atomic Energy of Canada Limited
Whiteshell Nuclear Research Establishment
Pinawa, Manitoba ROE 1L0

ABSTRACT

During crack propagation in Zr-2.5% Nb due to delayed hydride cracking (DHC), the
crack front moves forward in a stepwise manner when brittle hydride particles,
which have precipitated and grown at the crack tip, fracture under the applied
stress. An amplitude distribution analysis of the associated acoustic emission
(AE) signals was carried out at different stages of crack propagation in hydrided
Zr-2.5% Nb compact tension specimens. For DHC tests performed at temperatures
ranging from 423K to 523K, high amplitude events always occur at low stress
intensity factor (K) values. The amplitude of the events decreases significantly
as K increases. Also, the amplitude increases as the test temperature increases.
This behaviour is consistent with the spacing of striation markings observed on
the fracture surface after the DHC test is completed. The results suggest that AE
amplitude distribution corresponds to the size distribution of the fracturing
hydride particles at the crack tip. The observed stress and temperature
dependence of AE amplitude and the striation spacing can be rationalized in terms
of a critical normal stress required to fracture hydride particles at the crack
tip. The significance of these results in understanding the DHC mechanism is
discussed.

KEYWORDS

Hydride cracking, acoustic emission, amplitude distribution, striation spacing.

INTRODUCTION

Delayed hydride cracking (DHC) in cold worked Zr-2.5% Nb alloy is due to a
diffusion-controlled precipitation and growth of hydrides at a crack tip (under
the action of the stress gradient), followed by rapid crack propagation (Dutton
and co-workers, 1977). The crack-tip advances through the embrittled region and
then arrests in the ductile matrix until a new hydrided zone is formed at the
crack tip. The process repeats itself, in discrete steps, resulting in slow crack
growth. DHC is thus characterized by an incubation period, which represents the
time required for the hydrogen to collect at the crack tip and for hydride

precipitates to nucleate and grow to fulfill a critical condition for their
cracking. Because during DHC propagation the crack front moves forward in
discrete steps, the DHC growth rate can be expressed as (Dutton and co-workers,
1978)

$$v = \frac{da}{dt} = \frac{l_c}{\Delta t}$$

(1)

where l_c is the crack jump length and Δt is the incubation period between successive crack jumps. The latter is controlled by the hydrogen flux to the crack tip,
the kinetics of which has been extensively studied in the past (e.g., Dutton and
Puls, 1976; Simpson and Puls, 1979; Puls, 1984; Ambler, 1984), and thus, the
factors affecting hydrogen diffusion to the crack tip under the action of the
stress gradient are fairly well understood. In contrast, we have only a limited
understanding of the DHC propagation steps (l_c): consequently, the critical
condition that must exist at the crack tip for a propagation step to proceed is
not very well understood. As a result no sound model exists to predict the
threshold stress intensity factor below which no cracking takes place.

Investigations of the crack propagation step during DHC have been largely confined
to measurements of l_c from the regular striation markings, which typify the fracture surface (Simpson and Puls, 1979). In order to extend these observations, we
have undertaken a study of the acoustic emission (AE) that accompanies DHC.

The acoustic emission technique has been increasingly used for detecting and
monitoring cracks in materials either under test or in service. Among methods
used to analyze and quantify AE signals, ringdown counting, which consists of
counting the number of times the amplified AE signal crosses a preset trigger
level voltage (threshold voltage) is by far the most widely used. Despite the
simplicity of ringdown counting, it is difficult to relate it to a single physically meaningful parameter. Also, AE signals carry significant information that
cannot be extracted by this method alone. Nonetheless, this method has been
extensively used by Coleman and Ambler (1977,1978,1983) and by Tangri (1977) in
DHC studies, in particular for crack velocity measurements in zirconium alloys.
In the present work, we have attempted to obtain further information from the AE
data by anlayzing the distribution of signal amplitudes, based on sampling of AE
events (as opposed to ringdowns-see later).

Acoustic emission during DHC crack growth is associated with a rapid release of
stored elastic energy when hydride particles, accumulated and grown at the crack
tip, fracture under the applied stress. Recent studies by Waldley and Scruby
(1979) have shown a linear relationship between AE amplitude and incremental crack
growth. Because DHC proceeds in a stepwise manner, i.e. the crack front moves
forward in discrete variable amplitude events, it can be assumed that the amplitude distribution of events is indicative of the size distribution of fracturing
hydride particles producing the AE stress waves. The amplitude of each event
during DHC should thus be simply related to the distance of each incremental crack
advance step. Hence, acoustic emission amplitude distribution analysis is
expected to provide some insight into the crack-tip condition and crack
propagation mechanisms during DHC.

The work reported in this paper was conducted on Zr-2.5% Nb alloy that had been
quenched to produce a population of very fine hydrides. Crack velocity data and
the characteristic DHC kinetics of this particular material have been previously
described (Amouzouvi and Clegg, 1987). It was found that fine hydrides result in
a higher crack velocity than that measured in furnace-cooled material, containing
relatively coarse hydrides. In addition, whereas the crack velocity in the latter
material rapidly reached a constant plateau value, the crack velocity in the fine-
hydride material increased continuously with the stress intensity factor. An ob-

jective of the present work was to seek explanation for these differences in behaviour, using the AE analysis, supplemented by measurements of the striation spacings.

EXPERIMENTAL PROCEDURE

Material and DHC Testing

The material used in this investigation was a standard CANDU^{TM} pressure tube alloy, 28% cold-worked Zr-2.5% Nb. Compact tension (CT) specimens, 17 mm wide and 3.8 mm thick, were prepared from flattened pressure tube with the notch cut in the axial direction of the tube. The CT specimens were hydrided gaseously at 350°C for 4 hours to a concentration of about 50 μg/g (0.46 at%) and 100 μg/g (0.91at.%) hydrogen. These specimens were then homogenized at either 390°C (for material with 100 μg/g) or 320°C (for material with 50 μg/g) and the specimens were ice water quenched from these solution temperatures to produce a very fine distribution of hydride particles. Recent results (Amouzouvi and Clegg, 1987) have shown that DHC cracking readily occurs in such material. The CT specimens were fatigue precracked at low stress intensity factors, keeping the maximum stress intensity factor below 5 MPa√m for the last 50% of the precrack extension.

The specimens were stressed in a furnace, using the lever loading system shown in Fig. 1, at an initial stress intensity factor of 6 MPa√m, once the desired test temperature was reached. Isothermal DHC tests were performed at various temperatures, ranging from 150°C to 250°C. After initiation of crack growth, the load on the specimen was increased by small increments, generally of about 1 MPa√m. The crack length was continuously monitored, using the potential-drop technique.

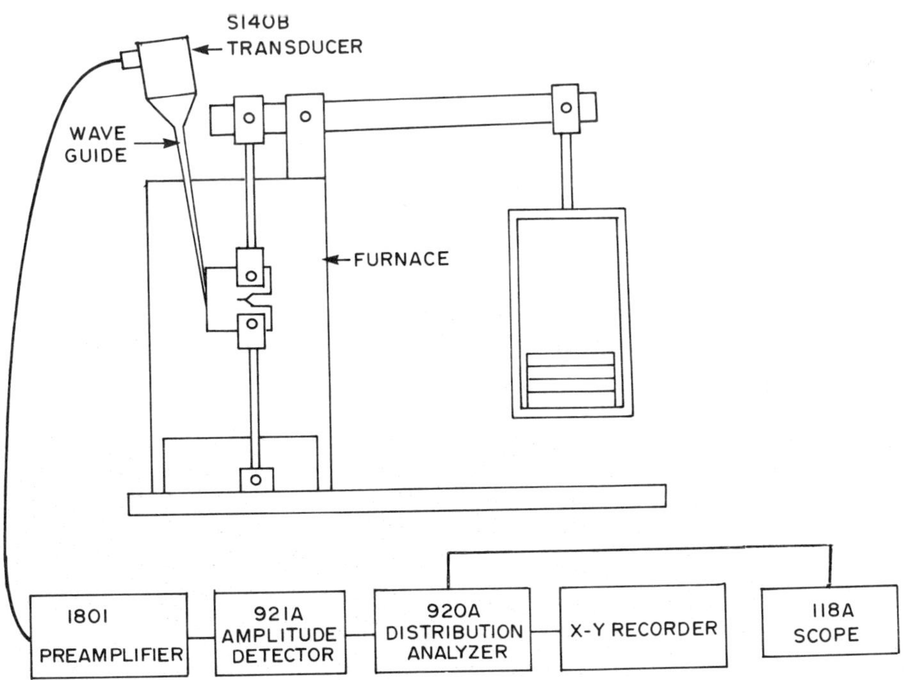

Fig. 1. AE amplitude distribution analysis experimental set-up.

AE Equipment, Signal Processing and Analysis

Acoustic emission equipment (Dunegan 3000) was used to characterize the fracture
events at different stages of the crack growth, using amplitude distribution
analysis. The acoustic emission set-up, shown in Fig. 1, includes a zirconium
wave guide, spot-welded to the CT specimen; a piezoelectric resonant transducer
(Model S140B); a preamplifier (Model 1800) with 40-dB gain; a peak amplitude
detector (Model 921A); a distribution analyser (Model 920A), which operates in
conjunction with the amplitude detector, performs a "per event" sorting and
storage of AE signals according to peak amplitude of the AE bursts. Figure 2 is a
schematic showing how the amplitude of <u>events</u> (three such "events" are depicted in
Fig. 2) is measured during DHC tests. Basically, the peak amplitude of each event
(V_i) is determined and each event is sorted and stored into 101 memory channels
with assigned values of 0,1,2---99, 100 dB. Output from the distribution analyser
provides an oscilloscope representation (histogram), or a plot of the amplitude
distribution of events. At any desired stage during the DHC test, the amplitude
distribution of the events can, therefore, be recorded photographically or plotted
on log paper, resetting the memory of the distribution analyser for an analysis of
subsequent stages of DHC. The usual display is a cumulative (Log-Sum)
representation of the amplitude distribution in which each channel shows the
number of events with peak amplitude appropriate to that channel and greater.

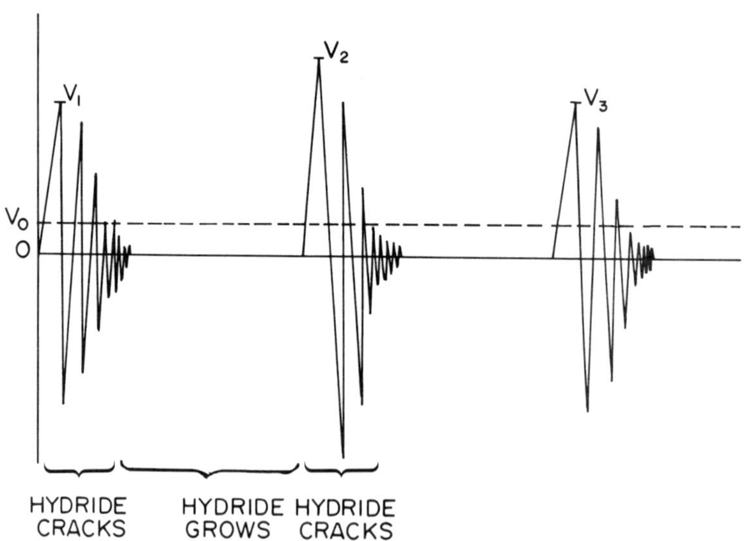

Fig. 2.   Schematic showing peak amplitude measurement
of three AE "events" during DHC.

It was found in previous materials testing (Scholz, 1968; Nakamura and co-workers,
1972; Pollock, 1973) that AE amplitude distributions often approximate to a power
law relationship developed by Gutenberg and Richter (1949) and Ishimoto and Iida
(1939):

$$N(V_i) = A\left(\frac{V_i}{V_o}\right)^{-b}$$

(2)

where $N(V_i)$ is the cumulative number of events with amplitude greater than V_i, V_o is the detector threshold, and b is the amplitude distribution parameter. The slope (b) of a log-log plot of $N(V_i)$ versus V_i is a highly mechanism-dependent constant (Scholz, 1968; Nakamura and co-workers, 1972, Pollock, 1981). For example, when acoustic emission is produced by plastic deformation that proceeds in a large number of small incremental steps, low-amplitude events are dominant and b is large. On the other hand, for mechanisms involving brittle crack growth, which proceeds in larger steps, a large proportion of high-amplitude emissions are produced and low b values result. It is thus possible to characterize the deformation or fracture mechanisms by the determination of a single parameter, b. When several mechanisms are operating simultaneously, the resulting amplitude distribution might be a superposition of individual mechanisms, as shown by Dilipkumar and co-workers (1979). This might be the case during tensile testing of a ductile metal containing brittle precipitates or inclusions. In this situation, one might expect to obtain a distribution consisting of high-amplitude events due to cracking of the brittle particles or inclusions, and numerous low-amplitude events corresponding to plastic deformation of the matrix.

Fractography

After the DHC tests, the fracture surfaces were examined using light microscopy. Oblique illumination, with the light directed in the crack propagation direction, was used to observe the intermittent nature of the crack growth process. In this way, the individual steps of crack propagation can be resolved on the fracture surface as striations, the spacing of which represent incremental crack growth (i.e., l_c in equation 1). The average striation spacing was estimated for different stress intensity factors and testing temperatures.

RESULTS

The amplitude distribution of AE events was recorded at different stages of the DHC tests. Figure 3 shows a logarithmic plot of $N(V_i)$, as a function of V_i (dB). A key feature of Fig. 3 is that the amplitude of the events changes significantly as the DHC test progresses. Thus, it can be seen that at the beginning of the DHC test (curve A), in the low stress intensity factor (K) range, very high amplitude ($V_i > 85$ dB) events are predominant. As the DHC progresses, the AE amplitude gradually decreases (curve B). Near the end of the test (curve C), in the high K range, the AE amplitude has considerably decreased (most of events having their amplitude below 50 dB). The AE amplitude distributions at low K (curve A) and at high K (curve C) can be approximated to straight lines, thus following the power law relationship (equation 2). The distribution parameter, b, was determined for both cases, being ~ 0.4 for curve A and ~ 1.78 for curve C. The significantly lower b value obtained for curve A at the beginning of DHC reflects the fact that most of the AE events are very energetic (high amplitude) at low K. In contrast, the relatively higher b value in the high-K regime is consistent with the occurrence of considerably lower amplitude events. We consider that the AE amplitude distribution obtained for the intermediate stage (curve B) represents a transition that combines features of curve A and curve C.

An important observation is that the amplitude of the AE signals generally increases with an increase in the test temperatures, at all stages of DHC propagation.

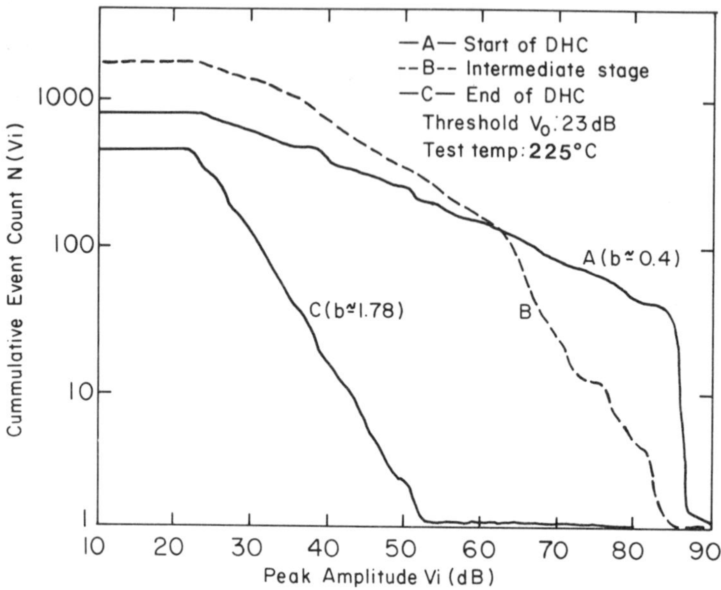

Fig. 3. AE Amplitude distribution of events at different stages of DHC.

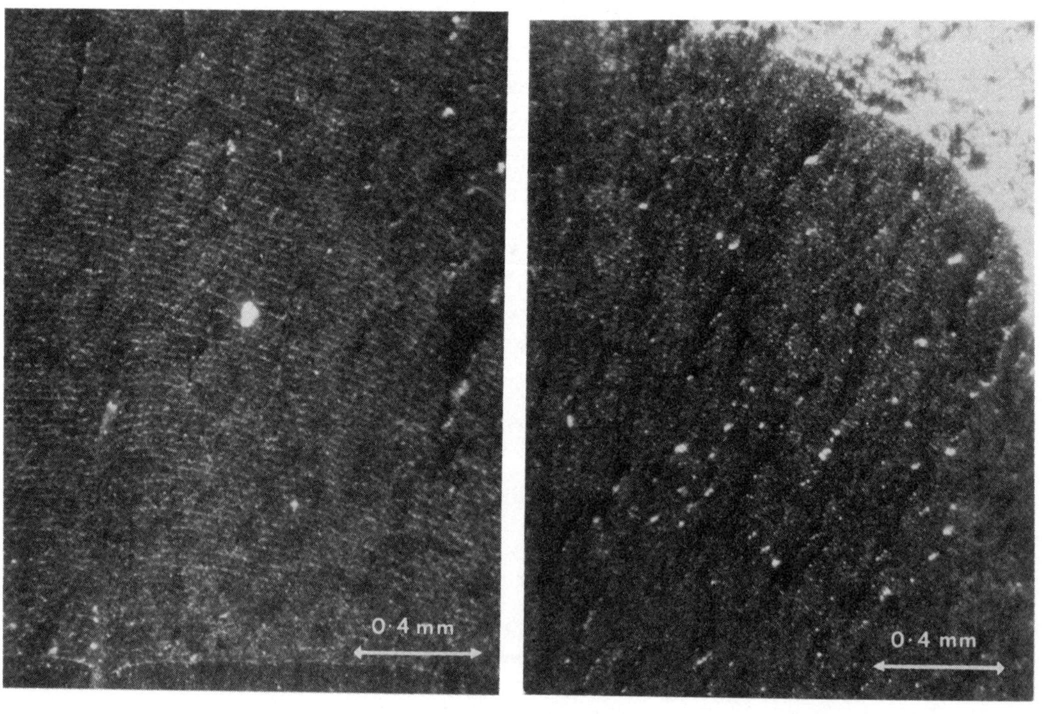

a) K = 8 MPa√m (start)   b) K = 30 MPa√m (end)

Fig. 4. Striation markings on the fracture surface at the beginning
(low K) and at the end (high K) of DHC propagation at 225°C.

The AE amplitude distribution analyses were complemented by examinations of
striation markings on the DHC fracture surface. In Fig. 4-a, which pertains to
the beginning of the DHC test, at low K, the striations are quite coarse (26 μm).
This corresponds to relatively large incremental crack growth events, producing
the high-amplitude AE signals of curve A in Fig. 3. In contrast, towards the end
of the DHC test (Fig. 4-b), in the high-K regime, much finer striations are
obtained (10 μm),corresponding to smaller incremental steps (and lower amplitude
AE signals) in the crack growth. The variation of the striation spacings with the
stress intensity factor is shown in Fig. 5 and their temperature dependence (at
low K) is shown in Fig. 6.

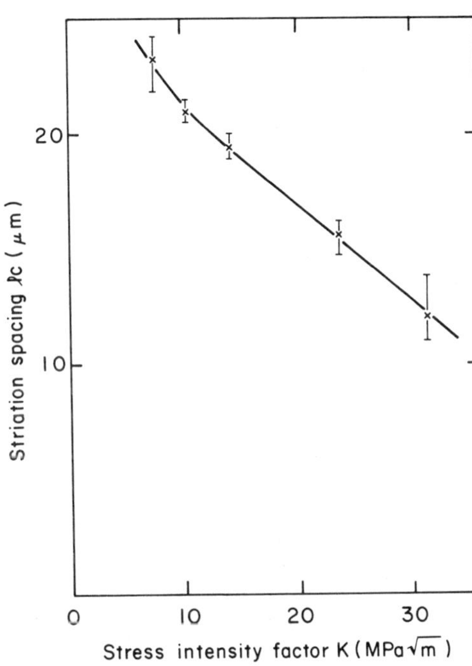

Fig. 5. Variation of striation spacings with applied
stress intensity factor, at 225°C.

DISCUSSION

As mentioned in the Introduction, previous work (Amouzouvi and Clegg, 1987) on
quenched Zr-2.5% Nb (hydrided to 50 μg/g) showed that the DHC crack velocity is
generally increased in material containing fine hydrides, compared with coarse-
hydride material. One of the postulated explanations for this was that the
distance of incremental crack advance (1_c) was greater in the presence of fine
hydride. Examination of Fig. 6 however shows that although the <u>maximum</u> striation
spacing for fine hydrides falls in the upper range of the coarse hydride data (as
reported by Simpson and Puls, 1979), there is insufficient evidence to support the
generality of the previous postulate. In addition, Amouzouvi and Clegg's observation that the crack velocity increases with K would require 1_c to increase with K
as well. This is completely contrary to the data shown in Fig. 5. Thus,
reference to equation 1 suggests that the higher crack velocities must require a
decrease in the incubation time, Δt. In addition, Δt must decrease further as K

increases. This is in complete accord with the alternative suggestion made by
Amouzouvi and Clegg (1987) that, in the presence of fine hydrides, dislocation
sweeping within the plastic zone can enhance the flux of hydrogen to the crack
tip, which increases with K.

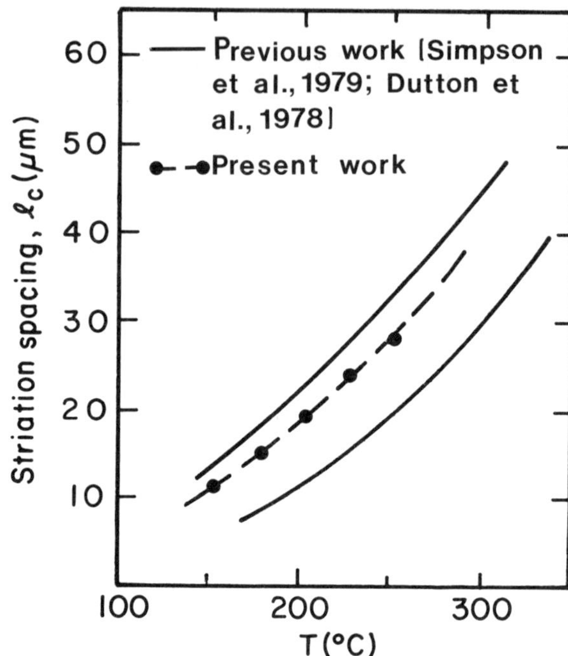

Fig. 6. Variation of striation spacings (at low K) with temperature.

In the original model (Dutton and co-workers, 1977), it was estimated that, at the
initial stage of DHC (stage I), i.e., at low K, the crack tip plastic zone size
(r_y) is small and l_c is larger than r_y so that the hydride would be growing beyond
the crack tip plastic zone, resulting in the crack velocity increasing with K.
However, in Stage II, (in which most of the cracking occurs) at higher K, r_y is
larger than l_c and the hydride platelet will be growing completely within the
crack tip plastic zone. In this case, the crack velocity would be almost independent of K (actually, a slight decrease with K is predicted), because the plastic
zone stress, which provides the major driving force for DHC, is fixed and limited
by the material yield stress. This is in agreement with experiments on furnace
cooled (coarse hydride) material (Simpson and Puls, 1979; Coleman and Ambler,
1977). A corollary of the model and experiments is that the striation spacing is
also independent of K (Dutton et al., 1978). Indeed, the limited experimental
evidence seemed to support this conclusion (Dutton et al., 1977). Clearly, the
behaviour of material containing fine hydrides is contradictory to this. In
particular, the striation spacing is inversely dependent on K (Fig. 5).

As shown in Fig. 3, the AE amplitude distribution changes significantly as K progressively increases. This is consistent with the K-dependence of the striation
spacing. Thus, it is reasonable to expect that the amplitude of the AE signals is
correlated with the incremental crack propagation distance, l_c. Consequently, at
low K, l_c is large and high-amplitude AE events dominate (i.e., low value of b).
In the higher K regime, l_c is smaller and, correspondingly, the AE amplitudes are
lower (i.e., high value of b). In addition, our experiments showed that as K

increases, quasi continuous low amplitude events were produced at a very high rate, indicating a very fast acoustic emission activity. This suggests that, at higher K, hydride platelets fracture soon after they form, in contrast to the beginning of the DHC test where hydride platelets, under a much lower K, grow to a larger size before fracturing, in agreement with Fig. 5. The effect of temperature also conforms to this premise. Thus, as reported earlier, the amplitude of the AE signals increases with temperature. This reflects the increase in l_c with temperature (Fig. 6).

It is important to note that we do not expect a single high-amplitude AE event to occur during the development of each individual striation: the number of striations per mm of crack growth is of the order of 10^2, far less than the number of AE events indicated in Fig. 3. Rather, we expect a very large number of events per striation, as many hydrides must rupture during the development of each unit area of fracture surface. In addition, it is probable that many AE events accompany the fracture of each hydride platelet (the rupture of the inter-hydride ductile ligaments will also produce many AE events, but most likely with low amplitude, as suggested by Pollock, 1979).

In an attempt to explain the stress dependence of the striation spacings and AE amplitude, we propose to use a crack propagation criterion based on a requirement of a critical stress at the crack tip for hydride platelet cracking. Thus, we assume that the critical hydride size, i.e., striation spacing, should be directly related to the stress required to fracture the platelet. We follow the energy balance concept proposed by Gurland and Plateau (1963) for fracturing precipitates (recently applied by Van Stone and co-workers (1974) to precipitates cracking at the crack tip of compact tension specimens). Thus, when the increase in the crack tip elastic strain energy is greater than the surface energy of the hydride particle, the particle will fail. The critical applied stress σ_f (note that the applied stress increases with K) necessary to fracture the hydride platelet of size l_c is then given by

$$\sigma_f = \frac{1}{q}\left(\frac{2E\gamma}{l_c}\right)^{1/2} \tag{3}$$

where q = the stress concentration factor at the particle,
E = hydride Young's modulus, and
γ = hydride fracture energy.

Equation 3 states that, prior to fracture, the increase in the crack-tip strain energy will be stored in the hydride that will fail when its length reaches l_c. This equation, written in the following form

$$l_c = \frac{2E\gamma}{q^2\sigma_f^2} \tag{4}$$

shows that the critical particle size (striation spacing) decreases when the applied stress (i.e., K) increases, as experimentally observed by both AE amplitude distribution analyses and striation spacing measurements.

Approximating γ with $K_{IC}^2/2E$ (where K_{IC} is the fracture toughness of the hydride), equation 4 becomes

$$l_c \simeq \left(\frac{K_{IC}}{q\sigma_f}\right)^2 \tag{5}$$

This equation, together with the experimental data, is plotted in Fig. 7, using the following parameters:

K_{IC} = 1.5 MPa√m (Simpson and Cann, 1979)
  q = 2.7 to 1.0, as discussed by McClintock (1968); and
  σ_f is calculated using the expression developed for a CT specimen by Tada et al. (1973).

Figure 7 shows that 1_c rapidly approach ∞ at low stresses. This is to be expected as the threshold stress intensity factor (estimated to be ∿ 6 MPa√m by Amouzouvi and Clegg, 1987) is approached. In the higher stress range (σ_f ≃ 200–700 MPa), equation 5 underestimates 1_c. Good agreement however with the experimental data is obtained if K_{IC} ≃ 5.0 MPa√m. Such a value is exactly what is expected (Puls et al., 1982) for the heavily hydrided zone (hydride plus matrix) at the crack tip.

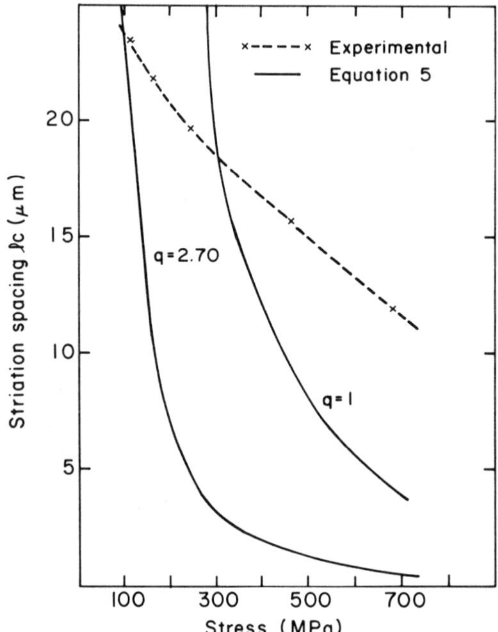

Fig. 7. Variation of striation spacings with the applied stress.

Equation 5 can also be used to explain the increase in 1_c with temperature (Fig. 6). Thus, it is expected that the fracture energy of the hydrided crack tip (i.e., K_{IC}) will increase with temperature. This could either be due to an increase in plasticity in the hydride itself (as discussed by Simpson and Puls, 1979), or the increase in ductility of the surrounding matrix (as suggested by Dutton et al., 1978).

A more detailed discussion of the role of 1_c in the DHC mechanism, including its stress and temperature dependence, will be presented in a future publication.

CONCLUSIONS

AE amplitude distribution analysis of DHC has been used to study DHC cracking
mechanisms in Zr-2.5% Nb containing a fine dispersion of hydrides. The amplitude
of AE signals produced by hydride platelets cracking during DHC was found to vary
with the applied stress intensity factor and the test temperature. This behaviour
was found to be consistent with striation spacing examinations on the DHC fracture
surface. An explanation of this behaviour, based on hydride platelets cracking at
the crack tip, was proposed. The inverse stress dependence of the striation
spacing is contrary to the expectations arising from the extant DHC model. Taking
this result into account, it is necessary to invoke an increase in the hydrogen
flux to the crack-tip to explain the observation of increasing crack velocity with
K. This supports an earlier conclusion that a K-dependent dislocation sweeping
mechanism occurs within the crack-tip plastic zone when fine hydrides are present.

A model based on the necessity of a critical stress criterion for the fracture of
hydride platelets has been used to explain the K and temperature dependence of the
striation spacing. The model also predicts a threshold value of K below which DHC
ceases. In contrast to the number of detailed modelling studies on the kinetics
of hydride accumulation at the crack tip, the details of the fracture event
associated with this locally hydrided zone has been neglected. The present study
has in part, addressed this need; however, further studies are required. We
believe that acoustic emission techniques and fractographic inspection of
striation spacings will play a strong role in future investigations.

ACKNOWLEDCMENT

We are grateful to Dr. R. Dutton for reading the manuscript, for fruitful
discussions and for making useful suggestions during the course of this study.

REFERENCES

Ambler, J.F.R. (1984). ASTM STP 824, 653-674.
Amouzouvi, K.F., and L.J. Clegg (1987). Metall. Trans., 18A, 1687-1694.
Coleman, C.E., and J.F.R. Ambler (1977). ASTM STP 633, 589-607.
Coleman, C.E., and J.F.R. Ambler (1978). Can. Met. Quart., 17, 81-84.
Coleman, C.E., and J.F.R. Ambler (1983). Scripta Metallurgica, 17, 77-82.
Dilipkumar, D., V.S.R. Gudimetla, and W.E. Wood (1979). Exp. Mech., 19, 438.
Dutton, R., and M.P. Puls (1976). Effect of Hydrogen on Behaviour of Materials,
    Proc. Inter. Conf., I.M. Bernstien, and A.W. Thompson (Eds.), Met. Soc. AIME,
    516-528.
Dutton, R., K. Nuttall, M. P. Puls, and L. A. Simpson (1977), Met. Trans., 8A,
    1553-62.
Dutton, R., C.H. Woo, K. Nuttall, L.A. Simpson, and M.P. Puls (1978). Hydrogen in
    Metals, Proc. of the 2nd Inter. Congress, Pergamon Press, Vol. 1, Paper 3C6,
    1-8.
Gurland, J., and J. Plateau (1963). Am. Soc. Metals, Trans. Quart. 56, 442-54.
Gutenberg, B., and C.R. Richter (1949). Seismicity of the Earth. Princeton
    University Press.
Ishimoto, M., and K. Iida (1939). Bull. Earthqu. Res. Inst., 17, 443.
McClintock, A.F. (1968). In Ductility, Americal Society for Metals, Metals Park,
    Ohio, 255-277.
Nakamura, Y., C.L. Veach, and B.O. McCauley (1972). ASTM STP 505, 164-186.
Pollock, A.A. (1979). Inter. Advances in Nondest. Test., Gordon and Breach
    Publishers, 6, 239-262.
Pollock, A.A. (1973). Nondest. Test., 6, 264-269.

Pollock, A.A. (1981). Inter. Advances in Nondest. Test., Gordon and Breach Science
  Publishers, 7, 215-239.
Puls, M.P. (1984), Acta Metall. 32A, 1259-1269.
Puls, M.P., L.A. Simpson, and R. Dutton (1982). Fracture Problems and Solutions in
  the Energy Industry, L.A. Simpson (Ed.), Pergamon Press, 13-25.
Scholz, C.H. (1968). Bull. Seis. Soc. Am., 58, 1, 399-415.
Simpson, L.A., and C.D. Cann (1979). J. Nucl. Mater., 87, 303-316.
Simpson, L.A., and M.P. Puls (1979). Met. Trans., 10A, 1093-1105.
Tada, H., P. Paris, and G. Irwin (1973). The Stress Analysis of Cracks Handbook,
  Del Research Corp.
Tangri, K. (1977). Proc. of the Institute of Acoustics, London, 4.17, 1-4.
Van Stone, R.H., R.H. Merchant, and J.R. Low Jr. (1974). ASTM STP 556, 93-124.
Wadley, H.N.G., and C.B. Scruby (1979). Acta Metall., 27, 613-625.

DETERMINATION OF THE FRACTURE TOUGHNESS OF TWO C–Mn STEELS FOR DIFFERENT TEMPERATURES AND LOADING RATES

Phuc Nguyên-Duy,   André Lapointe,   Jacques Flamand
Institut de Recherche d'Hydro-Québec (IREQ)
Varennes, Québec, Canada

ABSTRACT

The objective of this study was to develop and improve an inexpensive side-grooved precracked Charpy–V–notch technique for measuring the fracture toughness of material at crack initiation. Critical values of the J–integral were determined at different temperatures and different rates of loading. Results for two steels (SA–106B and SA–516 Gr. 70) show clearly that the addition of precracks and side grooves to small–size specimens allow a fracture toughness to be determined which proves highly comparable with that obtained by other standard methods.

Keywords: Fracture toughness; J–integral; slow–bending test; impact
testing; stretch–zone width; steels

1.   INTRODUCTION

Material fracture toughness is an important criterion for fracture–safe structural design. This laboratory determined criterion is assumed to describe the material's structural behavior, which depends on operating temperatures and the mode and rate of loading. Fracture toughness allows the ductile–brittle transition curve to be established for different loading rates. Usually these curves are determined by impact tests on Charpy–V–notch (CVN) specimens but the results express only the variation in the total energy absorbed by the specimen during impact as a function of the temperature and offer no information regarding the critical fracture parameter. On these curves, between the lower and upper shelves, the material presents a mixed behavior of crack initiation and crack propagation. The scatter in the experimental results, pronounced in this zone, is not statistical and generally leads to an overestimation of fracture toughness. The origin of the dispersion is metallurgical and/or mechanical. The total energy absorbed by a specimen during impact testing is composed of energy for crack onset, energy for crack propagation and energy for plastic deformation. In the case of precracked Charpy–V–notch (PCVN) and side–grooved PCVN specimens (SGPCVN) however, the deformation energy is reduced (Server, Oldfield and Wullaert, 1977; Nguyên–Duy and Bayard, 1982).

It is interesting to evaluate the part of the energy attributable to crack onset on these specimens since this permits determination of the critical fracture initiation parameters. The parameter based on Linear Elastic Fracture Mechanics (LEFM), K_{Ic}, is widely accepted for fracture–safe structural design but it cannot be measured validly according to the laboratory method (ASTM–E399, 1981) in

the case of Charpy-size specimens.   To overcome this limitation, the J-integral
concept enables the extension of LEFM, for large-scale plastic behavior in particular.

Determination of the critical value of the J-integral at different temperatures
allows the existence of a transition zone for crack-onset energy to be verified
and correlation with the variation in the total absorbed energy as a function of
the temperature determined by conventional impact testing to be established.   It
is also important to observe the variation in this initiation value and in the
transition zone as a function of the loading rate.

The principal objectives of this study are to develop and improve an inexpensive
SGPCVN specimen technique for evaluating the fracture toughness of materials at
crack initiation, quantified by critical values of the J-integral at different
temperatures and for different loading rates (impact to slow bend testing).   The
basis of this technique is the single-specimen method, which can be verified by
the multiple-specimen method.   This modified technique is capable of generating
toughness data which are both significant for fundamental design and reliable for
quality control, yet it retains the simplicity of the conventional test method and
specimen configuration.

2.   MATERIALS CHARACTERIZATION

Two plates of normalized hot-rolled SA-516 Gr. 70 and SA-106B, 5 cm in thickness,
were used in this study.   The results of the chemical analysis are reported in
Table 1.   In the case of SA-106B steel, the carbon content is lower than the specified carbon composition (0.275% as compared to 0.30%), whereas for SA-516 Gr. 70
the chemical composition is different from the specifications, the specified carbon content being 0.28% as compared to the actual carbon content of 0.221%.   It is
important to mention the copper and nickel contents (added to improve the low-
temperature properties) are respectively 0.32% and 0.12%.

Normalized steel microstructures consisting of a fine-grain structure of pearlite
in a matrix of ferrite were seen.   No preferential grain orientation in the case
of SA-106B was observed whereas a banded structure is present in the SA-516 Gr. 70
specimens (Fig. 1).

Tensile, Charpy-V-notch and four-point-bend specimens were cut from the steel
plates according to the scheme shown in Fig. 2.

SA - 516 gr 70   Steel   SA - 106 B   Steel

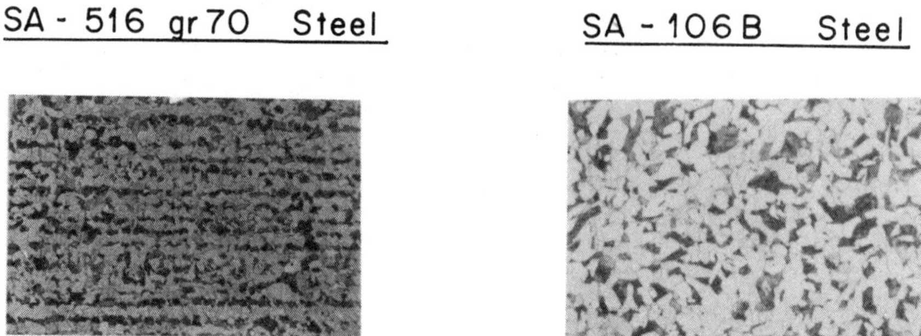

Fig. 1   Microstructure of SA-516 gr. 70 and SA-106B steels

As shown in this figure, the SA-516 Gr. 70 specimens were machined according to two directions TL(S) and TL(L) whereas the SA-106B specimens were machined in only one direction TL(L). The first letter of this identification indicates the direction of the larger dimension of the specimen, the second, the small dimension, whereas the letter in parentheses indicate the notch direction. Conventional mechanical testing results are reported in Table 2 and in Fig. 3.

All specimens used for facture toughness determination were precracked. Some of the precracked Charpy-V-notch specimens were side-grooved. Two side-groove profiles, namely round (RSGPCVN) and V-notch (VSGPCVN) were used.

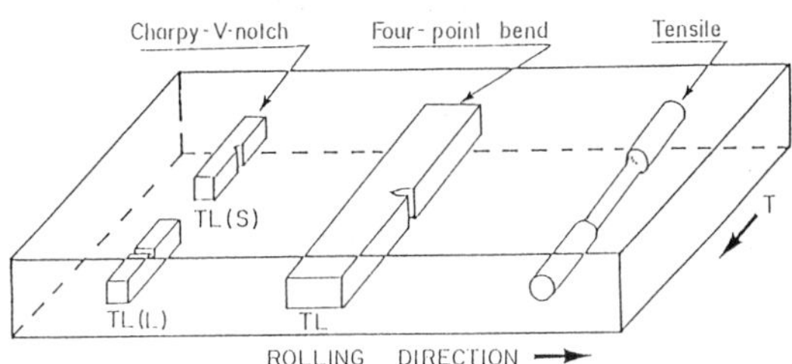

Fig. 2   Specimen location

3.   METHODOLOGY

3.1   CHARPY IMPACT TESTING METHOD:

The instrumented impact charpy tests give the same data as that obtained in notched slow bending and can be analyzed in the same way. Using a slender notch such as fatigue crack, the problem of overestimation of fracture-initiation energy is avoided but the primary advantage of the instrumented precracked charpy test is that it combines the best features of the energy-transition approach to fracture toughness. Recording loads during a precracked charpy test means that energy separation is possible and that the fracture toughness K_{Icd} or J_{Icd} can be measured under appropriate conditions. The only disadvantage of the test appears to be that it cannot measure full thickness behavior although this problem can be solved by using side-groove and precrack CVN specimens, which allows virtually full-scale specimen behavior. Side grooves constrain lateral deformation, thus creating triaxial-state stress at the lateral surfaces. For PCVN and SGPCVN specimens, frcture equation developed (Rice, Paris and Merkle, 1973) for bend bars can be applied:

$$J_{Icd} = \frac{2E_{sc}}{B_e b} \tag{1}$$

where J_{Icd} is the dynamic critical value of the J-integral, E_{sc} the critical strain energy corresponding to the crack-initiation energy, and B_e (B_e = B without side-groove) and b the effective specimen thickness and remaining ligament respectively.

The critical value of the strain energy is the area under the load-displacement curve at maximum load.

The experimental set up for impact testing is illustrated in Fig. 4. The software used for impact testing was developed in house; the machine compliance is substracted from the raw results to obtain E_{sc}.

3.2 SLOW BEND TESTING METHOD

Equation (1) was used to determine the critical value of the J-integral, except that in this case E_{sc} corresponds to the crack-onset point detected by electric potential measurements.

A typical set of results is presented in Fig. 5.

For determination of the crack-initiation point, a special step was included in the data treatment program allowing the operator to select four data points on the ΔV vs Load Line Displacement (δ) curve. Two of these points precede the region of crack initiation whereas the other two follow it. The program superimposes two straight lines on the ΔV vs δ curve (one passing through the first two points, the other through the last two points) and then reproduces the enlarged graph of ΔV vs δ between the first and last of the four selected points. The intersection of these two lines marks the crack-onset point (Nguyên-Duy and co-workers, 1986).

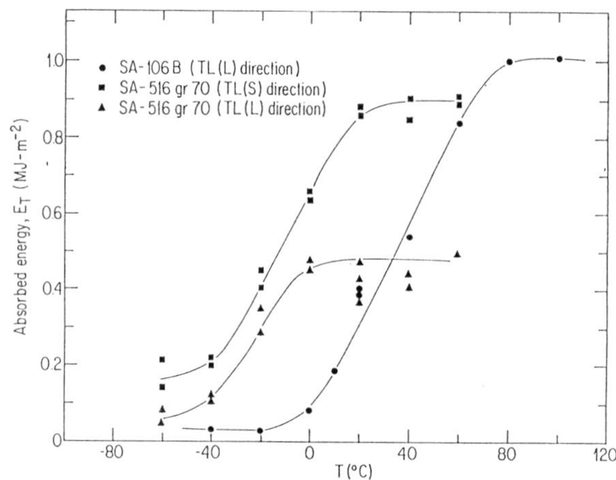

Fig. 3   Conventional CVN brittle-ductile transition curves obtained by impact testing of SA-516 gr. 70 and SA-106B steels

3.3 MODIFIED CRITICAL STRETCH ZONE MODEL

Theoretically, the stretch-zone width is related to the crack-tip opening displacement, CTOD, which can then be defined as the displacement of the original crack-tip position, namely the tip of the fatigue precrack in a COD specimen or a natural crack in a structure (Robinson and Tetelman, 1974; Landes and Begley, 1974). Considering a symetrical blunt crack relative to the oridinal line of fatigue precrack, it was developed elsewhere (Nguyên-Duy, 1981, Amouzouvi and Bassim, 1982) that W_{SZ_c} can be evaluated using the following equation

$$W_{SZc} = \frac{L \sec (\theta - \delta)}{G} \qquad (2)$$

where L is the measured length of the stretch zone on micrographs, θ angle formed by stretch zone and horizontal line, δ angle formed by incident and vertical line

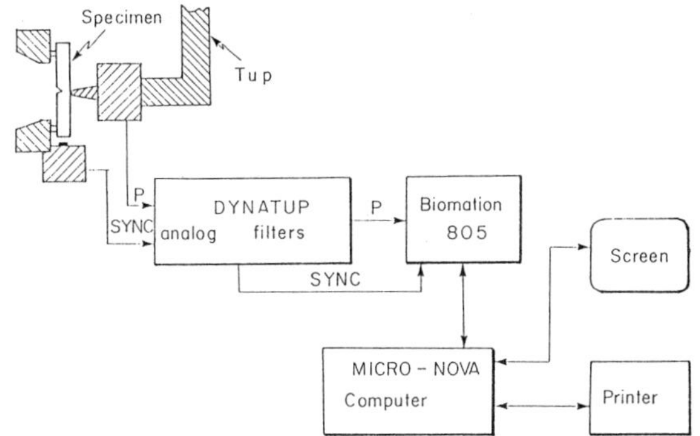

Fig. 4   Experimental set-up for impact testing

and G the magnification factor.   The scanning electron microscope is frequently used to observe the stretch zone of fractured specimens.

The only unknown angle on the right-hand side is the angle θ which represents the capacity of the material to resist to crack initiation.   This angle is assumed equal to 45°; this assumption can be removed by making two independent measurements of the same zone at different incident angles δ; Eq. (2) can be rewritten

$$W_{SZc} = L_1 \sec (\theta - \delta_1)/G = L_2 \sec (\theta - \delta_2)/G \qquad (3)$$

which after arrangement, becomes

$$\tan \theta = \frac{L_1 \cos \delta_2 - L_2 \cos \delta_1}{L_2 \sin \delta_1 - L_1 \sin \delta_2} \qquad (4)$$

all terms in the right-hand side member of this equation are known, angle θ and subsequently W_{SZc} can be evaluated (Nguyên-Duy and co-workers, 1986).

W_{SZc} is strictly a material property which can be used to evaluate the material's fracture toughness.   It is important to mention that its determination, although difficult, is direct and involves no assumptions.

4.   RESULTS

All specimens were precracked using a multi-specimen precracker designed by IREQ. The precrack length was limited to the range of $0.45 \leq a_0/w \leq 0.65$.   In all 200 specimens of SA-516 gr. 70 and 200 of SA-106B were precracked.   Side grooves were added using two side-groove geometries (1 mm radius round side groove and 1 mm deep V side groove).   For each steel, 60 round-side-groove specimens (RSGPCVN) and 60 V-side-groove specimens (VSGPCVN) were available.

4.1   IMPACT TEST RESULTS

4.1.1   Brittle-Ductile Transition Curves

Transition curves were determined on CVN, PCVN and SGPCVN specimens.   The tests were performed between -60°C and 120°C at intervals of 20°C using two specimens at

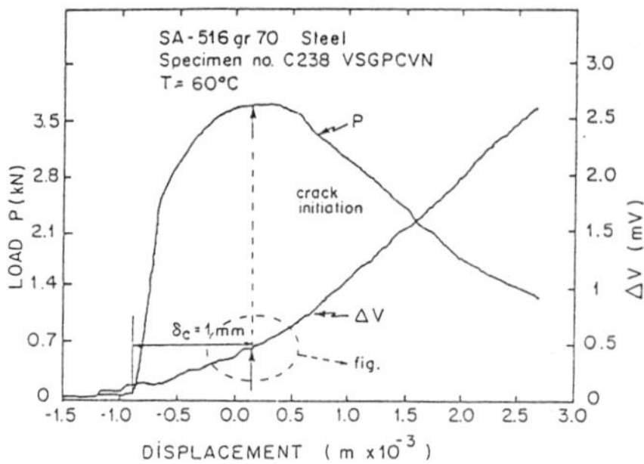 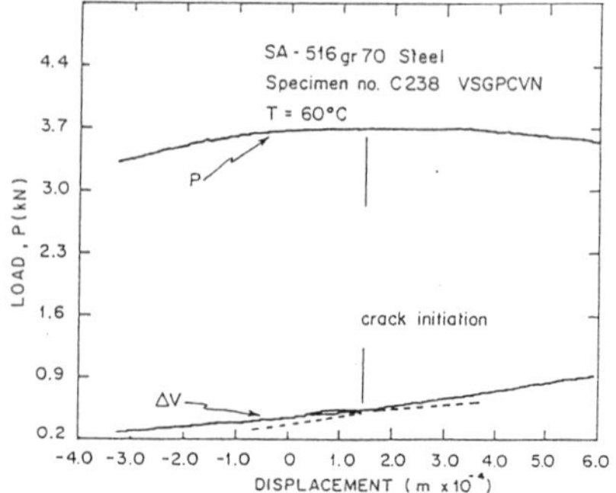

Fig. 5   Typical results from slow-bend testing

each temperature. The total absorbed energy per unit of area E_T was calculated from the measured absorbed energy E_{abs}.

The transition brittle-ductile zone can be seen to drift towards higher temperatures with the addition of precrack and side-grooves as shown in Fig. 6.

This indicates that cracks and side grooves act as embrittling factors, in the same way as the lowering of the temperature, thus enabling structural behavior to be approached when performing tests on small specimens (10 x 10 x 55 mm).

4.1.2 Dynamic Fracture-Toughness Determination

For each impact test, the total energy and the absorbed energy at maximum load were evaluated.

For both steels, the variations in J_{Icd} with respect to temperature for different geometries (PCVN, RSGPCVN AND vsgpcvn specimens) are illustrated in Fig. 7.

In the case of SA-516 gr. 70 steel, a brittle-ductile transition from $-40°C$ to $40°C$ was observed for all three specimen configurations. On the lower shelf, the J_{Icd} value is equal to 0.02 MJ-m^{-2} whereas on the upper shelf, the J_{Icd} for PCVN specimens is higher than that obtained for RSGPCVN and VSGPCVN specimens, the values ranging between 0.28 MJ-m^{-2} and 0.20 MJ-m^{-2}.

In the case of SA-106B steel, the variation in J_{Icd} as a function of the temperature presents a lower shelf up to $0°C$ for PCVN specimens and $+20°C$ for SGPCVN specimens; the brittle-ductile transition zone ranges from these temperatures to above $120°C$.

It was observed that for SA-516 gr. 70, all specimens (PCVN, SGPCVN) with cleavage initiation satisfy the validity criterion (Server, 1979).

$$a, \ b, \ B \geq 50J/\sigma_f$$

where σ_f is the flow stress and is equal to $(\sigma_u + \sigma_y)/2$. However, in cases where fibrous initiation occurs, few specimens satisfy this criterion. Nearly all specimens of SA-106B steel tested present a cleavage initiation and satisfy the validity criterion.

4.2 SLOW BEND TEST RESULTS

For each slow bend test, the initial displacement, δ_c, corresponding to the crack-onset point is evaluated by the variation in the specimen's electric poten-

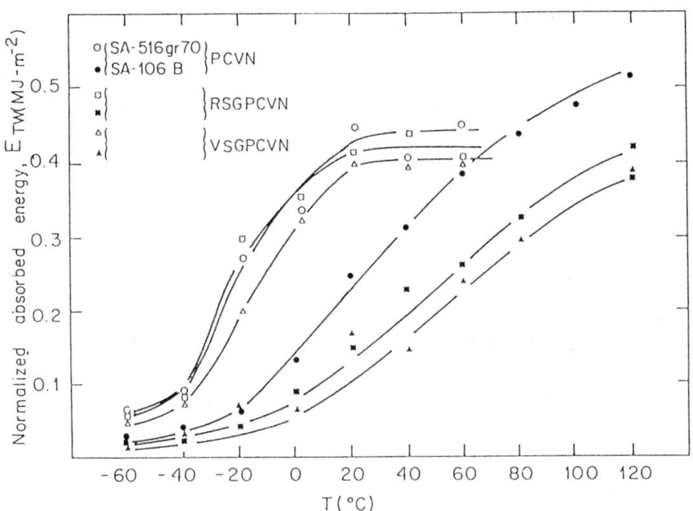

Fig. 6   Brittle-ductile transition curves for PCVN and SGPCVN specimens

tial as a function of the displacement. Then, from this displacement, the crack-onset energy, E_{Sc} and the critical value of the J-integral, J_{Ic}, were calculated. For each temperature and for a given specimen geometry, eight specimens were tested at different values of the total displacement δ on both sides of the critical point. The crack extension values obtained with different specimens allow calculation of J_{Ic} by means of the standardized multi-specimen method (ASTM-E-813, 1981) and subsequent comparison with the value of J_{Ic} determined for a single specimen by the electric-potential measurement. For both steels the values for J_{Ic} obtained on a single specimen by electric-potential measurements are close to those obtained for the same specimen using the standard multi-specimen method. An example of these results is shown in Fig. 8. The effect of

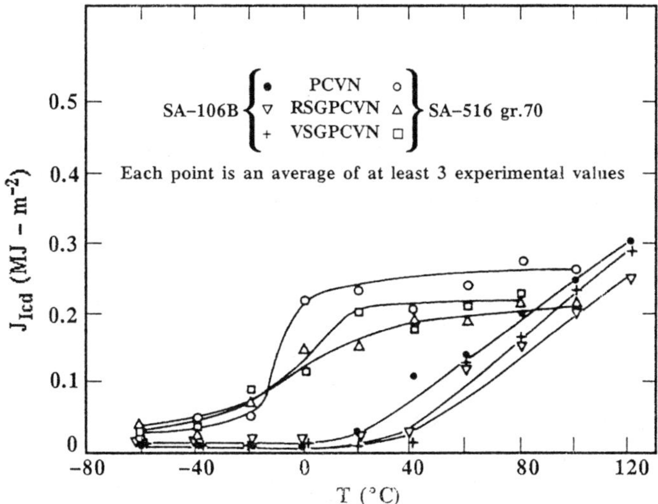

Fig. 7   Variation in J_{Icd} vs temperature for both steels

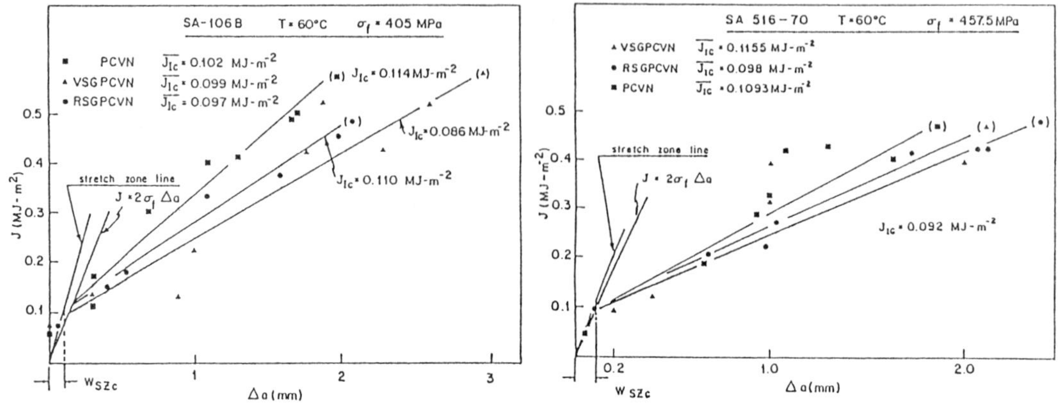

Fig. 8   Variation in J-Integral vs Δa for both steels at 60°C

the side groove is not clearly identified but the V-side groove geometry tends to result in lower J_{Ic} values.

Experimental results obtained for the two steels tested show clearly that the electric-potential measurement method for crack-onset detection is adequate for determining fracture toughness. However, at temperatures different from ambient, special care must be taken to avoid thermoelectric effects. One particular point observed was that the measured electric potential of the specimens decreases from the beginning of the test almost to the end of the elastic zone and then increases; this phenomenon is not yet well understood but for the time being, is associated with the crack-tip deformation.

4.3 STRETCH-ZONE WIDTH EVALUATION

Fracture surfaces, particularly the stretch zone located between the fatigue pre-crack and crack extension, of specimens of both steels were observed using a scanning electron microscope. The critital width of the stretch zone W_{SZc} is a characteristic of the material and is related to its fracture toughness. It is a directly measured value and its evaluation calls for no assumptions relative to the stress and strain rate in the near-crack-tip region as clearly shown in Eqs. 2 to 4. Typical results are shown in Fig. 9.

In the case of SA-516 gr. 70 steel, the critical values for the J-integral calculated from the stretch-zone model, assuming coefficient m to be equal to 2, are in agreement with those obtained from experiments. The blunting line established using the critical width of the stretch zone is very close to the blunting line established by the equation $J = 2\sigma_f \Delta a$. In the case of SA-106B steel, the J_{Ic} value evaluated by this model in this case is slightly lower than the J_{Ic} calculated from the electric-potential measurements and multi-specimen method (0.0915 MJ-m^{-2} compared to 0.118 MJ-m^{-2} (PCVN, 60°C)). The blunting line established with W_{SZc} in this case is steeper than the line $J = 2\sigma_f \Delta a$, resulting in lower values of J_{Ic}. As shown in Fig. 10, in the specimen C248 of SA-516 gr. 70 steel, a series of successive stretch zone was observed, which could explain why the existing crack was repeatedly initiated and propagated over a short distance then stopped.

TABLE 1: Chemical Compositions Of SA-516 gr. 70 and SA-106B Steels

Steel	C	Mn	P	S	Si	Cu	Ni	Cr	Mo	Nb	Al	N_2(PPM)
SA-516 gr. 70	.221	1.14	.018	.021	.25	.32	.12	.04	.030	.001	.042	95
SA-106B	.275	.89	.011	.018	.24	.03	.01	.05	.008	.007	0	55

TABLE 2: Conventional Mechanical Testing of Two Steels

	Hardness (R_B)	0.2% yield points σ_y (MPa) [a]	Tensile strength σ_u (MPa) [a]	% of elongation on 50 mm [a]	% of area reduction [a]
SA-516 gr. 70	80 - 82	365	547	28	55.0
SA-106B	75 - 78	243	518	43	56.6

(a) Average of 4 tests

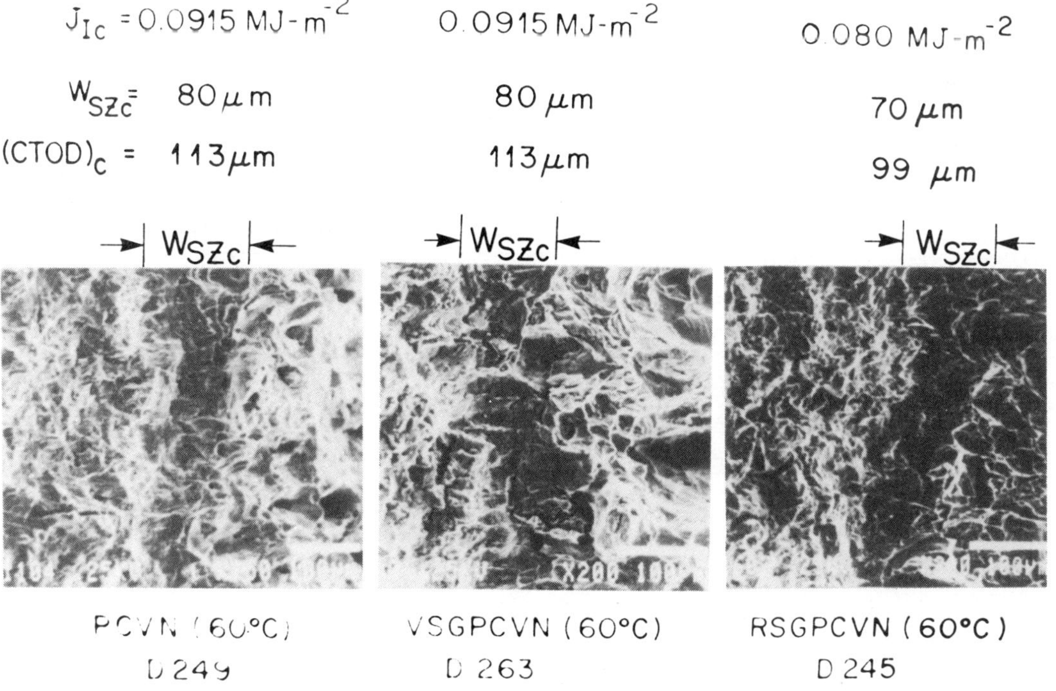

Fig. 9   Stretch-zone micrographs of SA-106B steel specimens

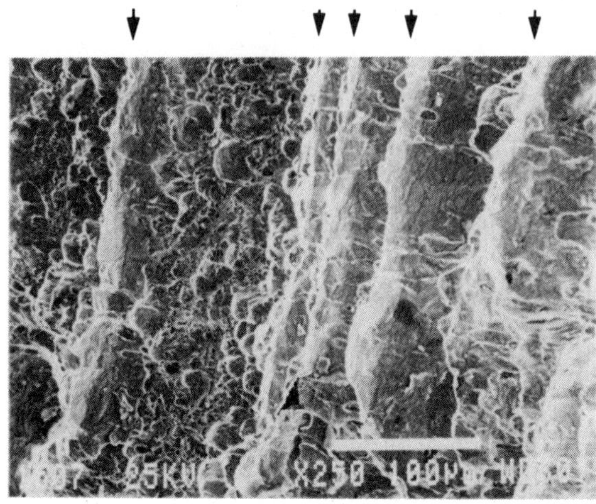

Fig. 10   Observations of a series of successive stretch zones
           (SA-516 gr. 70 steel)

5.   DISCUSSION

5.1 IMPACT TESTING OF TWO STEELS

5.1.1 Transition brittle-ductile zone

The two geometrical modifications (precrack and side-groove) to the specimens
generate plane strain conditions along the crack front through the specimen thickness. The precrack is a very efficient stress-raiser making crack onset brittle.
In the transition zone, this stress-raiser diminishes considerably, resulting in
scattering and the transition curve is then shifted to the right (Barsom and
Rolfe, 1970). In general, near the lateral surface of the specimens, the material
is not deformed in the same way as at the centre. A plane-strain zone observed at
the centre of the specimen thickness is replaced by a nearly plane-stress zone at
the lateral surfaces. The relative importance of this zone is inversely proportional to the thickness of the specimens. The addition of side grooves to PCVN
specimens helps to attenuate excessive lateral expansion due to triaxiality. This
expansion affects not only the initiation of a crack but also its propagation.

5.1.2 Dynamic Fracture toughness

For SA-516 gr. 70 steel, the J_{Icd} value determined on the lower shelf is the
same (0.02 MJ-m^{-2}) for PCVN and SGPCVN specimens, whereas on the upper shelf, the
value for the PCVN specimens is higher than that determined for the SGPCVN specimens. This can be explained by the fact that the addition of side grooves reduces
lateral expansion and has an embrittling effect on the crack-initiation process.
The same explanation is valid for SA-106B steel specimens.

As in the case of the total absorbed energy measured on conventional CVN specimens, a transition zone was observed for the total absorbed energy, E_T and
J_{Icd}, as determined on the modified CVN specimens (PCVN and SGPCVN), which is
shifted towards higher temperatures. This indicates that cracks and side grooves

behave as embrittling factors in the same way as lowering the temperature.

The results obtained from these two steels show a relatively large scatter, particularly in the upper-shelf region. Under conditions such as low temperatures and/or precracks and/or side grooves (embrittling factors), oscillations observed on Load-Displacement traces are important and are a result of a repetitive short separation between the specimen and the tip during impact. The determined values of J_{Icd} are believed to be overestimated owing to the fact that

- the onset of cracking is considered to be at maximum load;

- the dimensions of the Charpy specimens are insufficient for J determination.

This demension insufficiency is clearly reflected in the figures expressing the variation of J_{Icd} as a function of the remaining ligament b. For both steels, all specimens which present cleavage initiation meet the current specimen validity criterion

$$a, \ b, \ B \geq 50J/\sigma_f$$

whereas only some specimens of SA-516 gr. 70 steel with fibrous initiation satisfy this criterion.

5.2 SLOW BEND TEST

The set-up for the electric-potential measurements comprises two very important points where the current enters the specimen. For accurate measurements, conductors must be soldered to the specimen very carefully, keeping the contact resistance as low as possible. The electric-potential variation was measured as a function of the load-line displacement, any changes on this curve corresponding to changes in the specimen, such as theend of the elastic regime, plastic deformation, crack initiation and crack propagation. Experimental results obtained with SA-516 gr. 70 and SA-106B steels show clearly that this method of crack-onset detection is adequate for the determination of fracture toughness. At temperatures different from ambient, however, special care is essential in order to avoid the thermoelectric effects. It was observed that the measured electric-potential of the specimens decreases from the beginning of the test almost to the end of the elastic zone and then increases; this phenomenon is not yet well understood but is currently associated with crack-tip deformation (Tremblay, Nguyên-Duy and Dickson, 1987).

To compare the J_{Ic} value obtained on a single specimen using electric-potential measurements with that evaluated by the standard multiple-specimen, a program was developped to allow the calculation to be performed using different computer directories containing data of different tests (40 maximum) already performed by the single-specimen method. All single-specimen tests were done at different deflections creating different crack extension Δa. It was found that for both steels, the J_{Ic} values obtained from the single-specimen method using electric-potential measurements are in good agreement with those yielded by the standard multiple-specimen method.

The results obtained from the single-specimen method were verified by the simple method assumed in the following equation,

$$\frac{E_{Sc}}{WB} = -\left(\frac{J_{Ic}}{2}\right)\left(\frac{a_o}{W}\right) + \left(\frac{J_{Ic}}{2}\right) \tag{5}$$

the linearity of the variation in (E_{sc}/WB) as a function of (a_0/W) is an indication of the validity of the value of J_{Ic} determined on a single specimen in an appropriate range of (a_0/W) (Nguyên-Duy and Bayard, 1981). For both steels, this linearity was observed in the range of experimental accuracy. With regard to the quality of the results, the single specimen method using electric-potential measurements is a reliable method and is consistent with the J-integral definition which is a fracture-initiation concept.

The variation of J_{Ic} as a function of temperature is the same as in the case of the brittle-ductile transition curve obtained from impact tests on both steels. For both steels, the addition of side grooves in slow bending does not affect the value of J_{Ic}, probably because the stress raisers are less effective at low strain rate. It is therefore important to analyze the variation of J_{Ic} as a function of the strain rate. For SA-516 gr. 70 steel, in the upper shelf region (from 20°C), low strain rate J_{Ic} values are smaller than those obtained from impact testing, while the reverse trend is observed in the transition zone and lower-shelf region. This tendency is not clear in the case of SA-106B steel whose high-strain-rate (impact) J_{Ic} values are higher than its low-strain-rate J_{Ic} for temperature above 70°C. The explanation for these results is not simple, since the difference in fracture toughness is a combined effect of temperature and strain rate. It is known that lowering the temperature is generally equivalent to increasing the strain rate. At low temperatures, the movement of interstitial atoms, such as carbon, nitrogen, etc., responsible for the deformation of material is more difficult, resulting in a higher yield strength. The same explanation applies for the strain rate effect: at high strain rates, the interstitial atoms do not have enough time to move, resulting in an increase in the yield strength and a corresponding decrease both in ductility and fracture toughness. At higher temperatures, where the deformation mechanism is ductile regardless of the strain rate, it could be that the predominant factor affecting the fracture toughness is the work-hardening factor rather than the ductility. Since work-hardening is very sensitive to strain rate, it is plausible that at these temperatures, the high-strain-rate J_{Ic} values are higher than those obtained from slow-strain-rate testing. This difference could also be due to the fact that the heat-transfer process that occurred in the specimens during the tests is not the same at high as at low strain rate. If the plastic zone near the crack tip is considered as a thermodynamic system on which the energy balance is performed, at high strain rate this system is virtually adiabatic, whereas at low strain rate it is isothermal. In an adiabatic system, the temperature increases during the test, easing the flow of material and resulting in a higher fracture toughness. In light of this explanation, it is reasonable to consider the notion of an equivalent temperature for high-strain-rate testing, which is equal to the sum of the test temperature and the increase in temperature due to the adiabatic process occurring in the system constituted by the near-crack-tip plastic zone:

$$T_{eq} = T_{testing} + \Delta T(Q)$$

This term $\Delta T(Q)$ can be evaluated if the plastic-zone size, the strain energy and the heat capacity of the tested material are known.

It was found that for both steels, there exists a temperature, T_t, above which the high-strain-rate (HSR) fracture toughness if higher than the low-strain-rate (LSR) fracture toughness. This temperature seems to vary with the chemical composition and heat treatment conditions of the steels. The reverse tendency noted in the fracture-toughness variation as a function of the strain rate in the transition temperature range strongly suggests a serious consideration of the in-service temperature of components for design and quality-control purposes:

1) For in-service temperature lower than T_t, where the HSR

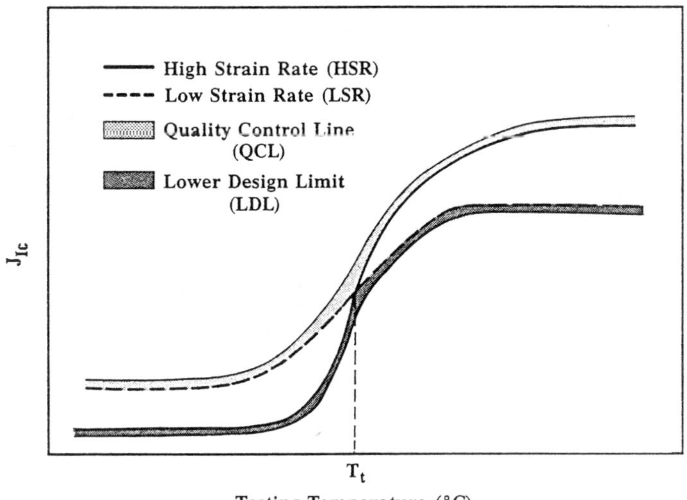

Fig. 11   Effect of strain rate of fracture toughness of steel

fracture toughness is lower than the LSR values, it is recommended
  - to choose the HSR fracture toughness as a lower limit for design purposes,
  - to use the LSR fracture toughness as a quality-control parameter to obtain good-quality final product.

2) For in-service temperature lower than T_t, where the HSR fracture toughness is higher, it is recommended
  - to select for quality control purposes the HSR values,
  - to select the LSR values as a design criterion.

This philosophy of quality control and design generally guarantees high quality products and a conservative safe design. Two envelope-curves whose upper curve is the Quality Control Line (QCL) and the lower is the Lower Design Limit (LDL) were illustrated in Fig. 11.

6.   CONCLUSION

In addition of precrack, side-grooves allow the behavior of small specimens (1 x 1 x 5 cm) to approach that of full thickness.

Impact testing performed on modified small specimens (PCVN and SGPCVN) gives reproducible fracture-toughness results, above a specific temperature, proper to a given material, the impact fracture toughness is high than the slow-bending fracture toughness.

Slow-bend testing conducted on modified small specimens (PWN and SGPWN) using electric-potential measurements allows measurement of the fracture toughness on a single specimen. These values are very comparable to those obtained using the standard ASTM multi-specimen method.

The critical stretch-zone width was confirmed as being a reliable parameter for characterizing material fracture toughness.

ACKNOWLEDGMENT

The authors thank the Canadian Electrical Association for the grant which made
this study possible (CEA 168G301).

REFERENCES

Amouzouvi, K.F. and Bassim, M.N. (1982). Determination of fracture toughness from
    stretch zone width measurement in predeformed 4340 steel. _Mat. Sci. Eng._,
    55, pp. 257–262.
ASTM (1981). Test for Plane–Strain–Fracture Toughness of Metallic Materials, _ASTM
    Standard E–399._
ASTM (1981). Test for J_{Ic}, a measure of Fracture Toughness, _ASTM Standard
    E–813._
Barsom, J.M. and Rolfe, S.T. (1970). Correlations between K_{Ic} and Charpy–V–
    Notch Test Results in the Transition–Temperature Range. _Impact Testing of
    Metals_, ASTM STP466, American Society for Testing and Material, pp. 281–302.
Landes, J.D. and Begley, J.A. (1974). Tests results from J–integral studies: an
    attempt to establish a J_{Ic} testing procedure. _Fracture Analysis_, ASTM
    STP560, American Society for Testing and Materials, pp. 170–186.
Nguyên–Duy, P. and Bayard, S. (1981). Fracture toughness of 4130 quenched and
    tempered steel. _J. of Eng. Mat. and Techno._, Trans. ASME, vol. 103, pp.
    55–61.
Nguyên–Duy, P. (1981). Relationship between critical stretch zone width, crack–
    tip–opening–displacement, and fracture–energy criterion: Application to
    SA–516 gr. 70 steel plates. _Fracture Mechanics: Thirteenth Conference_, ASTM
    STP743, American Society for Testing and Materials, pp. 543–552.
Nguyên–Duy, P. and co–workers (1986). Use of the side grooved precracked Charpy–
    V–Notch specimen technique for estimating fracture toughness. _Canadian
    Electrical Association_, CEA No. 168G301.
Rice, J.R., Paris, P.C. and Merkle, J.G. (1973). Some further results of
    J–integral analysis and estimates. _In Progress in Flow Growth and Fracture
    Toughness Testing_, ASTM STP536, American Society for Testing and Materials,
    pp. 231–245.
Robinson, J.N. and Tetelman, A.S. (1974). Measurement of K_{Ic} on small specimen
    using critical–crack–tip–opening displacement. In _Fracture Toughness and
    Slow–Stable Cracking_, ASTM STP559, American Society for Testing and Materials
    (Ed.), pp. 139–158.
Server, W.L. Oldfield, W. and Wullaert, R.A. (1977). Experimental and statistical
    requirements for developing a well–defined K_{IR} curve. _Electric Power
    Research Institute_, EPRI–NP–372.

Server, W.L. (1979). Static and dynamic fibrous initiation toughness results for
    nine pressure vessel materials. In _Elastic–Plastic Fracture_, ASTM STP668,
    J.D. Landes, J.A. Begley and G.A. Clarke (Eds.), American Society for Testing
    and Materials, pp. 493–514.
Tremblay, V., Nguyên–Duy, P. and Dickson J.I. (1987). Hydrogen embrittlement of
    high strength. _Twentieth National Symposium in Fracture Mechanics_, ASTM,
    Lehigh University Bethlehem, PA, U.S.A.

RELATIONSHIP BETWEEN STRAIN-RATE SENSITIVITY OF YIELD STRESS AND
TRANSITION TEMPERATURE FOR AN ARCTIC GRADE STEEL

B. Faucher, K.C. Wang and R. Bouchard
Physical Metallurgy Research Laboratories, CANMET, Ottawa

ABSTRACT

The yield strength of an Arctic grade steel has been measured over a wide range of
temperatures and strain rates. The fracture toughness of this steel in the cleavage
range is related to the yield strength through a micromechanical model. It is shown
how the temperature and strain rate dependence of fracture toughness can be
obtained from one set of experimental data and knowledge of the yield stress for
the appropriate conditions.

KEYWORDS

Fracture toughness; transition temperature; strain rate; yield strength; cleavage;
micromechanical model; variability.

INTRODUCTION

The ductile-brittle transition temperature of structural steels is strongly
dependent on the deformation rate. The effect of strain rate on brittle behaviour
has been related to the strain-rate dependence of yield properties(Corten and
Shoemaker, 1967; Curry, 1980;Hahn, Hoagland and Rosenfield, 1971; Henry and co-
workers, 1985; Krabiell and Dahl, 1981; Vlach, Holzmann and Man, 1981).The shift in
transition temperatures (defined at a reference value of toughness) measured at
different strain rates results from the effect of strain rate on the yield stress:
the same toughnesses are obtained at different strain rates and temperatures when
the associated yield strengths are the same.

Recently developed micromechanical models relate the probability of failure to the
fracture toughness and yield strength. For example, Tyson and Marandet (1987)
obtained the probability of failure by cleavage, ϕ, as a function of the stress
intensity factor, K, as:

$$\ln[1/(1-\phi)] = C \, B \, \sigma_y^{\alpha(N-1)/2} \, K^{\alpha} \tag{1}$$

where C is a constant for a specific material, B is the specimen thickness, σ_y and
N are, respectively, the yield stress and the stress exponent for the material and
α is a parameter that should be close to 4 from theoretical arguments. Faucher and
Tyson (1987) have shown that Eq.(1) adequately describes the thickness dependence
and the variability of fracture toughness in the brittle range.

It is the purpose of this paper to describe rate effects on the transition temperature in relation to the strain rate sensitivity of the yield stress for the Arctic grade steel used previously by Faucher and Tyson (1987). Yield properties, therefore, have been measured over a wide range of strain rates and temperatures. Dynamic values of fracture toughness have been evaluated with pre-cracked Charpy-V-notch specimens; they will be compared to fracture toughness values obtained at a low loading rate.

EXPERIMENTAL DETAILS

All the specimens were cut from the same 75 mm thick steel plate that was partially controlled-rolled and then normalized. The composition of this microalloyed steel is given in Table 1; it was experimentally developed by Algoma Steel to meet the stringent requirements of Lloyd's LT60 specifications for the Arctic (Ackert, McLean and Ray, 1984). Its Charpy V-notch transition temperature is about −80°C.

TABLE 1   Composition of Algoma LT60 steel (wt%)

Carbon	0.12	Chromium	0.17
Silicon	0.28	Copper	0.25
Manganese	0.39	Molybdenum	0.001
Phosphorus	0.017	Nickel	0.15
Sulphur	0.011	Niobium	0.041
Aluminum	0.027	Vanadium	0.053

Yield properties were evaluated from tensile and compression tests. Specimens were cylindrical with their axes in the longitudinal direction. Tensile tests were carried out with an MTS computer-controlled servo-hydraulic testing machine in the temperature range from −196°C to 25°C . Strain rates of 0.00002, 0.002 and 0.2/s were used on standard 25.4 mm gage, 6.35 mm diameter specimens. Figure 1 shows typical tensile stress-strain curves at several rates and temperatures. The strain was measured from the machine stroke below −160°C (Fig. 1b). The only condition when UTS was not observed was at the 0.2/s strain rate and below −160°C. Compression tests were carried out in a cam plastometer on specimens 19 mm in height and 6.35 mm in diameter at a strain rate of 20/s. The specimens were immersed in a cold liquid bath for at least 10 minutes after they had reached the

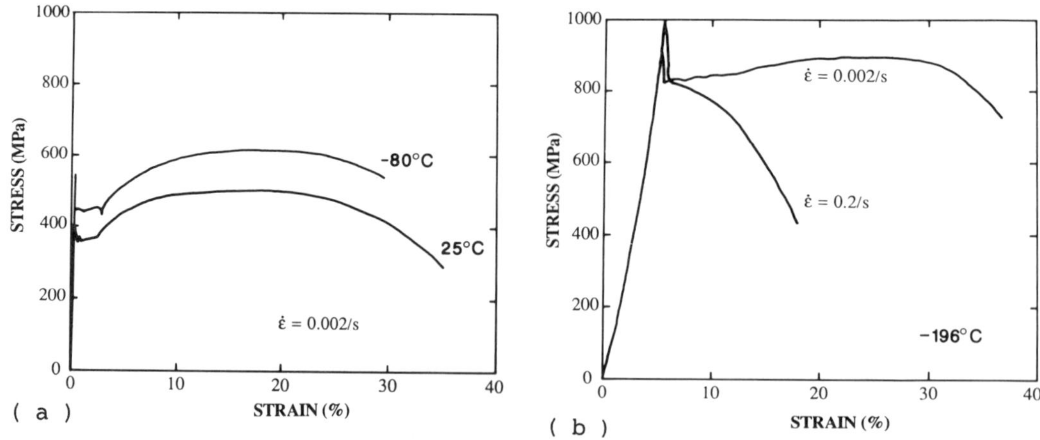

Fig. 1. Engineering stress-strain curves obtained at the indicated strain rates and temperatures.

bath temperature. They were then installed between the anvils of the plastometer and tests were initiated within 10 s of their removal from the bath. A thermocouple soldered at mid-height allowed measurement of temperature during the tests. These temperatures were about 15, -45 and -75°C.

The slow-rate fracture toughness tests were carried out on three-point-bend specimens 10 mm x 20 mm. The specimens were fatigue pre-cracked to a depth of about 12 mm. Some of them had V-grooves machined to 1 mm on each side. They were tested at a rate of 0.05 mm/s with the MTS machine in the temperature range from -160 to -110°C. Typical fracture surfaces are shown in Fig. 2.

(a)   (b)   (c)

Fig. 2. Typical fracture surfaces of slow rate specimens with (a)
and without side-grooves (b) and of impact specimens (c).

The dynamic tests were carried out on pre-cracked Charpy V-notch specimens 10 mm x 10 mm. The 2 mm V-notches of standard Charpy specimens were deepened by a 2 mm saw cut and fatigue pre-cracked to a depth of roughly 6 mm (Fig. 2c). The specimens were tested, after immersion for at least 15 minutes in a refrigerated bath at a temperature between -50 and -10°C, with a Dynatup ETI-630 instrumented impact system at an initial tup velocity of 3.1 m/s. The tests were initiated within 5 s after removal of the specimens from the bath. The load output from the transducer mounted on the tup was recorded for the full duration of the test (6 to 40 ms). Acceleration, velocity, displacement and energy were evaluated at each instant by taking into account the initial velocity, the specimen compliance, the machine compliance and the load. Typical curves are shown in Fig. 3.

EXPERIMENTAL RESULTS

Flow Properties

From tensile and compression test results, the lower yield strength has been obtained as the minimum stress value in the Lüders range (Fig. 1). Yield stresses are plotted as a function of temperature and strain rate in Fig. 4a. Work hardening

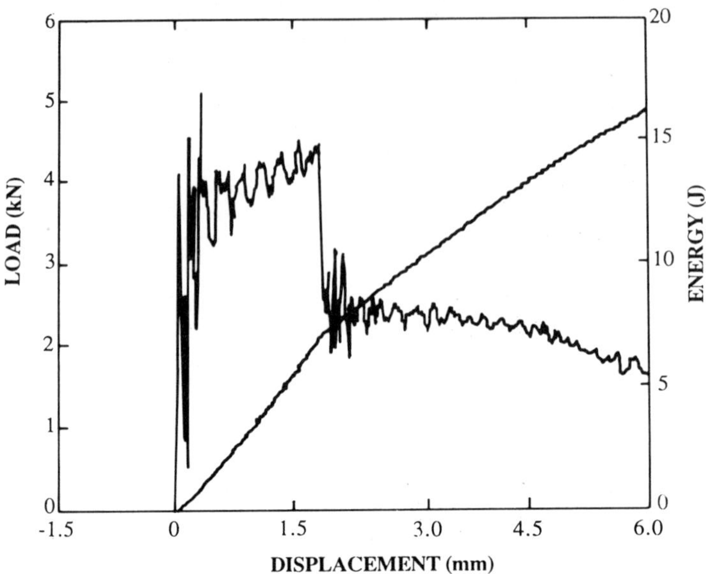

Fig. 3. Load and energy as a function of displacement
in an impact test at −30°C.

has been characterized by the strain-hardening exponent, n, defined from the true
stress–true strain relationship

$$\sigma = \sigma_o \varepsilon_p^{\,n} \tag{2}$$

where σ_o is the strength coefficient, ε_p is the plastic strain and n is the inverse
of the stress exponent N in Eq.(1). The strain hardening exponent is represented as
a function of temperature and strain rate in Fig. 4b.

The temperature and strain rate dependence of the yield stress are often
represented by a "rate parameter" $T \ln(\dot{\varepsilon}_o/\dot{\varepsilon})$ where T and $\dot{\varepsilon}$ are, respectively, the
temperature and the strain rate of the test and $\dot{\varepsilon}_o$ is a constant for a specific
material (Henry and co-workers, 1985; Krabiell and Dahl, 1981). This follows
because plastic deformation is a thermally activated process, and the rate
parameter is proportional to the apparent activation energy for the deformation
process (Krausz and Eyring, 1975). At high stress levels the apparent activation
energy corresponds to the forward activation over the controlling energy barrier
and its stress dependence is likely to be linear. Therefore, the constant $\dot{\varepsilon}_o$ was
chosen to give the minimum variance for the least squares analysis of the yield
stress as a function of the rate parameter. Only tensile stress values larger than
400 MPa have been considered because at low stress levels backward activation
becomes significant. Data obtained at liquid nitrogen temperature were also
disregarded because a different deformation mechanism may be operating at that
temperature. Results shown in Fig.5 demonstrate that the yield stress between 400
and 700 MPa can be accurately obtained from the relationship:

$$\sigma_y = 905.9 - 0.0942\, T \ln(\,2.7\ 10^8\,/\,\dot{\varepsilon}) \tag{3}$$

where σ_y is expressed in MPa, T in K and $\dot{\varepsilon}$ in s^{-1}.

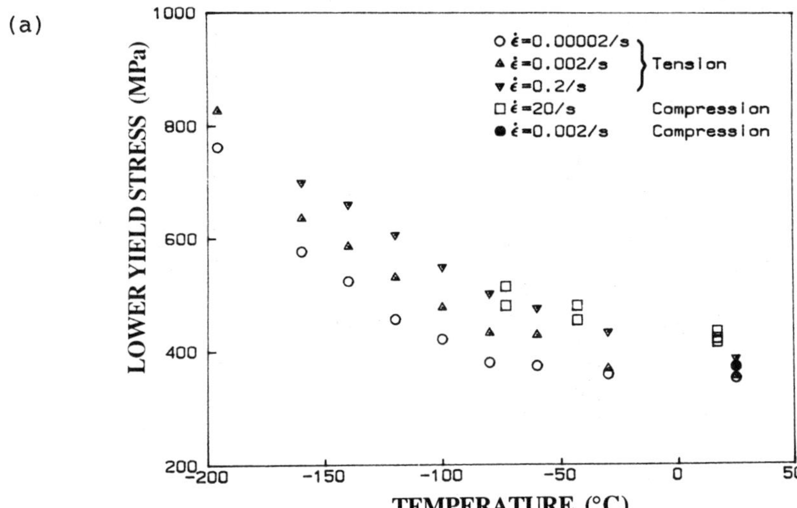

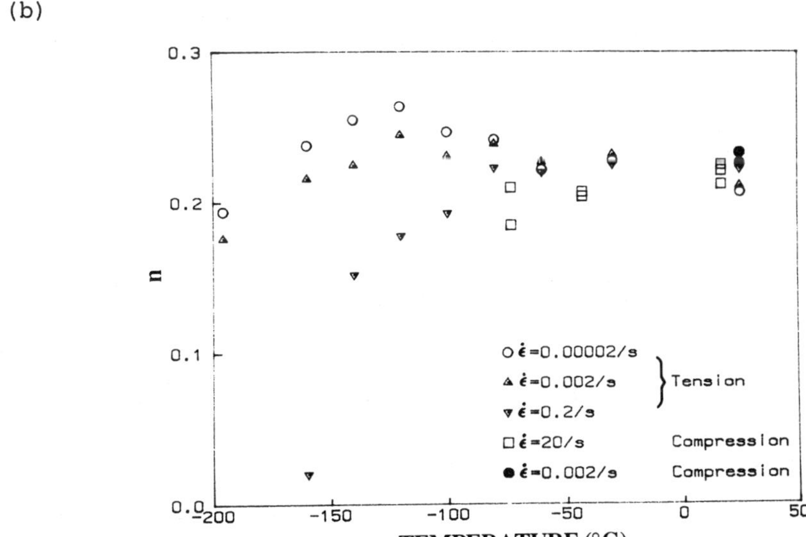

Fig. 4. Temperature and strain rate dependences of yield stress (a)
and strain-hardening exponent (b).

Figure 5 shows that the results obtained in compression tests at high loading rate do not fit the proposed relationship. This is likely a consequence of non-uniformity of the specimen temperature during these tests; the ends of the specimens were probably at a higher temperature than the middle of the specimen where the thermocouple was located.

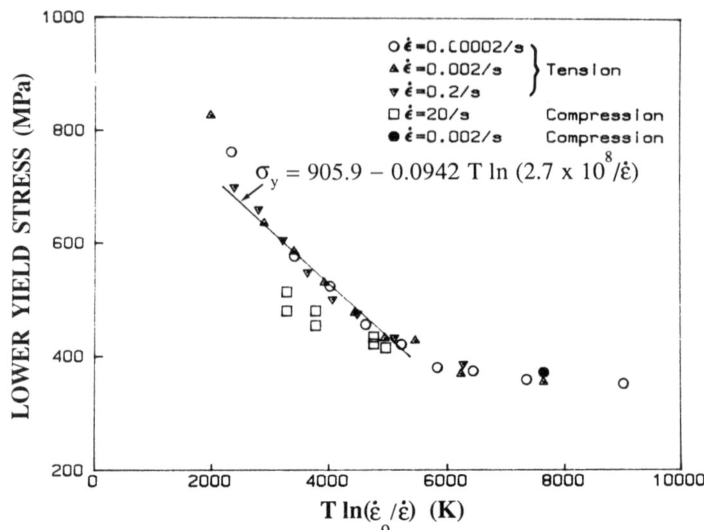

Fig. 5. Lower yield stress as a function of the rate parameter.

Fracture Toughness

From test results in the transition range, as shown in Fig. 3, fracture toughness can not be measured by either the plane strain fracture toughness, K_{IC}, because of excessive plasticity, or by the critical value of J-integral, J_{IC}, because there is no ductile tearing before the occurrence of cleavage. However, it is possible to evaluate the value of J at any point of the load-displacement curve of Fig. 3. Therefore, fracture toughness has been measured using K_J, obtained from J_c, the value of J at the occurrence of cleavage, defined as:

$$K_J = \sqrt{[EJ_c/(1-\nu^2)]} \tag{4}$$

where E and ν are, respectively Young's modulus and Poisson's ratio for the steel. When cleavage occurs, there is usually a large drop in load and a change in the slope of the energy curve (Fig. 3) which defines the critical event precisely. However, for two of the dynamic tests, cleavage produced only a small pop-in and it was followed by ductile tearing. Results from these two tests have been disregarded because it is difficult to ascertain the significance of pop-ins so small that brittle crack extension does not extend over the whole crack front making the load drop difficult to detect. At high temperatures, there is no cleavage, and the specimens fail by ductile tearing. Fracture toughness is then evaluated at the maximum load which corresponds roughly to the initiation of tearing. Results from both slow rate and dynamic tests are shown in Table 2.

DISCUSSION

Fracture Toughness-Yield Stress Relationship

Fracture toughness for cleavage is related to the yield strength of the material through Eq. (1). It is, therefore, critical to know the yield stress that should be associated with each value of K_J. This can be done using Eq. (3) if the proper

TABLE 2   Fracture Toughness, K_J, as a Function of Temperature, T, in Slow Rate and Dynamic Tests

Slow rate tests with side grooves		Dynamic tests	
T (K)	K_J (MPa$\sqrt{m}$)	T (K)	K_J (MPa$\sqrt{m}$)
113	30	223	65
113	34	233	53
123	27	233	56
123	48	233	64
133	37	243	82
133	40	243	84
133	101	243	267
133	124	253	169
143	98	253	336*
143	123	253	379*
153	71	Slow rate tests without side grooves	
153	78	T (K)	K_J (MPa$\sqrt{m}$)
153	106	113	40
153	137	133	40
163	146	143	124
163	201	143	141
163	221	153	112
163	226	153	179
163	258	163	152
		163	234

*Toughness at maximum load (no cleavage observed)

strain rate in the bend test is determined. Although there are large gradients of deformation and strain rate at the crack tip of a bend specimen, Irwin (1964) showed that at the elastic-plastic boundary the rate of deformation can be expressed as:

$$\dot{\varepsilon}=2(\sigma_y/E)(\dot{K}/K) \tag{5}$$

where $\dot{K}$ is the rate of increase of the stress intensity factor. The plastic strain rate should be evaluated at the general yield load, P_y, which for bend specimens with a span-to-width ratio of 4 is:

$$P_y=LBb^2\sigma_y/(4W) \tag{6}$$

where the constraint factor L is about 1.6, and b and W are, respectively, the ligament and the width of the specimen. Because for a specific geometry the stress intensity factor is proportional to the load, at general yielding the strain rate is:

$$\dot{\varepsilon}_y=2(\sigma_y/E)(\dot{P}/P_y) \tag{7}$$

It was shown by Faucher and Tyson (1985) that the compliance (ratio of load line displacement, δ, to load, P) of bend specimens can be expressed as:

$$EB(\delta/P)\sim13/(b/W)^2$$

for a wide range of crack lengths. Combining this expression with Eqs. (6) and (7), the plastic strain rate at general yielding is obtained as:

$$\dot{\varepsilon}\sim0.5\ \dot{\delta}/W \tag{8}$$

From Eq. (8) the strain rates were 0.00125/s and 155/s for the slow rate and the dynamic bend tests respectively. The yield stress values corresponding to the temperatures and strain rates in Table 2 can now be obtained from Eq. (3).

The yield stress dependence of the fracture toughness is represented in Fig. 6 for different tests used. The analysis of the data follows the procedure of Faucher and Tyson (1985) where Eq. (1) is rewritten as:

$$\ln K = -(1/\alpha)\ln(CB) - [(N-1)/2]\ln\sigma_y + (1/\alpha)\ln \ln[1/(1-\phi)] \qquad (9)$$

The two first terms of the right hand side of Eq. (9) represent the deterministic term of fracture toughness; they are calculated by linear regression of lnK on $\ln\sigma_y$. Experimental results that did not satisfy the ASTM E-813 size requirement, $b > 25J/\sigma_y$, were not included. The third term on the RHS of Eq.(9) is the "random" term that represents the variability of fracture toughness data; it is obtained from the difference between the measured value of lnK and the deterministic term. The cumulative failure probability, ϕ , is obtained by ranking the random terms. For the specimen with the i-th lowest random term among I test results (counting the invalid results),

$$\phi = (I-0.3)/(I+0.4)$$

Linear regression of the random term against $\ln \ln[1/(1-\phi)]$ gives the value of α. With these estimates of ϕ and α, the random term is subtracted from the experimental values of lnK, and a better estimate of the deterministic term is obtained. After a few iterations, convergence was obtained and yielded the following relation:

$$\ln K = 37.734 - 5.224\ln \sigma_y + 0.329\ln \ln[1/(1-\phi)] \qquad (10)$$

where the stress is expressed in MPa and the fracture toughness in MPa√m. The parameters obtained indicate that the stress exponent is N~11.4, and α in Eq.(1) is about 3 for the steel used. These values are close to those obtained previously by Faucher and Tyson (1987) who found on the average N=11.6 for the same steel and α=3.3 for 10 mm thick specimens. This value of N is much larger than the average

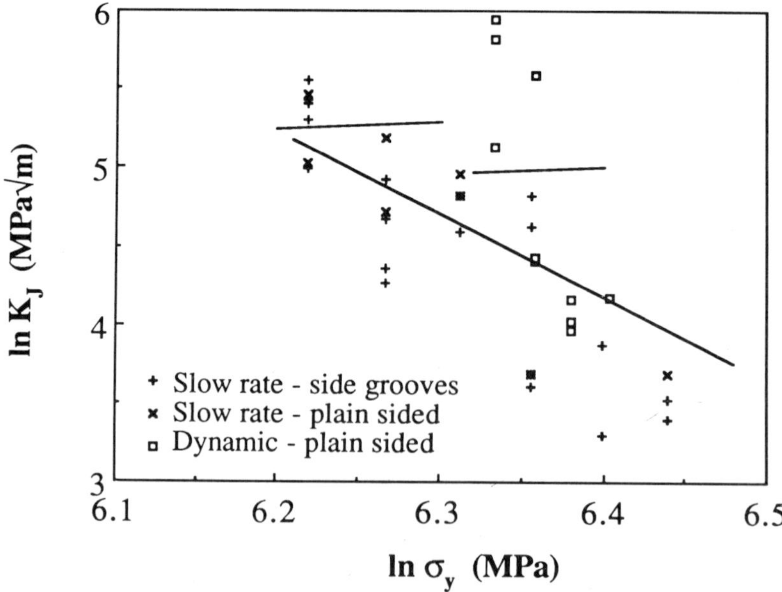

Fig. 6. Yield stress dependence of fracture toughness. Points above the horizontal lines do not satisfy the ASTM E-813 size requirement.

value of 4.5 (1/0.22) obtained from tensile test results (Fig.4b). However, a large value of N was shown by Faucher and Tyson (1987) to provide a better fit to bend test behavior.

The parameter α is usually obtained from a Weibull plot of Φ as a function of K. Because the existing results were measured under various conditions, an equivalent plot can be obtained by including the yield stress dependence of the fracture toughness in the variable. The resulting plot in Fig.7 shows that the data from different tests are well interspersed, which indicates that Eq.(10) provides an adequate representation of the variability of fracture toughness over a wide range of temperature and strain rates. There is a slight curvature at low toughnesses, which could result from a minimum toughness value K_{min} such that K in Eq.(1) should be replaced by $K-K_{min}$. However, the physical meaning of K_{min} is not yet well understood, and the apparent curvature may be a statistical effect of the limited number of data points.

<u>Effect of Deformation Rate on the Transition Temperature</u>

The ductile–brittle transition temperature is normally defined at a specific value of fracture toughness. At K= 80 MPa√m, according to Eq. (10), the yield stress is 580 MPa at a failure probability of 0.5. Because 580 MPa is well within the range of validity of Eq.(3), the temperature of transition T_t can be obtained as a function of the strain rate from Eq.(3) as:

$$T_t = 3460 \; / \; \ln(2.7 \; 10^8 \; / \; \dot\varepsilon) \tag{11}$$

Furthermore, by combining Eqs.(3) and (10), it is possible to obtain the temperature dependence of fracture toughness for a specific strain rate and

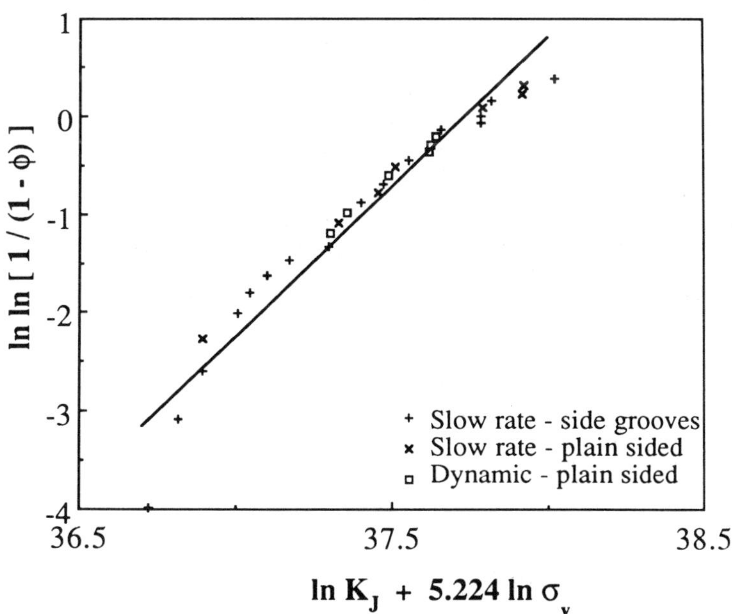

Fig. 7. Weibull plot of the fracture toughness corrected for the various test conditions.

probability of failure. For example, at a failure probability of 50% ($\phi=0.5$), the fracture toughness is given by the relationship:

$$\ln K_J = 37.613 - 5.224\ln[905.9 - 0.0942\ T\ \ln(2.7\ 10^8/\dot{\varepsilon})] \tag{12}$$

Calculated results are compared with experimental data in Fig.8. The transition temperatures at K=80 MPa√m, obtained from Eq.(11), are 133 and 241 K for slow rate and dynamic tests respectively, which is in excellent agreement with the experimental results. The model, however, predicts much lower toughnesses than the measured values in the dynamic tests above 250 K. Although the measured values are "invalid", some valid results would have been expected in the predicted range. These high observed values may be due to adiabatic heating at large plastic deformation and high loading rate, but further study will be necessary to clarify this point.

Several factors limit the applicability of the current study. Use of the toughness parameter K_J and of Eq.(1) imply that the stress ahead of the crack tip is well described by the "HRR" stress field. This is not true when the ASTM E-813 size requirement is not met, in which case the measured values may over-estimate the fracture toughness, as seen in Fig.8. This analysis also requires that the strain-hardening exponent shall be constant. As seen from Fig.4b, this is approximately true, except at extremely high strain rates and low temperatures, that is at large values of yield stress.

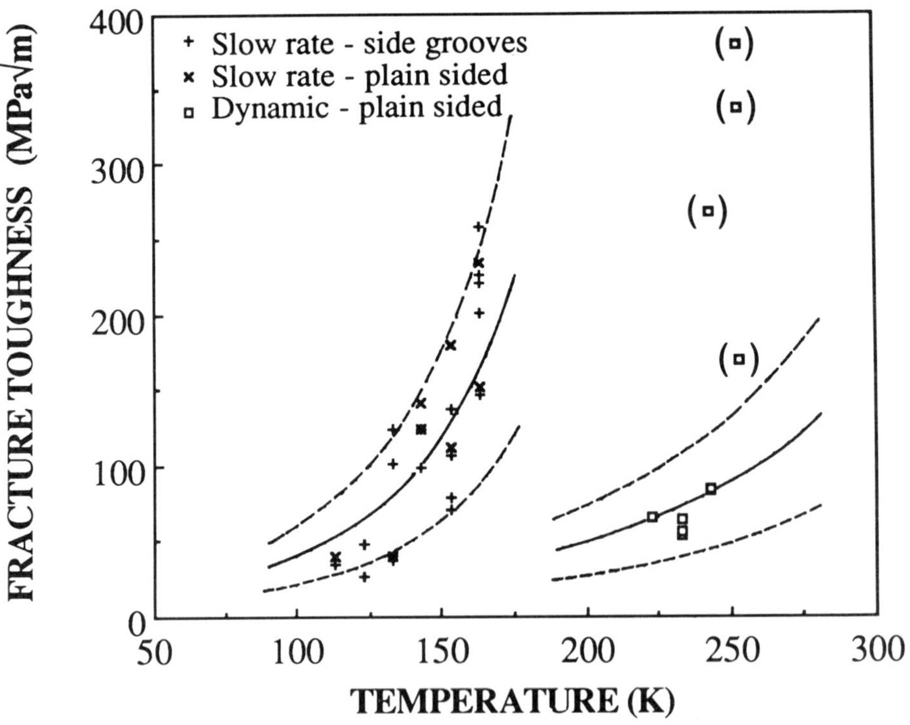

Fig. 8. Temperature and strain rate dependence of fracture toughness. The data points are from Table 2 and the curves were calculated from Eqs. (3) and (10) at failure probabilities of 10 and 90% (dashed curves) and 50% (continuous curve). For the data points in parenthesis, $b<25J/\sigma_y$.

Application of Eq.(12) is also limited by the range of validity of Eq.(3) that describes the temperature and strain rate dependence of the yield stress only for $400\text{MPa}<\sigma_y<700\text{MPa}$. A functional relationship that would apply over a larger stress range could be derived. However, it is this stress range that is critical for cleavage fracture. The validity of Eq.(3) at a high rate of loading should also be verified, even if the agreement between the experimental results and analysis is good.

A larger number of valid data points would be needed for a precise evaluation of the constants in Eq.(12). More data would also be needed to ascertain the effect of side grooves on the measured toughness. The experimental results in Figs.7 and 8 do not show any difference between plain-sided and side-grooved specimens. However, Eq.(1) predicts that side-grooved specimens, with a smaller net thickness than plain-sided specimens, should have a higher toughness. This has not been observed, probably because side grooving increases the constraint (stress level) at the crack tip, which should have the opposite effect on toughness, according to Eq.(1). Further experimental results are needed to quantify the effects of side-grooving.

CONCLUSIONS

The temperature and strain rate dependences of the yield stress of Algoma LT60 steel can be described by a single rate equation over the 400-700 MPa stress range.

On the basis of a micromechanical model, the toughness for cleavage fracture has been related directly to the yield strength of the steel.

The model provides a good description of the variability of fracture toughness at low strain rate, and of the effect of strain rate on the transition temperature.

When the various parameters of the micromechanical model are known, fracture toughness can be estimated in the cleavage range at any strain rate and temperature for which the yield stress is known.

ACKNOWLEDGEMENTS

This work forms a part of the PMRL/CANMET project on steels for offshore structures. The contributions of C.R. Trinque for the slow rate bend tests, B. Paquette for the dynamic tests, and Dr. D. Baragar and R. Malloy for the plastometer tests, is gratefully acknowledged.

REFERENCES

Ackert, R.J., McLean, M.K., and Ray, J.D. (1984). LT60 plate for Arctic application. 23rd Annual CIM Conference of Metallurgists, Quebec, August 1984.
Corten, H.C., and Shoemaker, A.K. (1967). Fracture toughness of structural steels as a function of the rate parameter T lnA/ε. Trans. of the ASME, J. of Basic Engineering, 86-92.
Curry, D.A. (1980). Predicting the temperature and strain rate dependences of the cleavage fracture toughness of ferritic steels. Materials Science and Engin., 43, 135-144.
Faucher, B., and Tyson, W. R. (1985). A comparison of crack-mouth opening and load-line displacement for J-integral evaluation using bend specimens. In E.T. Wessel and F.J. Loss (Eds.), Elastic-Plastic Fracture Test Methods: T Experience, ASTM STP 856, American Society for Testing and Materials, Philadelphia, pp.278-293.

Faucher, B., and Tyson, W.R. (1987). A study of variability, size and temperature effects on the fracture toughness of an Arctic grade steel plate. In Fracture Mechanics: Eighteenth Symposium, ASTM STP 945, American Society for testing and Materials, Philadelphia, to be published.

Hahn, G.T., Hoagland, R.G., and Rosenfield, A.R. (1971). The variation of K_{Ic} with temperature and loading rate. Metall. Trans., 2, 537–541.

Henry, M., Marandet, B., Mudry, F., and Pineau, A. (1985). Effets de la température et de la vitesse de chargement sur la tenacité à la rupture d'un acier faiblement allié. Interprétation par des critères locaux. J. de Mécanique théorique et appliquée, 4, 741– 68.

Irwin, G.R. (1964). Crack–Toughness Testing of Strain–Rate Sensitive Materials. Trans. of the ASME, J. Eng. Power, A86, 444–450.

Krabiell, A., and Dahl, W. (1981). Influence of strain rate and temperature on the tensile and fracture properties of structural steels. In D. François (Ed.), Advances in Fracture Research (Fracture 81). Proc. of the 5th Int. Conf. on Fracture (ICF5). Cannes, 1981, Vol. I. Pergamon Press, Oxford, pp. 393–400.

Krausz, A.S., and Eyring, H. (1975). Deformation Kinetics, John Wiley & Sons, New-York.

Tyson, W.R., and Marandet, B. (1987). Cleavage toughness variability and inclusion size distribution for a weld metal. In Fracture Mechanics: Eighteenth Symposium, ASTM STP 945, American Society for Testing and Materials, Philadelphia, to be published.

Vlach, B., Holzmann, M., and Man, J. (1981). The relationship between transition temperatures and fracture toughness parameters of steels. Kovove Materialy, 19, 601–611.

A COMPARISON OF FRACTURE TOUGHNESS MEASUREMENT TECHNIQUES APPLIED TO A QUENCHED AND TEMPERED LOW ALLOY STEEL

J. Morrison

Defence Research Establishment Pacific

Victoria, BC

ABSTRACT

Different experimental techniques have been used to determine the quasi-static fracture toughness of a low alloy steel tempered to produce a range of strengths and toughnesses. Linear-elastic and elastic-plastic procedures have been investigated using both compact and short bar specimens. Comparisons have been made between K and J values as appropriate, including the assessment of alternative measurement points on the crack growth resistance curves. CTOD and Charpy correlations have also been included. In general, consistent K values are obtained by the different methods, while J data can be differentiated on the basis of either ductile crack extension or limit load tearing.

KEYWORDS

Fracture toughness; critical stress intensity; elastic plastic fracture; J-integral; crack opening displacement; quenched and tempered steel.

INTRODUCTION

The increasing use of high strength, high toughness structural steels, coupled with greater performance requirements, results in increasing demands for meaningful and reproducible fracture toughness measurements, for example, for quality control. At the same time, improvements in structural integrity analyses based on flaw tolerant designs demand more sophisticated, but preferably less expensive experimental techniques for assessing crack tolerance, including critical defect size calculation. Fracture toughness determination is a key element in the material property measurements needed to support such evaluations.

An increasing number of standardized procedures for determining quasi-static toughness are being introduced. These offer the opportunity for quantitative assessment over a wider range of strength and toughness, using a variety of specimen types. The success of the fracture mechanics approach is judged in part by the extent to which data generated by small scale laboratory tests can be applied to large structures, rather than being restricted to a narrow range of application corresponding to the conditions under which the tests were performed.

Linear-elastic K_{Ic} measurements (ASTM, 1985a) on low alloy steels of limited thickness are frequently invalidated owing to inadequate specimen size, and/or to excessive local plasticity at the crack tip. Consequently their fracture characteristics are frequently quantified by means of the more complicated elastic-plastic J_{Ic} (ASTM, 1985b) and Crack Opening Displacement (BSI, 1979) techniques. Alternatively, because of the inherently greater crack tip constraint, chevron notched short bar specimens (SAE, 1981) can provide meaningful estimates of plane strain toughness using specimens of only half the thickness demanded by the K_{Ic} procedure. In the same way, for a given short bar specimen thickness, a valid plane strain toughness measurement can be made of twice the magnitude permitted using a through cracked compact or bend specimen.

This paper describes an experimental comparison of some toughness test methods on 25-mm thick low alloy steel specimens tempered to six different strength levels.

EXPERIMENTAL PROCEDURE

The material used in this study was a general purpose quenched and tempered low alloy steel equivalent in performance to AISI/SAE 4140 grade. The chemical composition is shown in Table 1. The steel was supplied in the form of 38-mm thick by 200-mm wide plates, from which were machined standard 25-mm thick compact specimens, and 25-mm square chevron notch short bar specimens.

TABLE 1   Chemical Composition (wt. %)

C	Mn	Si	S	P	Ni	Cr	Mo
0.42	1.2	0.2	0.02	0.03	0.4	0.9	0.06

In order to test material over a range of strength and toughness, six batches of specimens were used, each austenitized at 840 ^{o}C for one hour, oil quenched, and one batch each subsequently tempered for two hours at 370, 430, 480, 540, 590, and 650 ^{o}C. The mechanical properties quoted by the manufacturer for each temper are listed in Table 2. Figure 1 shows typical photomicrographs of the samples tempered at the lowest and highest temperatures. Both show tempered martensite, but, in the sample tempered at 650 ^{o}C, ferrite and carbide can be better resolved, and there are regions of coarser banded structure. Tensile properties and hardnesses were measured on each batch using standard ASTM procedures, as well as a limited number of room temperature Charpy impact tests. The results are shown in Table 3. The measured values closely follow those of Table 2, although at slightly reduced strength levels. Charpy impact results for the intermediate tempers are also somewhat lower than expected.

TABLE 2   Reported Variation in Mechanical Properties With Tempering Temperature

T ^{o}C	Yield Stress MPa	UTS MPa	Elongation %	Reduction of Area %	Hardness R_c	Izod J
370	1340	1600	10	42	47	12
430	1230	1440	10	43	43	16
480	1100	1310	11	45	39	27
540	970	1120	15	49	35	46
590	860	965	18	55	30	84
650	760	840	21	62	24	111

a) Tempered at 370 °C

b) Tempered at 650 °C

Fig. 1. Typical microstructures (x600)

Over the tempering temperature range investigated, the tensile strength and
hardness varied by a factor of two, and the impact energy by a factor of twenty. A
correspondingly wide range of fracture resistance was expected. Accordingly, for
compact specimens, both linear elastic K_{Ic} and elastic-plastic J_{Ic} techniques were
employed as appropriate. All of the tests described in the following sections were
carried out at room temperature on a 250 kN MTS servohydraulic machine equipped
with a DEC LSI 11/23 computer and MTS 468 interface.

TABLE 3   Average Measured Room Temperature Mechanical Properties

T °C	0.2% Proof Stress MPa	UTS MPa	Elongation %	Hardness R_c	Charpy V-notch at 20 °C J
370	1230	1445	9	45	6
430	1200	1310	11	42	14
480	1020	1145	14	38	23
540	935	1035	17	33	20
590	800	925	19	29	-
650	650	790	23	25	116

Compact Specimen Testing

The standard 25-mm thick compact specimens were equipped with integral machined
knife edges at the load line in order to accommodate a standard clip gauge.
Duplicate specimens were tested in accordance with the ASTM E813 procedure (ASTM,
1985b) for J_{Ic}. Specimens were sidegrooved a total of 10% of the thickness in
order to maintain crack front straightness, and fatigue precracked to an a/w of
0.55. Crack extension during both the precracking and the J test was monitored by
the unloading compliance technique. Typically between 20 and 30 unloadings were
performed, of which about one third fell within the exclusion limits for J_{Ic}
determination. Final crack extension was monitored by fatigue cycling to mark the
limit of the ductile crack, and the specimen subsequently was broken open after
immersion in liquid nitrogen.

In addition to the conventional linear extrapolation of the crack growth resistance curve to an intercept with the blunting line in order to determine J_{Ic}, a power law fit was made to the data, and the J integral value at an offset to the blunting line of 0.2 mm crack extension (designated $J_{0.2}$) used to define the critical J value (Fig.2). This procedure may be included in future standards, primarily because of the uncertainty and large scatter sometimes encountered using the blunting line intercept. The offset measurement is seen as strictly an engineering estimate of toughness without the theoretical significance as a crack initiation parameter sometimes invoked for the blunting line intercept J_{Ic}.

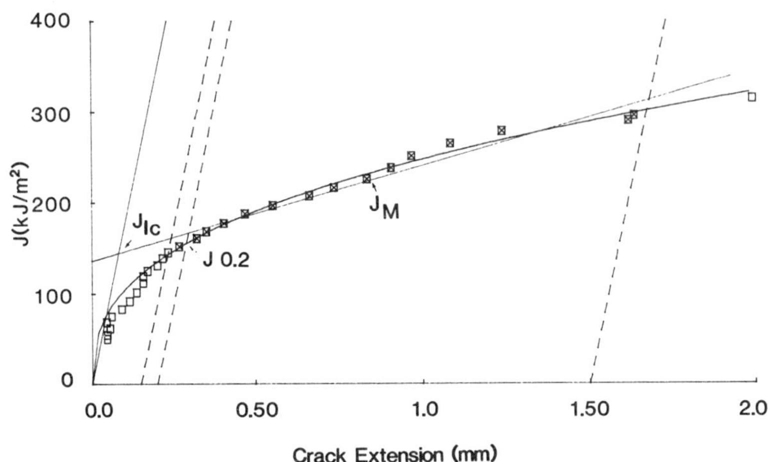

Fig. 2   J measurement points

Crack tip opening displacement (CTOD) was also calculated for each unloading point up to maximum load using the approach of standard BS5762 (BSI, 1979) (which only applies to bend specimens). For this calculation, the K solution from E399 was used, together with the unloading compliance slopes (for both Δa and the corresponding plastic component of displacement). Such an approach is likely to be included within a future ASTM standard. In a similar manner to $J_{0.2}$ testing, both a linear extrapolation to the ordinate (δ_i) and the 0.2 mm ($\delta_{0.2}$) crack extension point along a power law curve fit to the R-Curve were used as estimates of CTOD.

Chevron Notched Short Bar Testing

Toughness was also measured using square chevron notched short bar specimens. Specimen dimensions and a schematic view of the test arrangement is shown in Fig. 3. The test procedure has been described elsewhere (SAE, 1979; Barker,1983) and is also under serious consideration for ASTM standardization. This test has been proposed as being simpler, more widely applicable, and more reproducible as an estimate of plane strain fracture toughness than a standard K_{Ic} test. It is believed to generate a toughness, designated K_{IV} or K_{IVM}, which characterizes the resistance of a material to fracture in the presence of a growing crack, rather than to crack initiation from a preexisting flaw. The main advantages of this type of test are the elimination of the fatigue precrack and a relaxed specimen thickness requirement. In addition, for a linear-elastic material, as the crack extends from the tip of the chevron to a critical length, the crack tip stress intensity coefficient (which depends on specimen geometry) reaches a critical minimum value. This minimum value necessarily occurs at maximum load, a very desirable feature of a toughness test. Where significant deviations from linear-elastic behaviour occur, the critical stress intensity will be displaced from the maximum load point. In this case, unloading slopes can be used to determine the point on the load extension curve at which the critical crack length is reached, and the corresponding applied load and stress intensity K_{IV}. Earlier work (Morrison and Gough, 1986) indicated that the maximum load toughness (designated K_{IvM} in cases where the unloading slope analysis is omitted) correlated reasonably closely but non-conservatively with K_{IC} and J_{IC}, and that the unloading slope procedure was not very useful for quenched and tempered steel. Nevertheless, 3 or 4 unloadings were performed during the current experiments in order to permit a comparison to be made between the two short bar K values.

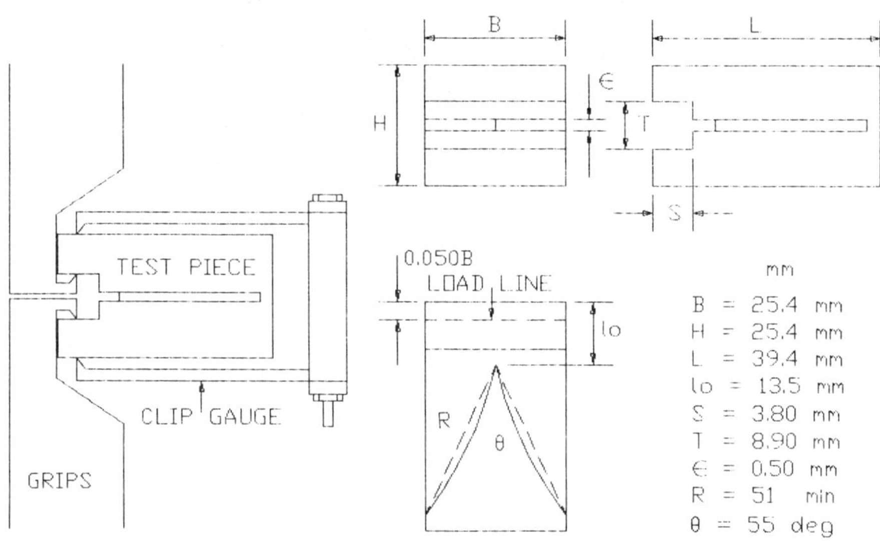

Fig. 3.   Schematic short bar testing arrangement.

EXPERIMENTAL RESULTS

<u>Tensile Properties</u>

Figure 4 shows the true stress-strain curves for the six different heat treatment conditions, derived from the engineering stress-strain data on the assumption of constant volume. The samples tempered at 650 °C show a distinct yield point; the remaining curves are smooth. These data were fitted to the strain hardening equation:

$$\sigma = K\varepsilon^n \tag{1}$$

where ε is the true plastic strain. The results are shown in Fig. 5. The fit is very sensitive to the strain range over which it is made, in this case from 0.005 to maximum load strain (increasing with tempering temperature from about 0.02 to 0.1). Overall, however, the results are consistent with previously reported studies (Schwalbe and Backfisch, 1977; Winter and Woodward, 1986). The strength coefficient K shows a typical correlation with hardness. There is no hardness plateau indicative of the delayed tempering effect sometimes observed in low alloy steels. Delayed tempering results from the precipitation of alloy carbides at tempering temperatures above 500 °C, arising from such additions as vanadium and molybdenum, which are not present in this steel in significant amounts. The work hardening index n has a minimum in the region of a 500 °C temper, which is a somewhat higher temperature than reported elsewhere on similar

material (Winter and Woodward, 1986). Fracture toughness can be expected to
correlate with strain hardening index in situations where crack tip behaviour is
controlled by local plasticity, since n is related to strain localization, and
hence to void formation. Such a correlation will not apply in cases where fracture
results from brittle cleavage or intergranular separation.

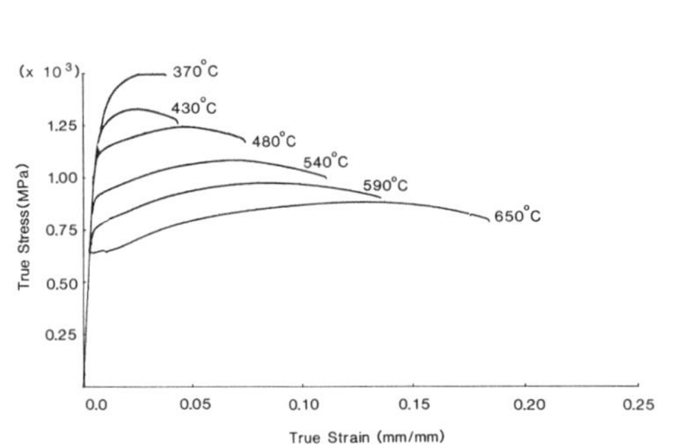

Fig. 4. True stress-strain curves.

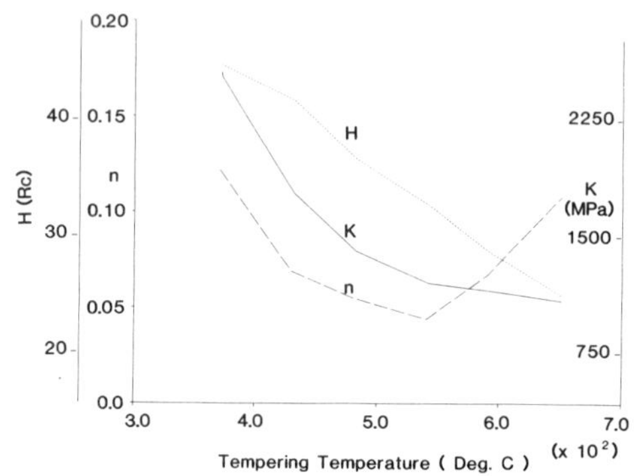

Fig. 5 Strain hardening coefficients.

Toughness Test Results

Charpy impact energies. The maximum K_{Ic} toughness expected from these tests can be
estimated from the upper shelf Charpy energy, for which there is a well
established correlation (Barsom and Rolfe, 1970) of the form:

$$\left[\frac{K_{Ic}}{\sigma_y} \right]^2 = 0.66 \left[\frac{C_v}{\sigma_y} - 9.6 \times 10^{-3} \right] \qquad (2)$$

where C_v in Joules is the standard Charpy energy at 27 °C; σ_y is the yield strength
in MPa, and K_{Ic} is in MPa.√m. On this basis, the K_{Ic} toughness estimate from the
data in Table 3 would be about 220 MPa.√m for specimens tempered at 650 °C, assuming
that a room temperature Charpy value for this treatment is representative of the
upper shelf energy. At this toughness level, insufficient material thickness is
available to permit a valid plane strain toughness using either compact or short bar
specimens. At lower tempering temperatures, the relatively low impact energies are
unlikely to be representative of the upper shelf so that the correlation is not
considered applicable.

Compact specimen tests. Figure 6 shows the load-clip gauge displacement curves. The
desired wide range of toughness is apparant in the fourfold change in load bearing
capacity of these specimens, and also in the extremes of crack growth resistance.
The corresponding J-Δa curves are shown in Fig. 7. For specimens tempered above 430°
C, the test resulted in an extended region of stable crack extension, whereas
specimens tempered at the two lowest temperatures exhibited sudden failure at
maximum load without observable crack extension. In the latter cases, a K_Q
calculation was made based on ASTM Standard E399 (ASTM 1985a) for plane strain
fracture toughness measurement. The results are summarized in Table 4, which also
includes, for the case of slow stable crack extension, a comparison of J_{Ic} and $J_{0.2}$

values determined using the two different criteria described above. It can be seen in the Table that the two J methods give essentially the same results. It should be noted, however, that when applied to a linear fit rather than to a curve fit, the offset method gives a value which is 15-20 % higher than the intercept values. As far as could be ascertained from this limited number of tests, the reproducibility was about the same for both techniques, variability increasing with toughness to about ±10%. K_{Ic} reproducibility was approximately ±5%.

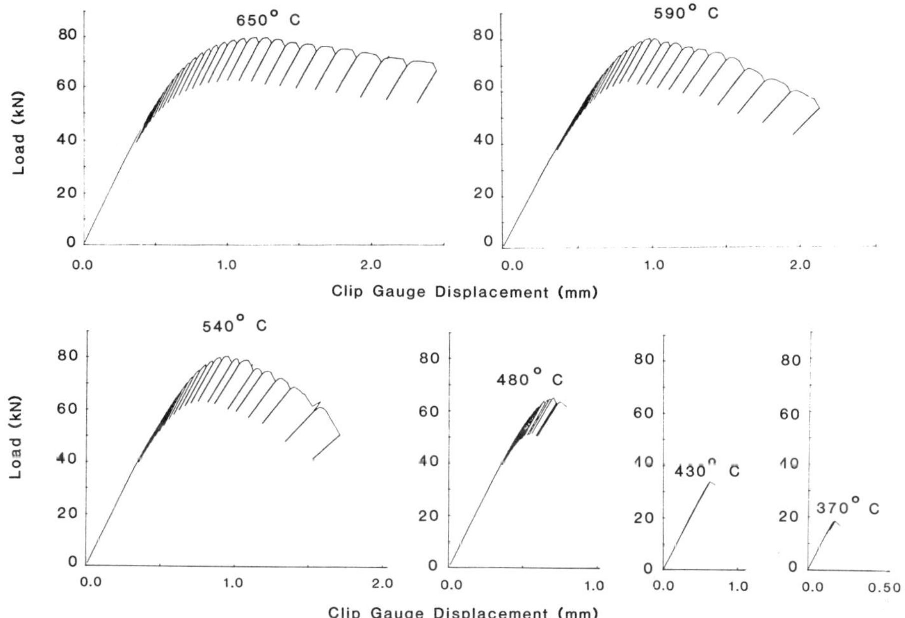

Fig. 6 Load-clip gauge displacement curves from compact specimens.

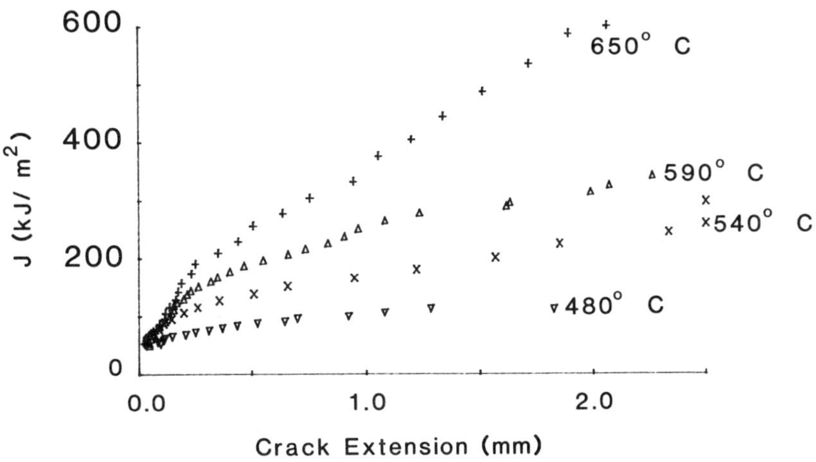

Fig. 7. J-Δ_a curves from compact specimens.

Also shown in Table 4 are the maximum load J values, and also the ratio of the
maximum load measured during the test to the theoretical limit load based on the
flow stresses (the average of yield and ultimate stresses). The same maximum load is
obtained in these tests for the three toughest steels even though in only one of
case is the the theoretical limit load exceeded, and the maximum load J values are
very different. There is also a small but systematic variation between the measured
fatigue precrack and that calculated from compliance which slightly overestimates
crack length for the lower toughness specimens. Typical compact specimen fracture
surfaces are shown in Fig. 8.

TABLE 4   Results of Compact Specimen K_{Ic} and J_{Ic} Tests

T	K_{Ic}	J_{Ic} a)Linear Intercept	b)0.2 mm Offset	J at Maximum Load	Tearing Modulus	P_{Max}/P_L
°C	MPa.√m	kJ/m^2	kJ/m^2	kJ/m^2		
370	40	–	–	–	–	0.15
430	77	–	–	–	–	0.28
480	(138)	63	59	92	8	0.61
540	–	98	93	179	15	0.82
590	–	137	125	217	27	0.95
650	–	168	182	258	97	1.13

( ) - invalid

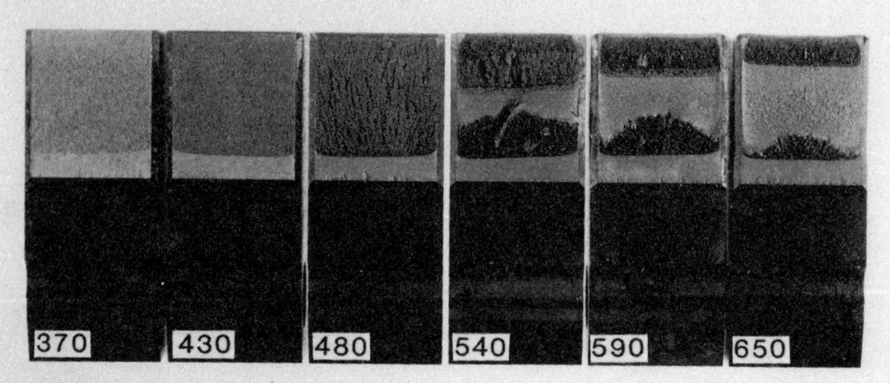

Fig 8. Compact specimen fracture surfaces.

Short bar tests. Figure 9 shows the load deflection curves for each temper. A
similar range of behaviour to that of the compact specimens is evident. At high
toughness, the curves are rounded without any apparant instabilities up to maximum
load. As the toughness decreases, an arrested instability is observed prior to
maximum load. In the case of the 430 °C, temper fracture occurs by a series of such
instabilities, and, in the case of the two lowest tempers, final fracture occurs
abruptly. In the terminology used to describe short bar test records, samples
tempered below 540°C would be classified as having crack jump test records, in that
the initial instability usually resulted in an unloading of greater than 5%. The
remaining curves would be described as smooth. Tearing in the toughest specimens
eventually gives way to bending of the specimen arms, whilst in the intermediate
tempers separation sometimes occurs by one arm breaking off, leaving a fracture
normal to the original crack plane. Figure 10 shows typical fractured half
specimens.

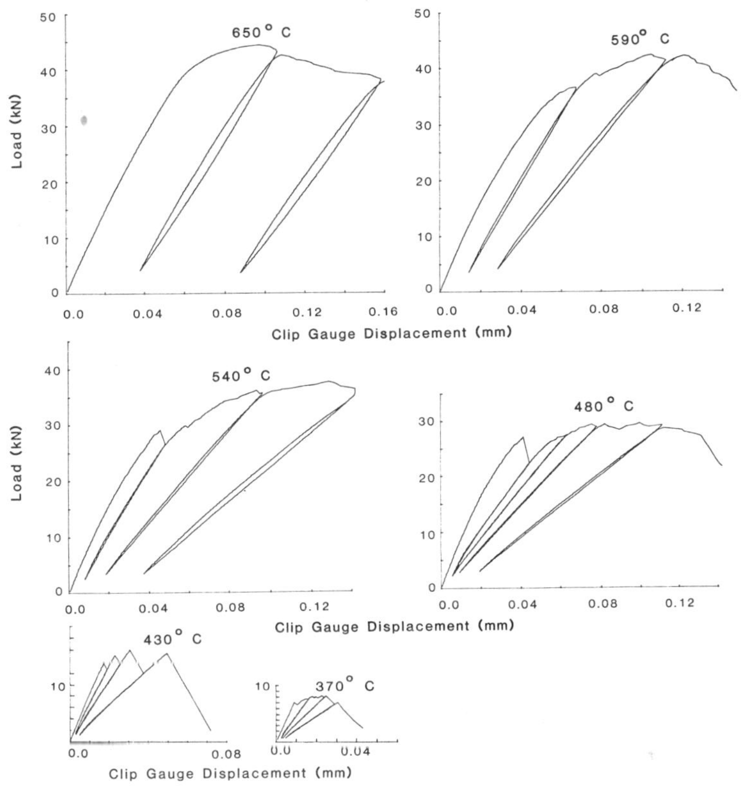

Fig. 9. Load-displacement curves from short bar specimens.

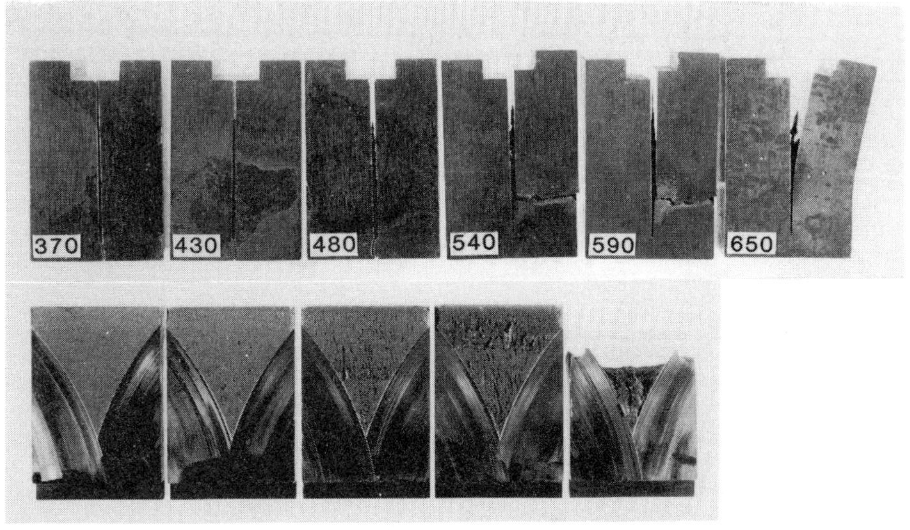

Fig. 10. Broken short bar specimens.

The results are summarized in Table 5, which includes K values based on both maximum load (K_{IvM}) and on a value determined at the critical unloading slope (K_{Iv}) . For the type of specimen used in this study K_{Iv} is estimated by extrapolation to the point corresponding to an unloading slope of 0.3 of that of the initial linear part of the test record, which corresponds to the critical crack length for maximum load in a linear-elastic test. Also shown in Table 5 is the maximum K capacity at each yield strength level; a measured K higher than this value would be invalid by virtue of insufficient specimen thickness.

TABLE 5   Short Bar Test Results

T °C	K_{Iv} MPa.√m	K_{IvM} MPa.√m	Maximum Measurable K MPa.√m
370	42	44	191
430	(69)	83	175
480	151	155	157
540	(172)	(198)	138
590	(210)	(217)	122
650	(219)	(232)	108

( ) - invalid

Overall, the results for the three lower tempers provide valid plane strain toughness measurements, in that they meet the minimum thickness and plasticity restrictions. Most of the data from specimens tempered above 480 °C, on the other hand, are invalid, primarily because of insufficient specimen thickness. The limit load behaviour exhibited by the 650 °C temper, for example, results in either too much plasticity, or insufficient crack extension to permit a valid K_{Iv} estimation.

Crack opening displacement. Table 6 shows the crack tip opening displacement results from the compact specimen J_{Ic} tests. Included are δ_m (maximum load) values, and δ_i (initiation) values based on a linear extrapolation of the CTOD-Δa curve to the ordinate, in accordance with BS5762 for multispecimen tests. Also shown for comparison are the $\delta_{0.2}$ values derived from a simple curve fit. Specimens tempered below 480 °C, being brittle, do not give useful CTOD data. It can be seen that δ_i is always small, and is approximately constant for the three highest tempers, reflecting limit load behaviour, whereas $\delta_{0.2}$ and δ_m increase throughout.

TABLE 6   Crack Tip Opening Displacements

T °C	δ_i mm	$\delta_{0.2}$ mm	δ_m mm
480	0.005	0.008	0.032
540	0.009	0.015	0.081
590	0.011	0.024	0.098
650	0.008	0.052	0.143

TOUGHNESS COMPARISONS

Critical stress intensity values of varying validity with respect to test standards are obtainable from specimens tempered at up to 480 °C. These results are shown in

Table 7(a), which includes K_{Ic}, K_{IvM}, and, in the case of the 480 °C temper, a K derived from $J_{0.2}$ using the linear-elastic relation:

$$K_{Ic} = \frac{\sqrt{JE}}{(1-\nu^2)} \qquad\qquad (3)$$

For specimens tempered above 480 °C (i.e. having an equivalent plane strain toughness above about 150 MPa.√m), test techniques have been compared on the basis of J estimates. Table 7(b) includes J measured at both the 0.2 mm crack extension offset and maximum load. Also included are values calculated from K_{IvM} data using $J = K^2/E$, the plane stress equivalent of equation (3), and from Charpy using equations (2) and (3).

TABLE 7   Toughness Comparisons

a) Critical stress intensities MPa.√m

T °C	K_{Ic}	$K_J{}^1$	K_{IvM}
370	40	–	44
430	77	–	83
480	(138)	121	155

1) $K = \sqrt{JE}/(1-\nu^2)$

b) J values kJ/m²

T °C	$J_{0.2}$	at P_{Max}	from $K_{IvM}{}^1$	from Charpy²
540	93	179	189	131
590	125	217	227	136
650	182	258	260	161

Notes: 1) $J = K^2/E$
         2) equation (2)

There is good agreement between K_{Ic} and K_{IvM} for the two lowest tempers. In the other tempers, estimates from K_{IvM} are very close to maximum load J values from compact specimens. Maximum load values are often of interest in instability analyses of high toughness steels, depending on the compliance of the structure relative to the testing facility. The correlation between the two specimen types is somewhat surprising given the dissimilarity in mode of plastic tearing at maximum load in some of the short bar specimens. This is usually not the remaining ligament tearing found in compact specimens, but bending and fracture of a specimen arm.

The generalised relationship between J and CTOD is given by:

$$J = m\sigma_y\delta_m \qquad\qquad (4)$$

where, for a LEFM plastic zone correction analysis, the theoretical constraint factor m is of the order of √3 (Turner, 1984), and experimental estimates usually fall between 1 and 3. A comparison of J and CTOD at maximum load from Tables 4 and 6 results in values of m between 2 and 2.5. There is no systematic correlation with work hardening capacity.

CONCLUSIONS

Several different experimental and computational techniques are available to provide either direct measurements of, or correlations with, quasi-static room temperature toughness. A quenched and tempered low alloy steel has been heat treated to provide a wide range of toughness for which no single parameter will provide a material-characterizing estimate of crack resistance. The results indicate that, in selecting a suitable procedure, care must be exercised in that the different tests provide only partial correlations within different toughness ranges.

J estimates based on a linear interception to the blunting line were similar to those determined from a 0.2-mm offset along a power law fit to the crack growth resistance curve. A CTOD estimate made at 0.2-mm crack extension along a curve fit to the R-Curve is about double that measured from a linear extrapolation to zero crack extension, though in these samples all are very small.

There was good agreement between valid plane strain toughness from both compact and short bar specimens. In cases where thickness limitations invalidated K_{IvM} short bar data, a J estimate based on the linear-elastic J-K relation gave the same result as that determined from a maximum (limit) load J from a compact specimen. The short bar test result is always non-conservative with respect to compact specimen results. Upper shelf Charpy energy also provided a fair estimate of maximum load toughness.

REFERENCES

ASTM (1985a). Standard test method for plane-strain fracture toughness of metallic materials. Designation E399-83. In _Annual Book of ASTM Standards_, Vol. 03.01. ASTM, Philadelphia.

ASTM (1985b). Standard test for J_{Ic}, a measure of fracture toughness. Designation E813-81. In _Annual Book of ASTM Standards_, Vol. 03.01. ASTM, Philadelphia.

Barker L.M. (1983). Compliance calibration of a family of short rod and short bar fracture toughness specimens. _Engrg. Fracture Mech._, _17_, 289-312.

Barsom, J.W., and S.T. Rolfe (1970). Correlation between K_{Ic} and Charpy V-notch test results in the transition temperature range. In D.E. Driscoll (Ed.), _Impact Testing of Metals_, ASTM, Philadelphia. pp 281-302.

BSI (1979). Methods for crack opening displacement (COD) testing. Designation BS5763:1979. The British Standards Institution, London.

Morrison, J. and J.P. Gough (1986). Chevron notched short bar testing of high toughness steels. _Theoret. Appl. Fracture Mech._, _5_(2), 109-116.

SAE (1981). Determination of short bar fracture toughness of metallic materials. Aerospace Recommended Practice ARP 1704. SAE, Warrendale, PA.

Schwalbe, K.H., and W. Backfisch (1977). Correlations between fracture toughness, tensile properties, fracture morphology, and microstructure of a high-strength steel. _Fracture 1977_, Proceedings ICF4, University of Waterloo Press, Canada. pp 73-78.

Turner, C.E. (1984). Methods for Post-yield Fracture Safety Assessment. In _Post-yield Fracture Mechanics_, by D.G.H. Latzko, C.E. Turner, J.D. Landes, D.E. McCabe, and T.K. Hellen, Elsevier Applied Science Publishers, pp 25-221.

Winter, P.L., and R.L. Woodward (1986). Effect of tempering temperature on the work-hardening rate of five HSLA steels. _Metall. Trans. A_, _17A_, 307-313.

DEVELOPMENT OF A PROPOSED ASTM TEST METHOD
FOR CRACK ARREST FRACTURE TOUGHNESS OF FERRITIC MATERIALS

B. Mukherjee* and D.M. McCluskey*

*Ontario Hydro Research Laboratory, Toronto, Ontario, Canada

ABSTRACT

This paper documents the experience gained by Ontario Hydro from a Round Robin
Program that was conducted to evaluate a proposed ASTM Standard Test Method for
Determining the Crack Arrest Fracture Toughness of Ferritic Steel. Twenty–seven
laboratories from the United States, Canada, Europe and Japan participated in the test
program. Two bridge steels, A514 and A588, and a pressure vessel steel, A533B, were
selected for the test program. Specimen blanks were cut from large plates, identified
and shipped to each participant who was responsible for the fabrication of specimens
and test fixtures. Each participant agreed to test 3 specimens of A514 at -30°C, 3
specimens of A588 at -30°C, and 6 specimens of A533B, 3 at 10°C and 3 at 25°C. A
brittle weld crack starter was used to promote run–arrest events. Ontario Hydro's test
results were consistent with the overall Round Robin Program results. In light of the
experience gained from the Round Robin, the test procedure is currently being revised,
prior to its submission to ASTM Committee E–24 as a Proposed Standard Test Method.

INTRODUCTION

Published literature of the 1970's on the development of crack arrest methodology is
complex and appears to be contradictory to the uninitiated. However, two recent
reviews go a long way toward explaining the issues clearly (Kanninen, 1985; Ravi-
Chandra and Knauss, 1984). Two opposing views for the arrest of rapidly propagating
cracks have been advanced (Kanninen, 1985). In one view, crack arrest is considered as
the reverse of crack initiation. In this approach, global experimental parameters are
measured when a crack is stationary, that is, before crack initiation and just after
crack arrest. It is not necessary to consider the rapid crack propagation process
preceding crack arrest. The opposing view is the dynamic approach.

As indicated in (Kanninen, 1985), the dynamic point of view gives direct consideration
to crack propagation. In the linear elastic condition, a dynamically computed value of
stress intensity K, drives a moving crack, while a material and crack speed dependent
critical value of K, conventionally denoted by K_{ID}, is the resistance parameter. Crack
arrest occurs when the driving force K becomes less than the material resistance K_{ID}.
In this approach, the K_{ID} versus crack velocity, $(K_{ID} - \dot{a})$, relationship is regarded as a
temperature dependent material property. Some investigators (but not all (Dally, Fourney and Irwin, 1985)), have found K_{ID} values to be reasonably geometry independent.
According to the dynamic analysis, K_{ID} can depend on K_o, the stress intensity at in-

itiation of rapid crack propagation. The nature of this relationship is governed by the $(K_{ID} - a)$ material property and by the size of the crack jump relative to the size of the specimen and structure.

According to the quasi-static approach, a crack arrests at a toughness value, K_{Ia}, during a run-arrest event. K_{Ia} is considered as a material property (Crosley and Ripling, 1979; Ravi-Chandra and Knauss, 1984). In ferritic steel K_{Ia} is relatively independent of K_o, the stress intensity at initiation of rapid crack propagation. K_{Ia} is a plain strain linear elastic fracture mechanics parameter and is characterized in terms of static value of crack driving force at the arrested crack length a few milliseconds after the arrest event.

Kanninen (1985) has indicated where the static and dynamic approaches are compatible and where they are not. K_{ID} at the instant of crack arrest ($\dot{a} = 0$) or at low crack speed region ($a = 100$ m/s) is denoted by K_{IA}. It can be shown from the analysis of a crack in an infinite medium that the statically measured K_{Ia} is a good approximation of K_{IA} provided the crack jump lengths are short relative to the specimen geometry so that reflected stress waves do not return to interfere with the moving crack tip. However, he indicates that dynamic treatment may be necessary when the crack propagation path parallels a nearby free surface as in a double cantilever beam or compact crack arrest specimen.

A brief history of the development of crack arrest toughness tests of ferritic steel is given by Barker and co-workers (1988). Early work in this area was done by Ripling and Crosley (1971) and Hoagland and co-workers (1977). A large Cooperative Test Program using a test method that employed compact specimens was conducted during 1977-1979 (Crosley and co-workers, 1983). Static analysis K_{Ia} values, obtained from a test method similar to that of the Cooperative Test Program, were used to assist in the planning of large cylinder thermal shock fracture experiments at Oak Ridge National Laboratory. The K_{Ia} values from the large cylinder tests were comparable to those obtained using the static analysis small-specimen tests (Cheverton and co-workers, 1985). Therefore, there was confirmatory experimental evidence that K_{Ia} values from relatively simple laboratory experiments could provide useful estimates of the actual K_{IA} for run-arrest events in large structures. An ASTM Task Group E-24.01.06 subsequently examined the experiences gained from the Cooperative Test Program and produced a draft document describing a Proposed Test Procedure for Determining the Crack Arrest Fracture Toughness of Ferritic Steel. This test procedure was followed in the Round Robin Program. This paper describes Ontario Hydro's experience in conducting crack-arrest tests according to the proposed procedure. It documents results obtained by Ontario Hydro and reproduces some Round Robin test results for comparison.

<u>Round Robin Test Program</u>

The objective of the Round Robin Program was to simultaneously review and validate a proposed ASTM Standard Test Method for Determining the Plane-Strain Crack-Arrest Fracture Toughness K_{Ia} of Ferritic Steels. The initial testing procedure was based on the experience gained from the Co-op Program (Crosley and co-workers, 1983) and the suggestions of a group of individuals who have been extensively involved with crack arrest testing over the past several years (Rosenfield and co-workers, 1984). The participants were encouraged to send in their comments and suggestions for modifications.

The round robin test program was managed by Professors W.L. Fourney and R. Chona of the University of Maryland. The project was sponsored by the U.S. Nuclear Regulatory Commission. Twenty-seven leading laboratories from the United States, Canada, Europe and Japan participated in this Round Robin test program. Two bridge steels (A514 and A588) and a pressure vessel steel (A533B) were selected for the test program.

The specimen blanks for the two bridge steels were flame cut from large 5 cm thick plates, such that each specimen was approximately 21.5 x 25.5 cm in size, with the shorter dimension parallel to the rolling direction of the parent plate. The direction of crack propagation in these specimens was transverse to the rolling direction of the parent plate. The specimen blanks of the A533B material were cut from large blocks which were, in turn, cut from a 254 x 305 cm section of a 25.4 cm thick rolled plate. Each specimen blank was marked with an alphanumeric code, which identified the block location within the parent plate and the location of the particular specimen blank within the block. The direction of crack propagation in these specimens was also transverse to the rolling direction of the parent plate.

Test temperature for A514 and A588 bridge steels was -30°C. It is the lowest temperature at which results would be useful with regard to practical applications to steel structures. It was suggested that A514 specimens (σ_{ys} = 890 MPa) should be tested first, since these were most likely to pose problems in achieving successful crack arrest tests. These tests, therefore, served as trial tests in identifying problems with test setup and instrumentation. These were followed by tests on the A588 specimens. Because of the low yield stress of the material (σ_{ys} = 330 MPa), the crack opening necessary to initiate a run–arrest event was larger than that allowed by the proposed test procedure for a single cycle loading. Therefore, a sequential load–unload cycling procedure was required for these tests. The final tests were done using A533B (σ_{ys} = 480 MPa) specimens at 10°C (3 specimens) and 25°C (3 specimens).

Experimental Procedure

A proposed ASTM Standard Test Method for Determining the Plane–Strain Crack–Arrest Fracture Toughness, K_{Ia}, of Ferritic Steels was utilized in the test program. The procedure involves testing of modified compact specimens that have been notched by machining. A wedge and a split–pin assembly are used to apply a load at the crack line. The wedge is forced into a split pin, which applies an opening force, causing a run–arrest segment of crack extension. The loading system was designed to minimize the introduction of additional energy into the specimen during the run–arrest event.

The test fixture is shown in Fig. 1a. It was designed for use in a 250 kN servohydraulic testing machine. A hold down plate was included to facilitate wedge extraction during incremental load cycling. The wedge, split pin, base plate and specimen hole were lubricated using thin sheets of teflon. A displacement gauge was used to measure the crack–mouth opening displacement at 0.25 W from the load–line, where W is the specimen width.

A starter notch was produced in all specimens by depositing a brittle weld across the specimen thickness. The Murex–Hardex–N electrodes specified in ASTM E208, Standard Method for Conducting Drop–Weight Test, were used. The weld bead was notched by spark machining. The function of the starter notch was to sustain sufficiently high crack opening displacement to permit an appropriate length of crack jump prior to arrest. Notched brittle weld starter notches were necessary for conducting crack arrest tests for low and medium strength steels. For high strength steels a simple machined notch has been used successfully. A crack arrest test specimen is shown in Fig. 1b.

Tests at subambient temperatures were conducted using cooling coils embedded in the specimen support base. The specimens were thermally insulated to minimize temperature gradient across the specimen thickness. The specimen temperature was measured on the top and bottom surfaces with thermocouples.

A cyclic loading technique, which has been described in detail in the test procedure, was used to apply load to the wedge. The specimen applied load was inferred from the crack mouth displacement. The cyclic loading technique was designed to estimate and separate out components of the opening displacement which did not contribute to the opening load.   The occurrence of unstable crack extension was clearly identified both audibly, and as an abrupt load drop on the wedge load versus crack mouth displacement test record.

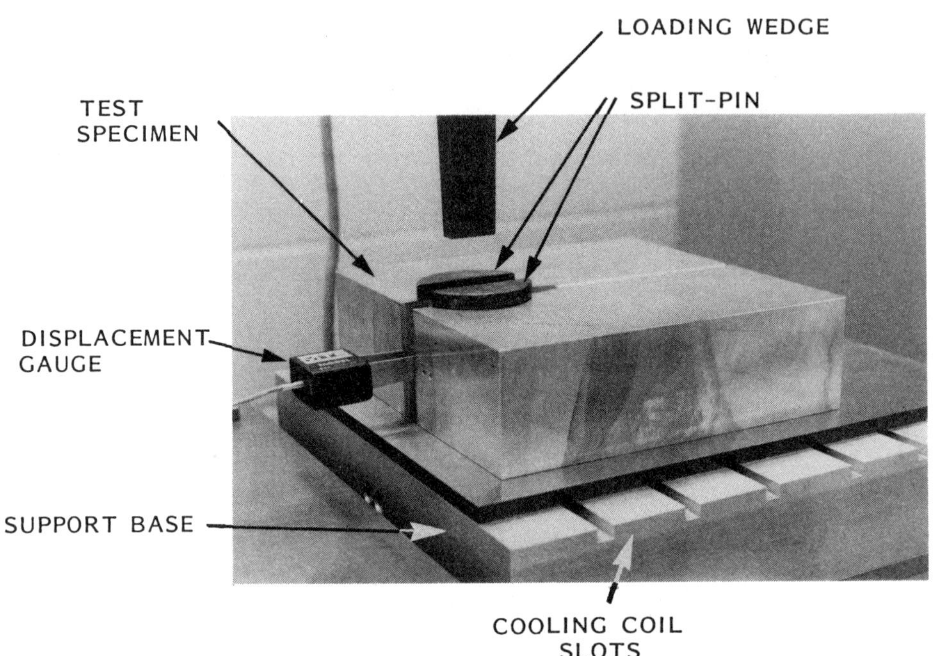

Fig. 1a.   Crack arrest test fixture.

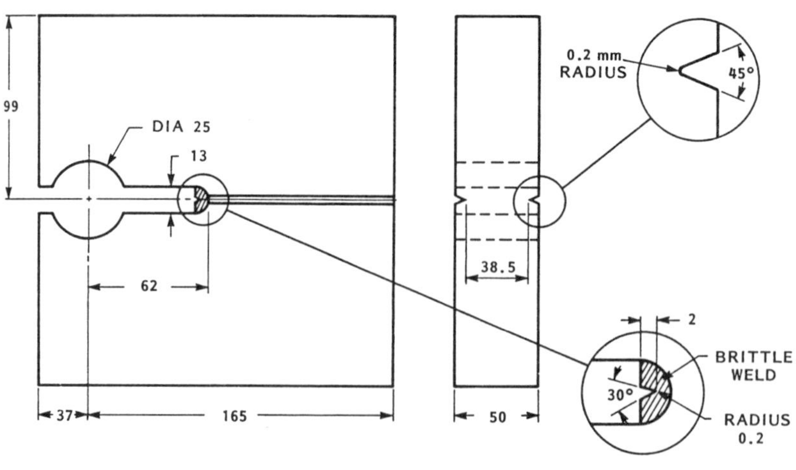

Fig. 1b.   Crack arrest test specimen.

The arrested crack front was marked by heat tinting before breaking the specimen open. The arrested crack length was defined as the average of three measurements taken at the centre of a specimen and midway between the centre and the bottom of the sidegroove on each side.

Calculation of the effective stress intensity at initiation, K_o, is based on measurements of the machined notch length and opening displacement (δ_o) at the onset of unstable crack growth. The value of crack arrest toughness, K_a, is based on the measurement of crack length and opening displacements (δ_a), after arrest. K_o and K_a were calculated from:

$$K = E\delta\ (x)\ \sqrt{B/B_N}\ /\ \sqrt{W}\quad MPa\sqrt{m} \tag{1}$$

where

$$(x) = \frac{2.24(1.72 - 0.9x + x^2)\ \sqrt{1-x}}{(9.85 - 0.17x + 11x^2)}$$

W	= specimen width;
B	= specimen thickness;
B_N	= net specimen thickness after side grooving.

and $x = a/w$

The value of K_a calculated from equation (1) can be considered a linear–elastic plane-strain value, K_{Ia}, provided the unbroken ligament, thickness, and crack–jump length, satisfy a set of criteria. These criteria are based on specimen geometry and dynamic yield strength estimate at the test temperature (see Appendix A). A complete test record for a crack arrest test is also shown in Appendix A.

RESULTS AND DISCUSSION

Wedge load versus crack mouth displacement records from 4 tests are shown in Fig. 2 to 5. Figure 2 is the test record for specimen number 1 (A514 at -30°C). In this test

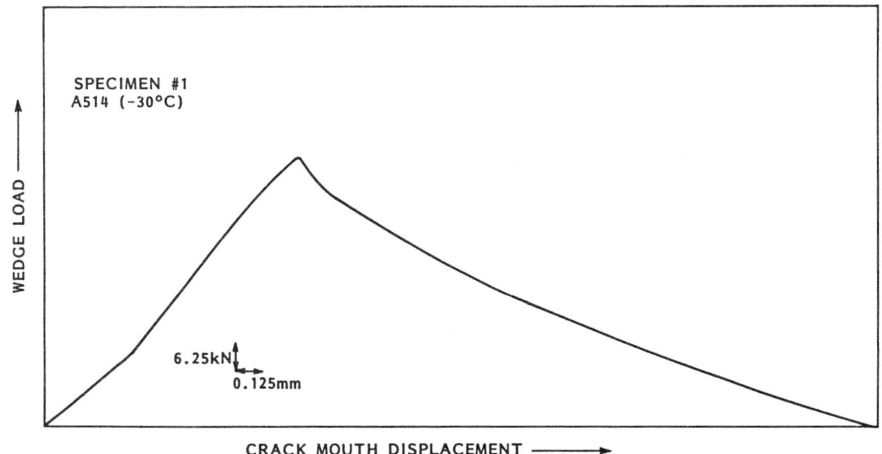

Fig. 2. Weld load versus crack mouth displacement of Sp. No. 1 (A514) at -30°C

a rapid dynamic crack propagated from the starter notch, but it did not arrest and the specimen split in two halves. Figure 3 is the test record for specimen number 53, where a run–arrest event was observed in the first loading cycle. Small load drops on the loading trace were possibly due to cracking of the brittle weld bead.

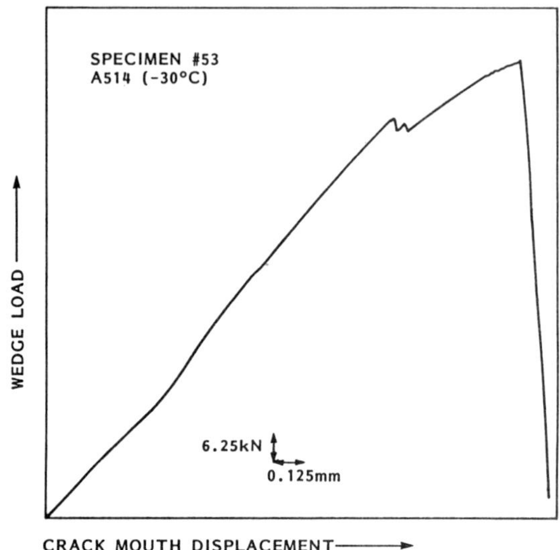

Fig. 3. Wedge load versus crack mouth displacement of Sp. No. 53 (A514) at –30°C

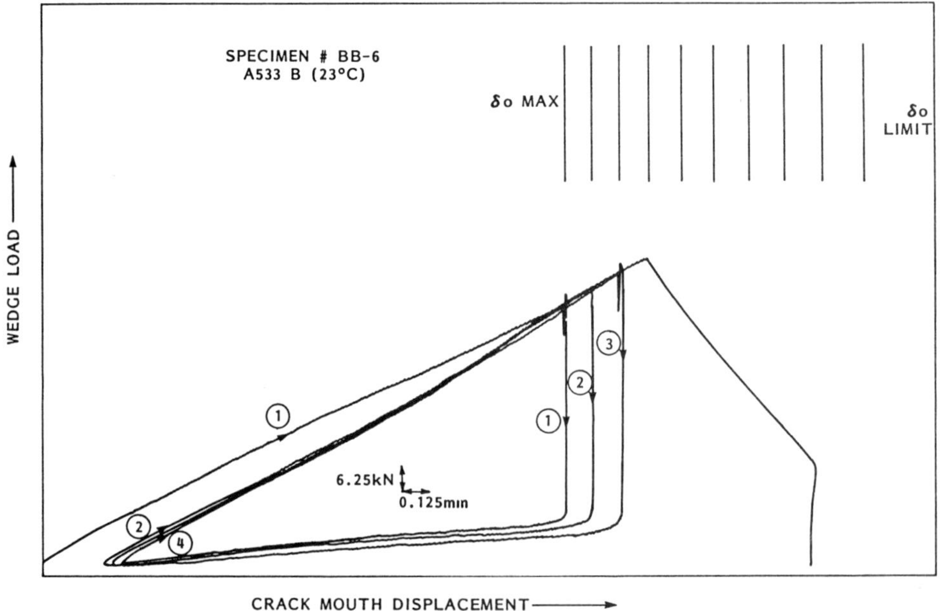

Fig. 4. Wedge load versus crack mouth displacement of Sp. No. BB6 (A533B) at 23°C

A cyclic loading procedure was required for A533B specimens at 10°C and 23°C. Figure 4 shows how displacement was increased by a given increment per cycle. A dynamic crack had initiated and arrested in this test during the fourth cycle. Figure 5 shows the cyclic load trace for specimen A1, where a dynamic crack initiated during the third loading cycle but it did not arrest.

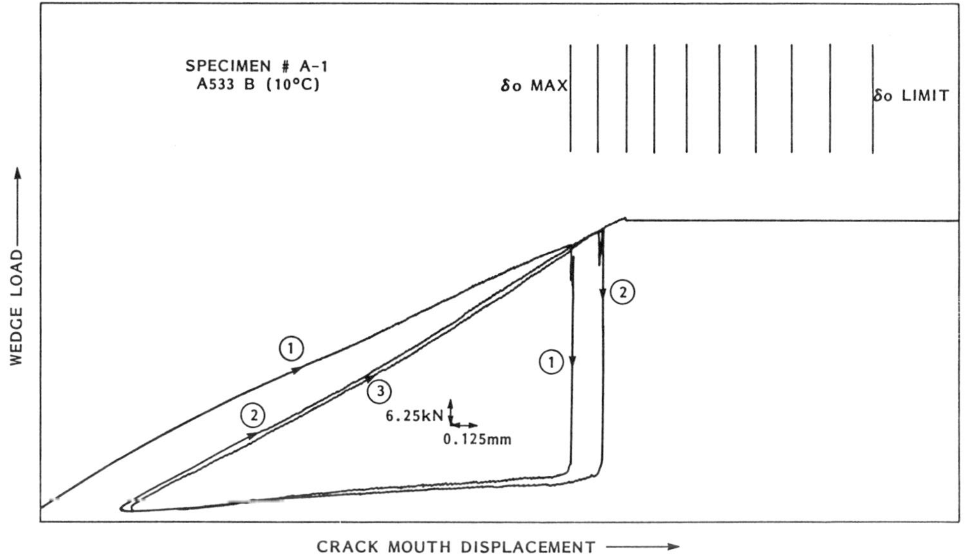

Fig. 5. Wedge load versus crack mouth displacement of Sp. No. A1 (A533B) at 10°C

TABLE 1 Ontario Hydro Test Results

ASTM Round Robin on K_{Ia} Testing Results

Specimen ID	Temp (°C)	Material	a/W	K_o	K_a	
A–1	10	A533B	–	156.6	N/A	Did not arrest
P–6	10	A533B	.77	171.8	84.3	Valid
W W–5	10	A533B	.85	154.1	63.7	Valid
Z–8	23	A533B	.81	187.0	83.1	Valid
H–3	23	A533B	.96	172.9	71.2	Crack jump
B B–6	23	A533B	.94	167.2	46.1	too long
1	–30	A514	–	164.5	N/A	Did not arrest
30	–30	A514	–	198.5	N/A	Did not arrest
53	–30	A514	.79	297.3	131.2	Slant crack front
3	–30	A588	.74	102.7	57.9	Valid
31	–30	A588	.82	129.4	57.5	Valid
65	–30	A588	.84	134.6	56.6	Valid

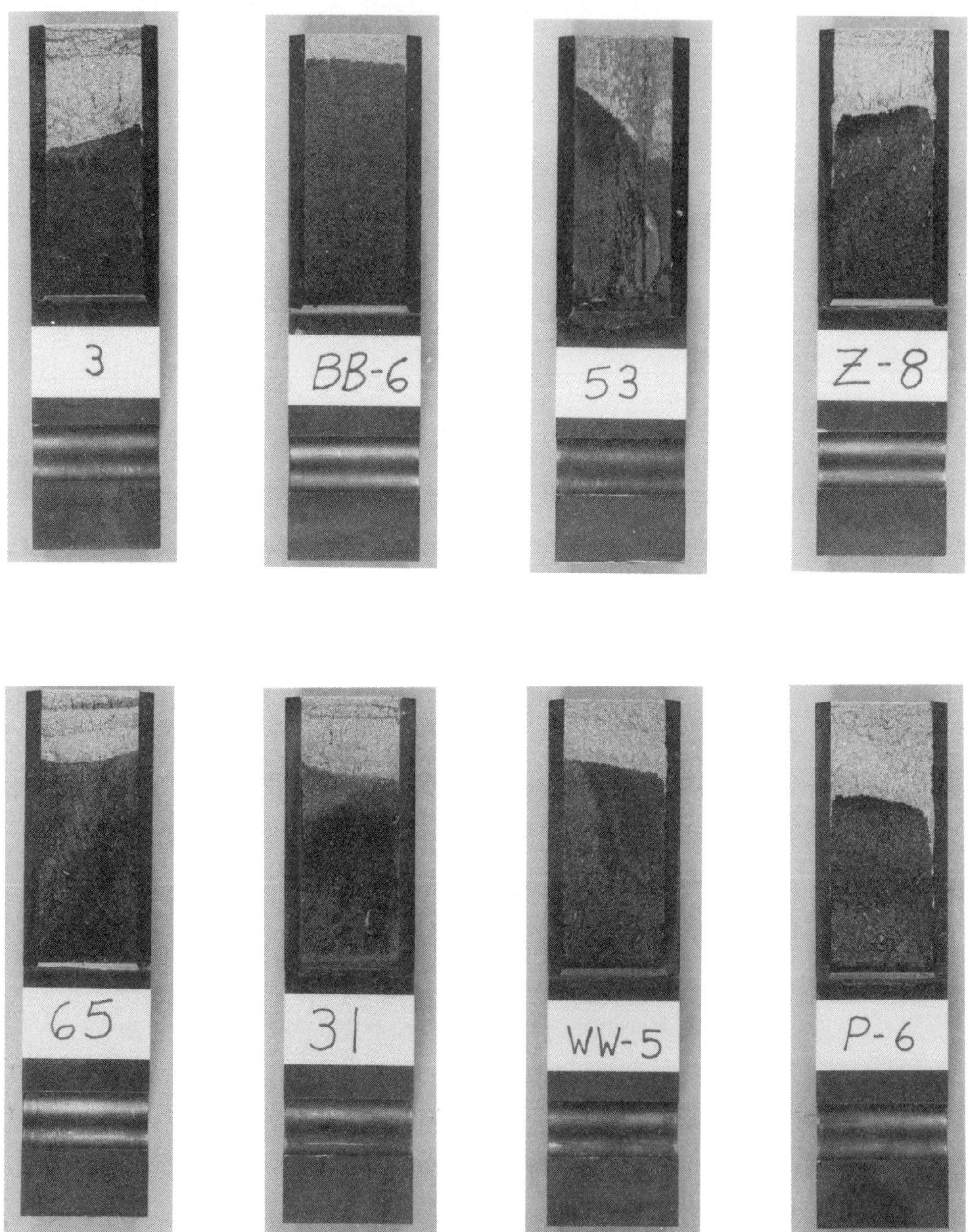

Fig. 6.   Fracture surface of compact crack arrest specimens.

Results from all Ontario Hydro tests are tabulated in Table 1 and several fracture surfaces are shown in Fig. 6. Valid tests are identified in Table 1 and reasons for non validity are also indicated. All three tests on A514 at -30°C were not valid. Dynamic cracks initiated, but did not arrest in two specimens. The third specimen showed excessive crack front curvature and transverse cracks on the fracture surface. It seems that a machined starter notch may have been more appropriate than the brittle welds for this high strength bridge steel. All three tests on A588 bridge steel at -30°C were valid.

Results on A533B steel were mixed. Two 23°C tests were invalid due to excessive crack jump (a /W>0.85) and in one 10°C test, the crack did not arrest.

Barker and co-workers (1988) summarized the test results from 21 participants. Consistent with Ontario Hydro experience, different success rates were achieved for different steels tested. Test results were summarized as follows:

A588 @ -30°C: 54 specimens tested; all initiated; none failed to arrest
40 specimens arrested in the specified range
Mean K_a = 61.5 MPa$\sqrt{m}$; Standard Deviation = 6.4 MPa$\sqrt{m}$
(10%)

A533B @ 10°C: 70 specimens tested; 10 failed to initiate; 13 did not arrest
30 specimens arrested in the specified range
Mean K_a = 78.2 MPa$\sqrt{m}$; Standard Deviation = 9.7 MPa$\sqrt{m}$
(12%)

A533B @ 25°C: 65 specimens tested; 9 failed to initiate; 4 did not arrest
38 specimens arrested in the specified range
Mean K_a = 91.2 MPa$\sqrt{m}$; Standard Deviation = 16.6 MPa$\sqrt{m}$
(18%)

A514 @ -30°C: 54 specimens tested; 4 failed to initiate, 21 did not arrest
18 specimens arrested in the specified range
Mean K_a = 102 MPa$\sqrt{m}$; Standard Deviation = 21.0 MPa$\sqrt{m}$
(20%)

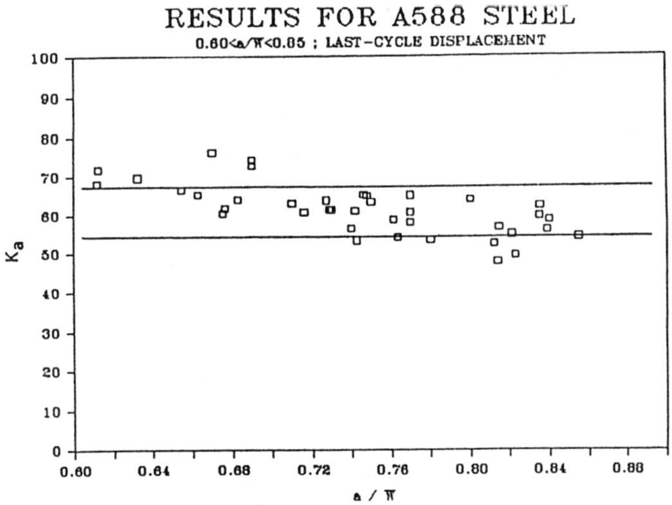

Fig. 7. K_a versus K_o for A588 steel at -30°C

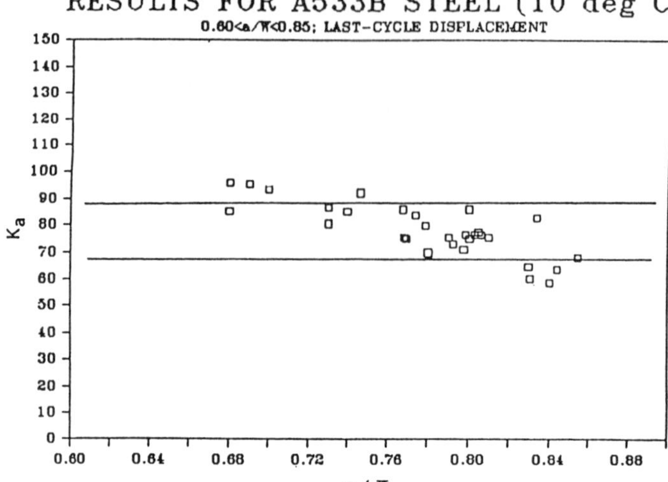

Fig. 8. K_a versus K_o for A533B steel at 10°C

Several of the specimens tested displayed fracture surface irregularities such as tunnelled cracks, cracks that ran out of the side grooves, large unbroken ligaments that clearly affected the fracture propagation behaviour, etc. These problems were almost entirely in the group of A533B specimens tested at 25°C and the A514 specimens. Elimination of these test results led to:

A533B @ 25°C: 28 valid test results; Mean K_a = 83.4 MPa m; Standard Deviation = 10.6 MPa m (13%)

A514 @ –30°C: 12 Valid test results; Mean K_a = 88.4 MPa m; Standard Deviation = 10.2 MPa m (12%)

Figures 7 and 8, which are reproduced from Barker and co-workers' (1988) paper, show K_a as a function of K_o for the A588 and A533B (10°C) specimens. Also shown on Fig. 7 and Fig. 8 is the scatter band for K_a corresponding to the mean K_a value ± one standard deviation. It can be seen from these figures that there is no apparent dependence of K_a on the K_o value calculated for initiation. It can also be seen that valid Ontario Hydro test results lie within the band of test data produced by the round robin program.

SUMMARY AND CONCLUSIONS

Results obtained by Ontario Hydro during the Round Robin Test Program to evaluate a proposed ASTM Test Method for Crack Arrest Fracture Toughness of Ferritic Steels have been presented.

Ontario Hydro's test results were consistent with the overall Round Robin Program results.

In light of the experience gained from the Round Robin, the test procedure is currently being revised, prior to its submission to ASTM Committee E–24 as a Proposed Standard Test Method.

REFERENCES

Barker, D.B. and Co-workers (1988). A Method for Determining the Crack Arrest Fracture Toughness of Ferritic Materials. To appear in T.A. Cruse (Ed.), Fracture Mechanics: Nineteenth Symposium, ASTM STP. ASTM, Philadelphia.

Cheverton, R.D. and Co-workers (1985). Pressure Vessel Fracture Studies Pertaining to the PWR Thermal Shock Issue: Experiments TSE-5, TSE-5A and TSE-6. NUREG/CR - 4249 (ORNL-6163), Oak Ridge National Laboratory, Oak Ridge, Tennessee.

Crosley, P.B. and E.J. Ripling (1971). Crack Arrest Toughness of Pressure Vessel Steels. J. Nucl. Eng. and Design, 17, 32–45.

Crosley, P.B. and E.J. Ripling (1979). Crack Arrest Studies. EPRI Report NP-1225, Final Report. EPRI, Palo Alto.

Crosley, P.B. and Co-workers (1983). Final Report on Cooperative Test Program on Crack Arrest Toughness Measurement. USNRC Report NUREG/CR-3261, Nuclear Regulatory Commission, Washington, DC.

Dally, J.W., W.L. Fourney and G.R. Irwin (1985). On the Uniqueness of Stress Intensity Factor – Crack Velocity Relationship. Int. J. Fracture, 27, 159–168.

Hoagland, R.G. and Co-workers (1977). A Crack Arrest Measurement Procedure for K_{Im}, K_{Id} and K_{Ia} Properties. In G.T. Hahn and M.F. Kanninen (Ed.), Fast Fracture and Crack Arrest, ASTM STP 627, pp. 177–202.

Kanninen, M.F. (1985). Application of Dynamic Fracture Mechanics for the Prediction of Crack Arrest in Engineering Structures. Int. J. Fracture, 27, 299–312.

Ravi-Chandra, K. and W.G. Knauss (1984). An Experimental Investigation into Dynamic Fracture: I. Crack Initiation and Arrest. Int. J. Fracture, 25, 247–262.

Rosenfield, A.R. and Co-workers (1984). Recent Advances in Crack-Arrest Technology. In R.J. Sanford (Ed.), Fracture Mechanics: Fifteenth Symposium, ASTM STP 833, pp 149–164.

APPENDIX A

Test Report – Crack Arrest

Crack Plane Orientation: L–S
Specimen Number: Z–8
Test Temperature (C): 23°C

Material Identification

Material Type:	A533B	Young's Modulus (σ_{ys}, MPa):	206800
Yield Strength (MPa)	480	Dynamic Yield (σ_{ys}, MPa)	605

Starter Notch

Type of Notch:	Weld Embrittled	Notch Width, N(mm):	12.7
Notch Root Radius, ρ(mm):	.16		
Hardness of Weld–Front (RC):	48	Hardness of Weld–Back (RC):	40
Hardness of H.A.Z.–Front (RC):	27	Hardness of H.A.Z.–Back (RC):	22

Convention for front/back of specimen is: Weld bead is put down going from back to front across the thickness.

Specimen Dimensions

Thickness, B(mm):	50.2
Net Thickness, B_N(mm):	38.4
Width, W(mm):	164.9

<u>Crack Length Measurements</u>

Initial Crack Length, a_o(mm):	61.9
Arrested Cr. Length at 1/4 Pt., a_1(mm):	135.6
Arrested Cr. Length at 1/2 Pt., a_2(mm):	133.1
Arrested Cr. Length at 3/4 Pt., a_3(mm):	131.7
Avg. Arrested Cr. Length, a_a(mm):	133.5

<u>Load–Displacement Summary</u>

Max. permissable displacement on first cycle, (mm):	1.210
Type of Loading Used:	Load/Unload Cycling
Total Number of Cycles:	7
Accumulated Displacement at Start of Cycle No. 2 (mm):	.14
Accumulated Displacement at Start of Cycle No. 3 (mm):	.16
Accumulated Displacement at Start of Cycle No. 4 (mm):	.18
Accumulated Displacement at Start of Cycle No. 5 (mm):	.21
Accumulated Displacement at Start of Cycle No. 6 (mm):	.24
Total Accumulated Displ. at Start of Last Cycle (mm):	.27
Displ. on Last Cycle Corresponding to Cr. Initiation d_o(mm):	1.35
Displ. on Last Cycle Corresponding to Cr. Arrest d_a(mm):	1.5
Maximum Load, P_{max}(kN):	77.3
Load Drop as % of P_{max}:	100

<u>Calculated Values of K_o and K_a</u>

K_o MPa$\sqrt{m}$:	186.96
K_a MPa$\sqrt{m}$:	83.08

<u>Validity Requirements</u>

$$W - a_o \geq 0.15\ W$$
$$W - a_a \geq 1.25\ (K_a/\sigma_{YD})^2$$
$$B \geq 1.0\ (K_a/\sigma_{yd})^2$$
$$a_a - a_o \geq 2\ N$$
$$a_a - a_o \geq (K_o/\sigma_{ys})^2/2\pi$$

SESSION 2:   TESTING AND MECHANICS - NONMETALS

CHAIRMAN: P. S. NICHOLSON
McMaster   University,
Hamilton

FRACTURE TOUGHNESS OF VERY HARD TOOL STEELS

D.F. Watt[1], P.A. Jones[1], Guo Shi Wen[1] and N. Zamani[2]

[1]Dept. of Engineering Materials & [2]Dept. of
Mathematics, University of Windsor, Windsor, Ontario,
Canada N9B 3P4

ABSTRACT

It is common practice for tool designers to temper back tool steels to add toughness at the expense of wear resistance. However, there previously have been very few measured fracture toughness values for the hardness levels at which these steels are commonly used. A modified K_{IC} specimen has been developed which allows measurement of the fracture toughness at hardness levels of Rockwell C48 and higher. At these hardness levels, exceeding the fatigue threshold stress intensity factor (ΔK_{th}) only slightly leads to very high rates of crack growth. This makes fatigue precracking chancy if not impossible. Therefore, the fatigue precracking has been replaced by a slow crack growth process made possible by the design of the specimen. Because of the high hardness levels, the crack tip plastic zone size is small, and so a sharp precrack can be produced without fatigue.

Crack lengths are measured by compliance measurements, and by heat tinting. The fracture toughness of 01 (Atlas Keewatin) steel has been measured as a function of tempering temperature.

The test specimen geometry, which was developed empirically, has been analyzed using the finite element method to clarify the nature of the slow crack growth process, and to calculate stress intensity values. The results have been used to calculate critical crack sizes for various cracks in practical applications with some interesting results.

KEYWORDS

Tool steels; fracture toughness, brittle materials, fracture, fracture mechanics, steels, mechanical properties, fracture testing.

TOUGHNESS TESTING IN BRITTLE MATERIALS

For many applications where wear is the primary consideration, it would be desirable to use a tool steel in its hardest possible condition. However, these steels are brittle, so they are tempered back to gain fracture toughness. This logic has been followed for over a century, yet there are almost no data available to

quantify just how much toughness is gained as the steel hardness decreases. A
review of previous work is given by Watt, Nadin and Biner (1987).

There are two reasons for this information deficiency. The typical designer using
tool steels needs to be made aware of the principles of the fracture mechanics
design methodology, and must also be convinced ot its utility for practical application. Secondly, good working data is required. Modernizing the design procedure
first requires that the data be generated. This can then be used as ammunition in
the campaign to implement modifications of the design procedure.

The perspectives and basic principles of fracture mechanics methodology were well
defined twenty years ago in, for example, ASTM-STP 410. Because hard tool steels
have very small plastic zones at the crack tip, they would appear to be ideal
candidate materials for fracture mechanics type of calculation. Why has the data
not been made available?

The standard test for the measurement of fracture toughness is delineated in detail
in the ASTM-E399 standard. In order to get reproducible values for the fracture
toughness, K_{IC}, this test specifies that fatigue precracking be performed to produce a sharp crack tip. This requirement is a necessity for ductile materials
where plasticity leads to crack tip blunting. In hard tool steels, sharp cracks
are the only type observed regardless of the growth mode. It is therefore easy to
generate sharp cracks but as all the references listed confirm, it is difficult to
conform to ASTM-E399 because fatigue cracks are highly unstable in these brittle
materials. Controlling the crack growth rate at low K_I values is difficult, so
that a high fraction of these expensive specimens is destroyed in the precracking
stage. To overcome this difficulty, we have successfully developed a new method
for precracking fracture toughness specimens.

The standard ASTM specimen has been altered by adding to it a "paddle handle" as in
Fig. 1. To precrack the specimen, dowel pins are put in the outer holes (right-

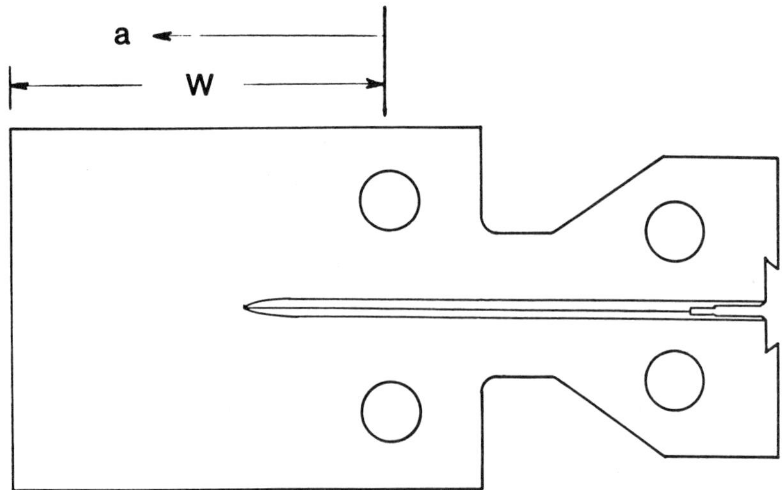

Fig. 1. The paddle-type specimen designed for testing the
fracture toughness of brittle materials.

hand side, Fig. 1) and a stiff wedge is driven slowly between these pins while the
left-hand side of the specimen is being squeezed in the vertical direction in a
custom designed viselike grip. In practice the wedge is attached to the load cell
of a screw-driven universal tensile tester while the vise rests on the slowly rising cross-head. When the specimen is heard to crack, the cross-head is stopped.
At this stage the crack will propagate into the narrow portion of the paddle han-
dles, guided by side grooves on the specimen. To conduct the K_{IC} test, the crack
tip must have previously advanced to a position close to a/w = 0.5. How can we get
the crack to grow under stable conditions? If, using the standard ASTM loading
clevises, we apply a force through the outer paddle-handle loading holes, then it
is possible to grow the crack into the specimen in a controlled fashion. In practice, the test machine cross-head is slowly lowered while recording the load vs.
time. When the crack advances, the cross-head motion is stopped. Crack propagation is manifested by an instantaneous drop in load (often accompanied by an
audible pop), followed by a very gradual further decrease in load representing
further slow crack growth. After several minutes, this slow growth ceases and the
load remains constant. The current crack tip position can be found by measuring
the specimen compliance and comparing it to the standard values in Fig. 2. The
crack arrest positions are highly visible on the final fracture surfaces as clearly
demarcated lines when viewed with the naked eye. The number of these lines corresponds with the number of load drops during the precracking procedure, and they are
typically spaced 2-10 mm apart. Ironically, they are very difficult to discern on
an SEM at any magnification.

Why do propagating cracks arrest during the precracking stage? Consider Figs. 2
and 3. The latter figure shows the change in the stress intensity factor K_I, as a
function of crack tip position for a constant load of 1000 lb. applied at the outer
hole position. These data were calculated using a modification of the APES finite
element program developed at the David Taylor Naval Shipyard in Washington, D.C.
A typical 2d mesh, shown in Fig. 4, utilizes 12 noded isoparametric elements with
enriched elements at the crack tip. The singularity is incorporated in the shape
functions used near the crack tip. The close correspondence in Fig. 2 between
measured compliance values and those calculated using the APES program gives cre-
dence to the fidelity of the FEM simulation.

The salient feature of Fig. 3 is that for constant load. K_I passes through a minimum at about a/w = 0.2. Beyond a/w = 0.3, K_I for a fixed load increases but the
compliance also increases with crack length. Evidently the increase in compliance
results in a large enough load drop during crack growth to allow the crack to re-
arrest when these specimens are loaded through the paddle handle loading holes
using a screw-driven Instron machine. The design configuration was arrived at
intuitively. We have commenced a CAD sequence to optimize the paddle-handle size
and shape.

Fig. 5 shows some typical values measured for 01 steel using this type of specimen,
and Fig. 6 shows the measured variation in yield strength for the same steel and
heat treat conditions. Fig. 7 shows the critical crack depth for a semi-elliptical
surface crack as a function of hardness and applied load for 01 steel. What are
the consequences of this data in terms of a choice of suitable heat treatment?

Fig. 5 tells us that, for a given size of defect, the allowable stress level (which
is proportional to K_{IC}) is increased by about 40% on tempering back from R_c62 to
R_c55, and it is doubled on tempering to R_c50 from R_c62. Fig. 7 gives us a good
sense of the absolute size of critical defects in these materials. If one is to
take advantage of the very high strengths of these materials, then it is important
to produce material defect-free to very fine levels. Even when tempered back to

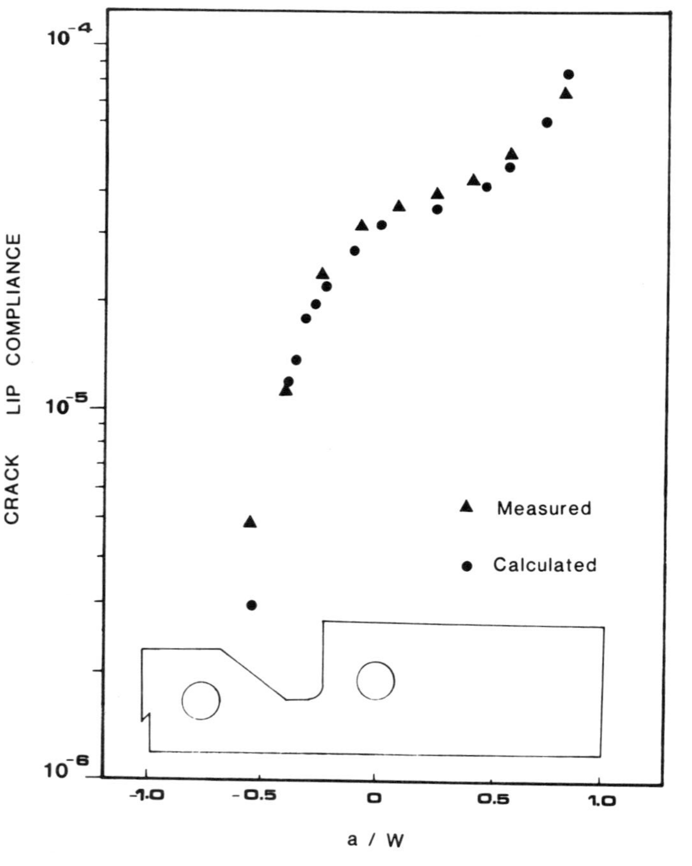

Fig. 2. The crack lip compliance for outer hole loading calculated using finite element analysis compared with measurements made on a successively sawcut specimen.

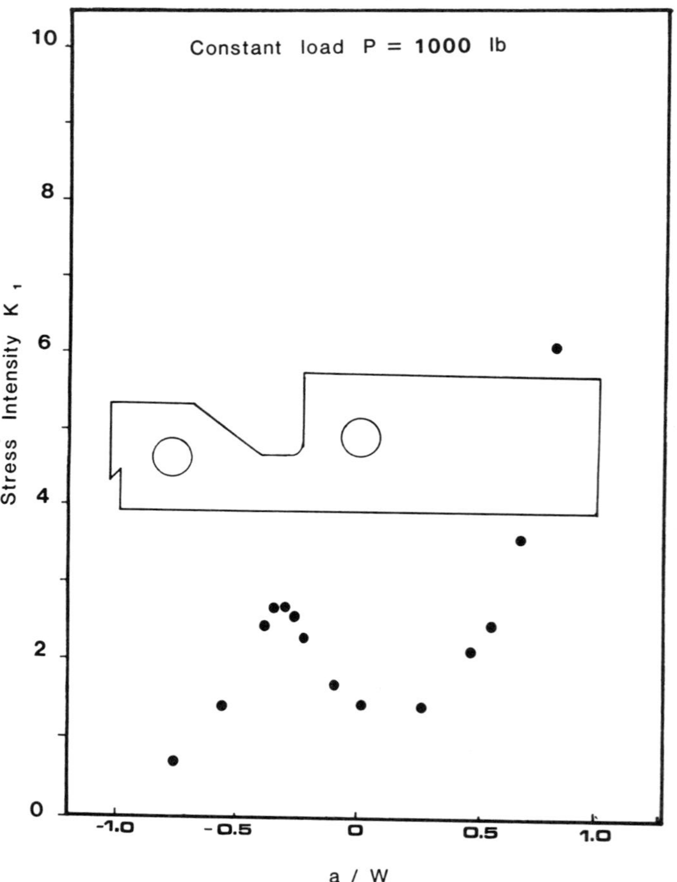

Fig. 3. The variation of K_I with crack length when a constant load is applied at the outer hole position.

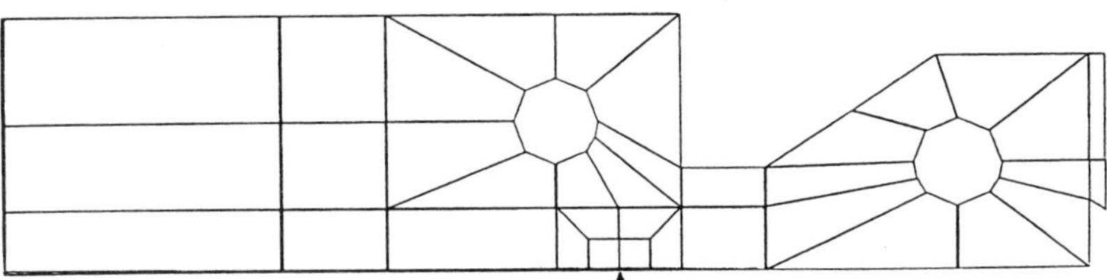

Fig. 4. A typical finite element mesh used for the calculations in Figs. 2 and 3. The crack tip is at the
position indicated by the arrow.

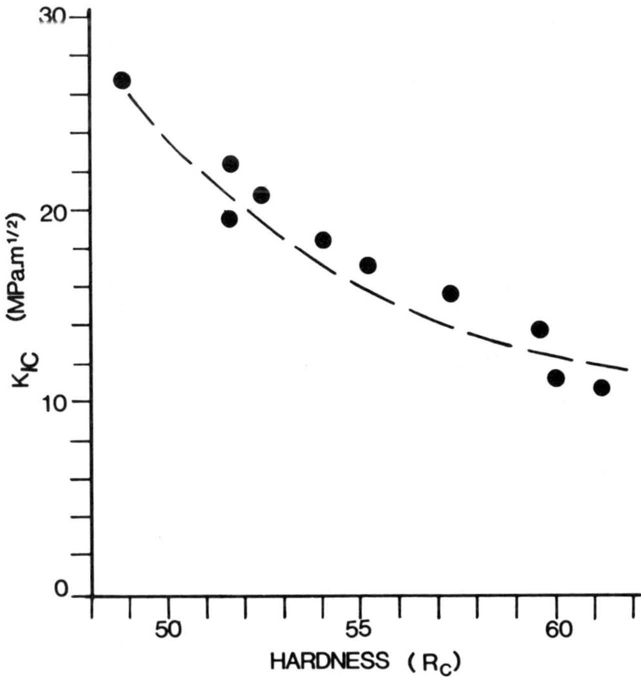

Fig. 5. The fracture toughness of 01 tool steel as a
function of hardness.

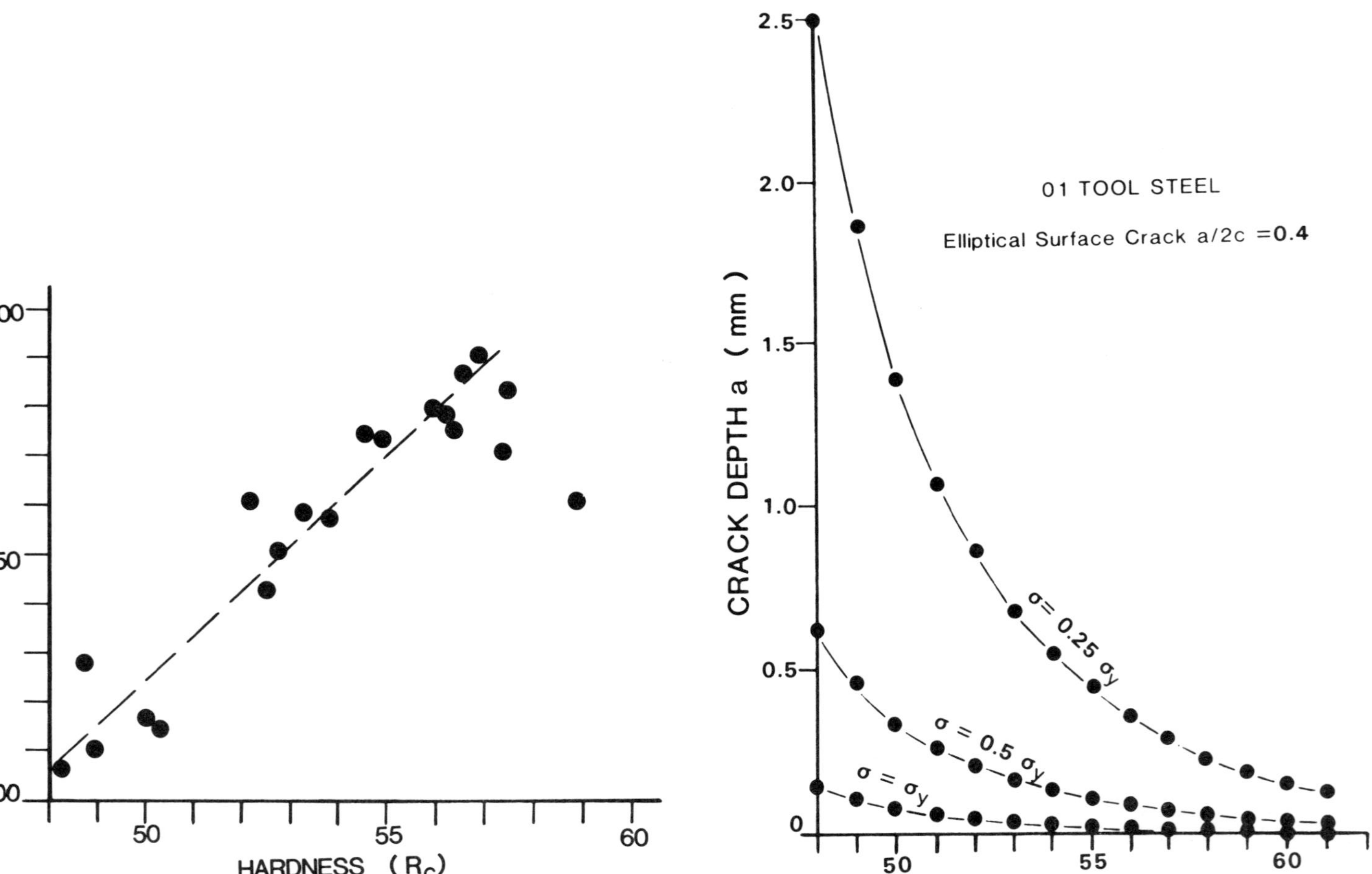

Fig. 6. The yield strength of 01 steel as a function of
hardness. None of these specimens was ductile
but all showed perceptible yielding before
fracture.

Fig. 7. Critical crack depth for semi-elliptical
surface cracks in 01 tool steel calculated
from the data in Figs. 5 and 6.

R_c48, this material when loaded to half its yield strength has a critical size of $a = 0.6$ mm by $2c = 1.5$ mm. At R_c62, these dimensions shrink to the range of tens microns. Such defects are impossible to detect reliably by any standard non-destructive evaluation procedure. The critical crack sizes are of the order of the carbide spacings of these materials.

The very small critical crack sizes have another interesting consequence. These brittle materials have very small plastic zones at the crack tip, but are they negligible compared with the critical crack depths? At R_c57, using the Irwin approximation of the plastic zone size $r = (K_{IC}/\sigma_y)^2/6\pi$ we find that $r = 2\mu$m which is considerably smaller than the critical crack size. It is, in fact, about one-eighth the critical crack depth for $\sigma = \sigma_y$, and about one-fortieth of a_{cr} for $\sigma = 0.5\sigma_y$. At $R_c = 49$, $r = 15\mu$m which is one-seventh the critical crack size at $\sigma = \sigma_y$ and about one-thirtieth of a_{cr} for $\sigma = 0.5\sigma_y$.

If plane stress is assumed rather than plane strain, then the plastic zones are estimated to be three times as large. Is the size of these plastic zones of large enough magnitude compared to the crack depth to invalidate the basic assumptions of linear elasticity and of the nature of the crack tip singularity? How inaccurate is the Irwin approximation for estimating the plastic zone size under these conditions? The implications of the plastic zone size calculations are that small critical cracks may require a higher stress to propagate than that implied in Fig. 7. The significant question is whether Fig. 7 can be used as a guide to the toughness of these steels for small critical crack sizes. It is clear that experimental work is required to see if the critical stress intensity factor for propagating very small cracks can be estimated from data acquired from the long cracks used in the test specimens described in the main body of this paper. The test data will still be of use for cracks of sufficient size to be detectable by standard non-destructive testing methods. For 01 steel at hardness R_c49, the crack length to plastic zone size requirement (from ASTM-E399) implied in a 2.5 $(K_{IC}/\sigma_y)^2$ is fulfilled for $a = 0.7$ mm, while at R_c57 it is 94μm.

CONCLUSIONS

A specimen has been developed which facilitates fracture toughness testing of brittle materials. Results have been presented for 01 tool steel tempered back to various hardness levels. This data is suitable for applications for cracks larger than 1 mm in depth.

ACKNOWLEDGEMENTS

This research has been supported by an NSERC of Canada (Grant A1289). We would also like to thank Atlas Steel of Welland, Ontario for the test material. Useful discussions with J.D. Embury and Luo Li Geng of McMaster University are also gratefully acknowledged.

REFERENCES

Knott, J.R. (1973). Fundamentals of Fracture Mechanics, John Wiley and Sons, 254-256.
Rescalvo, J.F., and Averback, B.L. (1979). Fracture and Fatigue in M50 and 18-4-1 High Speed Steels. _Metallurgical Trans._, _10A_, 1265-1271.

Watt, D.F., Nadin, P.A., and Biner, S.B., (1987). The Fracture Toughness of
   Hardened Tool Steels, to be published in <u>ASME J. of Engineering Materials and
   Technology</u>.

FABRICATION OF LANXIDE^{TM} CERAMIC COMPOSITES

Richard S. Webb

Alcan International Limited, 1188 Sherbrooke St. West,
Montreal, Quebec H3A 3G2

ABSTRACT

Lanxide Corporation (Newark, Delaware) is developing a novel technology for the
fabrication of ceramic composites. Alcan has a major stake, as a partner in the
development and commercialisation of this technology, which includes the incorpo-
ration of composite filler materials in a ceramic/metal matrix obtained by the
directed oxidation reaction of a molten metal and a gaseous oxidant. Information
published by Lanxide Corp will be reviewed, and Alcan's role explained.

KEYWORDS

Ceramics; composites

REFERENCES

Newkirk, M.S., Urquhart, A.W., Zwicker, H.R. "Formation of Lanxide^{IM} Ceramic
    Composite Materials," J. Mater. Res. 1 [1] 81 (1986).
Newkirk, M.S., Lesher, H.D., White, D.R., Kennedy, C.R., Urquhart, A.W.,
    Claar, T.D. "Preparation of Lanxide^{TM} Ceramic Matrix Composites: Matrix
    Formation by the Directed Oxidation of Molten Metals," to be published in
    Ceramic Engineering and Science Proceedings

THE ANALYSIS OF THE FRACTURE ORIGIN TYPES AND SEVERITY IN Y$_2$O$_3$–PARTIALLY STABILIZED ZrO$_2$

Jason Sung and Patrick S. Nicholson

Ceramic Engineering Research Group
Department of Materials Science and Engineering
McMaster University
Hamilton, Ontario, Canada

ABSTRACT

A fractographic study was conducted on 4.5 wt% Y$_2$O$_3$-partially-stabilized ZrO$_2$ four-point bend bars densified by dry-pressing/isostatic pressing/pressureless sintering. Several fracture origin types were identified by SEM/EPMA and classified by their responses to various elimination techniques. An elliptical crack model was used to characterize the correlation between fracture stress and fracture origin size. Consistent results showed that an order-of-severity existed, which enables a definition of "fracture origin severity parameter" for the different origin types. The existence of a relative fracture-origin severity is related to the residual stress fields at the defect-matrix interfaces. A spherical-defect model yielded satisfactory predictions of the numerical values of the fracture origin severity parameters

KEYWORDS

flaw elimination; flaw-strength relationship; severity of defect types.

INTRODUCTION

The excellent mechanical properties of Y$_2$O$_3$-partially-stabilized ZrO$_2$ (Y-PSZ) at ambient temperature has recently attracted extensive research (Lange, 1986a; Matsui, 1984; Masaki, 1986; Tsukuma, 1985a, 1985b, 1985c). It has been found that elastic-brittle materials like Y-PSZ will have fracture stresses correlatable with the fracture origin sizes. Several defect elimination techniques have been employed to remove or reduce the size of the defects reponsible for fracture (Lange, 1986a; Rhodes, 1981; Taguchi, 1985; Tsukuma, 1985a, 1985b). In a recent study by the authors (Sung, Nicholson, 1987), a systematic improvement of the bend strength (MOR) of 4.5 wt% Y-PSZ was realized by conventional dry-pressing/isostatic-pressing/pressureless sintering combined with sedimentation and burnout techniques. The behaviour of several identified fracture origin types were studied over a wide range of fracture stresses and fracture origin sizes.

The relative severity of the defects in ceramics has been little studied. Previous investigators (Evans, 1977; Evans, 1980; Lange, 1986b) showed that there existed a tendency of severity among different defect types but quantitative analysis was not done. Attempts (Evans, 1980; Kirchner, 1976) have been made to correlate fracture stresses with fracture origin sizes through crack models and so identify the nature of the different types of defect. Various crack models of similar kind have been developed (Barratta, 1978; Evans, 1972; Irwin, 1962), refined (Bansal, 1976; Evans, 1979) and applied (Mecholsky, 1976; Seshadri, 1981) and a part-through-elliptical-crack model (Bansal, 1976; Irwin, 1962; Seshadri, 1981) is thought to be the most realistic.

The present study examines the different types of fracture origin in 4.5 wt% Y-PSZ and characterizes the correlation between the fracture stress and the fracture origin size so as to identify the severity of the fracture origin types.

ANALYTICAL PROCEDURE

A part-through elliptical-crack-extension model, first developed by Irwin (Irwin, 1962) and later refined by Bansal (Bansal, 1976), was used in the present study. This model, shown schematically in Fig. 1, calculates

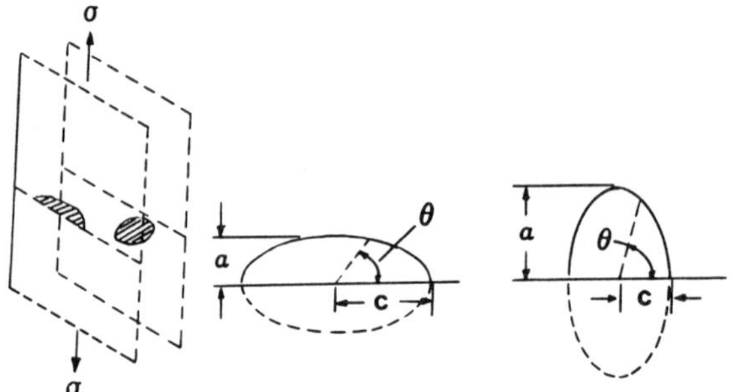

Fig. 1. Part-through elliptical crack extension model.

the stress intensity factor for an elliptical crack in the interior (or a semi-elliptical crack at the surface) of a finite plate subjected to a uniform tensile stress perpendicular to one of its elliptical axes. The plane-strain critical stress intensity factor for the opening mode (K_{IC}) along the periphery of a crack is given by;

$$K_{IC} = \frac{Y\sqrt{1-v^2}}{\Phi}\,\sigma_f\,\sqrt{a} \quad \text{for } a \le c \tag{1a}$$

$$K_{IC} = \frac{Y\sqrt{1-v^2}}{\Phi}\,\sigma_f\,\sqrt{c} \quad \text{for } a \ge c \tag{1b}$$

where v is the Poisson's ratio, σ_f the fracture stress, c the semi-axis perpendicular to the applied stress, a is the other semi-axis and Y is a geometrical parameter describing the position of a crack relative to the tensile surface. For a surface semi-elliptical crack, Y is taken as 1.94, and for a subsurface elliptical crack Y = 1.77. Φ is a "shape parameter", describing the shape variation of the crack. It is an elliptical integral which varies with the ratio a/c, i.e;

$$\Phi = \int_0^{\pi/2} \left[1-\left(1-\frac{a^2}{c^2}\right)\sin^2\theta\right]^{1/2} d\theta \tag{2}$$

where θ is an angle defined in Fig. 1. From equation (1) it is clear that the smaller crack dimension controls the fracture. If "a" denotes the smaller crack dimension, equation (1) becomes;

$$\sigma_f \cdot \frac{Y\sqrt{a}}{\Phi} = \frac{K_{IC}}{\sqrt{1-v^2}} = \text{constant} \tag{3}$$

and a plot of σ_f vs "$Y\sqrt{a}/\Phi$" should give a curve characterized by a constant K_{IC} value. $Y\sqrt{a}/\Phi$ will be termed the "equivalent fracture origin size".

EXPERIMENTAL PROCEDURE

4.5 wt% Y_2O_3-partially-stabilized ZrO_2 submicron powder* was used as the starting material. Three processing routes were employed (Fig. 2). Route 1 was the as-received powder (without powder beneficia-

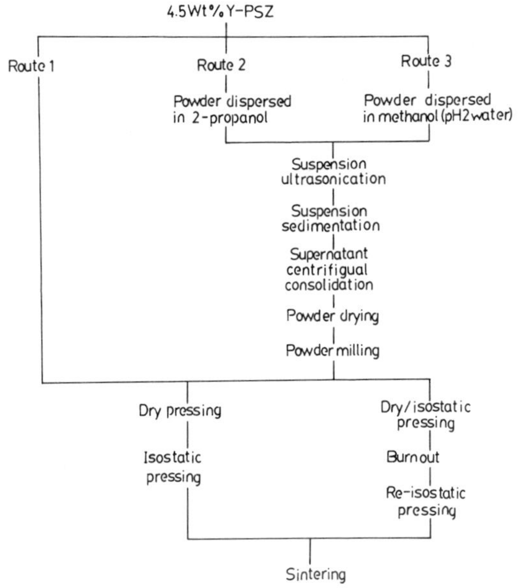

Fig. 2. Various processing routes for 4.5 wt% Y-PSZ.

tion), Route 2 the as-received powder 2-propanol sedimented and Route 3, the as-received powder methanol (or pH-2 water) sedimented followed by a burnout/re-isostatic pressing step. In route 1 (also the common procedure) the powders were dry-pressed at 40 MPa and isostatically-pressed at 300 MPa. The compacted plates were pressurelessly sintered in air at 1400°C for 3 hours, cut into bars and ground and edge-chamfered to final dimensons of $25 \times 2 \times 1.5$ mm. The finished bars were annealed at 1400°C for 0.5 hour to relieve surface residual stresses. Bars were broken in four-point-bend with an inner span of 10 mm and an outer span of 20 mm. The machine cross-head speed used was 0.2 mm/min. Test procedures conformed with American Standard MIL-STD-1942 (MR) (DOD, 1983). Both fracture surfaces of each bar were examined by optical microscope at 80 X to approximately locate the fracture origins and representative half specimens were then examined by SEM. Inclusion-type fracture origins were analyzed by EPMA to identify their composition. Route 2 involved dispersing and sedimenting the as-received powders in 2-propanol prior to pressing and sintering. The sedimentation technique (Rhodes, 1981; Parish, 1984) was introduced to eliminate powder agglomerates and inclusion-type defects. Route 3 introduced the additional burn-out step (Lange, 1986a) prior to final sintering. The as-received powders were first sedimented in methanol (or pH-2 water), compacted and fired at 600°C for 24 hours. They were then re-isostatic pressed at 300 Mpa and sintered.

RESULTS AND DISCUSSION

<u>Experimental Results</u>

The bend strengths of Route-1 synthesized samples were 650-1020 MPa, average 880 MPa. Five types of fracture origin were identified, i.e. fiber inclusions, agglomerates, iron inclusions, alumina inclusions, and

---

*Zircar Inc., Florida, N.Y., USA.

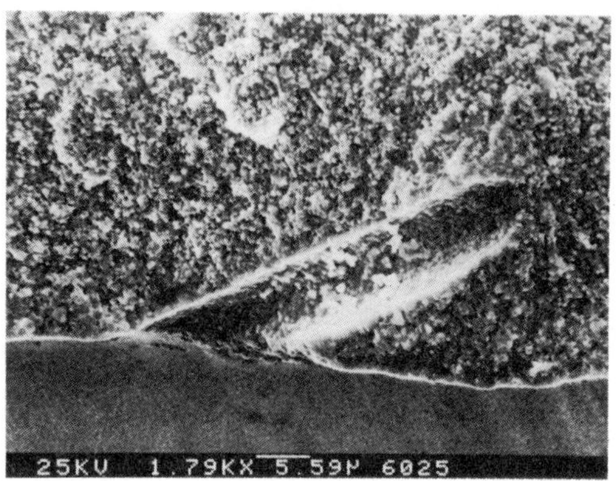

Fig. 3. Type A fracture origin-fiber inclusion.

Fig. 4. Type B fracture origin-agglomerate.

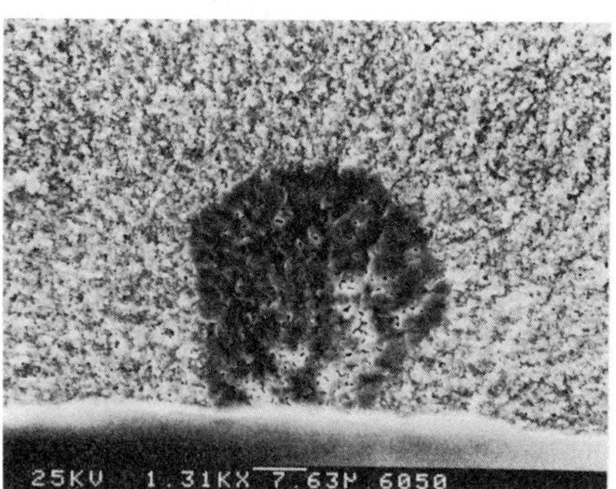

Fig. 5. Type C fracture origin-iron inclusion.

Fig. 6. Type D fracture origin-alumina inclusion.

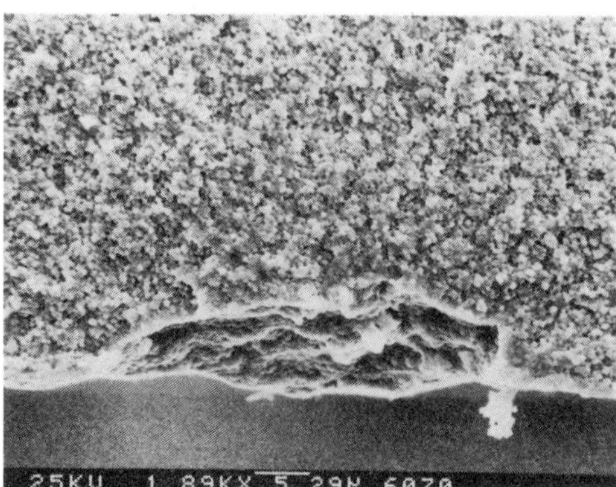

Fig. 7. Type E fracture origin-pore.

pores (Figs. 3-7). Fiber inclusions (Fig. 3) are evidenced by regularly-shaped elongated pores. They are left by some kind of fiber-shaped inclusions after burn-off. Agglomerates (Fig. 4) show as aggregates of large grains centered in a cavity surrounded by smaller grains. The large grains and surrounding porosity result from the differential sintering rate of the agglomerate and the matrix (Lange, 1983). Iron inclusions (Fig. 5) show as round or elliptical dark regions. EPMA detected iron. Iron oxide impregnated ZrO_2 grains sinter abnormally generating a porous region. Alumina inclusions (Fig. 6) are large regions of irregular shape elongated along the surface direction. EPMA detected aluminum. Pores (Fig. 7) are irregular-shaped cavities left after pressing or developed during sintering. The bend strengths for Route-2-synthesized samples were 740-1280 MPa, average 995 MPa. The agglomerates and iron inclusions were reduced in size by sedimentation but the fiber inclusions were unaffected. The resulting strengths for Route-3 bars were 950-1470 MPa, average 1150 MPa. Agglomerates were further reduced in size and the iron inclusions totally eliminated. The fiber inclusions were also removed. Detailed experimental results are described elsewhere (Sung, Nicholson, 1987).

The Existence of the Severity of Defects

The experimental data were plotted as fracture stress (σ_f) vs. calculated equivalent fracture origin size ($Y\sqrt{a}/\phi$) (Fig. 8). The fracture origin types follow the part-through crack extension model (equation (3)). The agglomerate-type fracture origins deviate below an equivalent size $\sim4\times10^{-3}\,m^{1/2}$ (fracture stress ~900 MPa) and finally merge with the curve for iron inclusions and pores (these are in one curve).

The fracture origin types differ with respect to their responsibility for fracture, i.e. fiber inclusions are the most severe, then agglomerates, iron inclusions and pores, and alumina inclusions. The part-through crack model applies to a pore with traction-free periphery. Therefore the fiber inclusion and the agglomerate fracture origins will be more severe than predicted, whereas, the iron inclusion and pore fracture origins will be as predicted. The alumina inclusion fracture origins will be less severe than predicted.

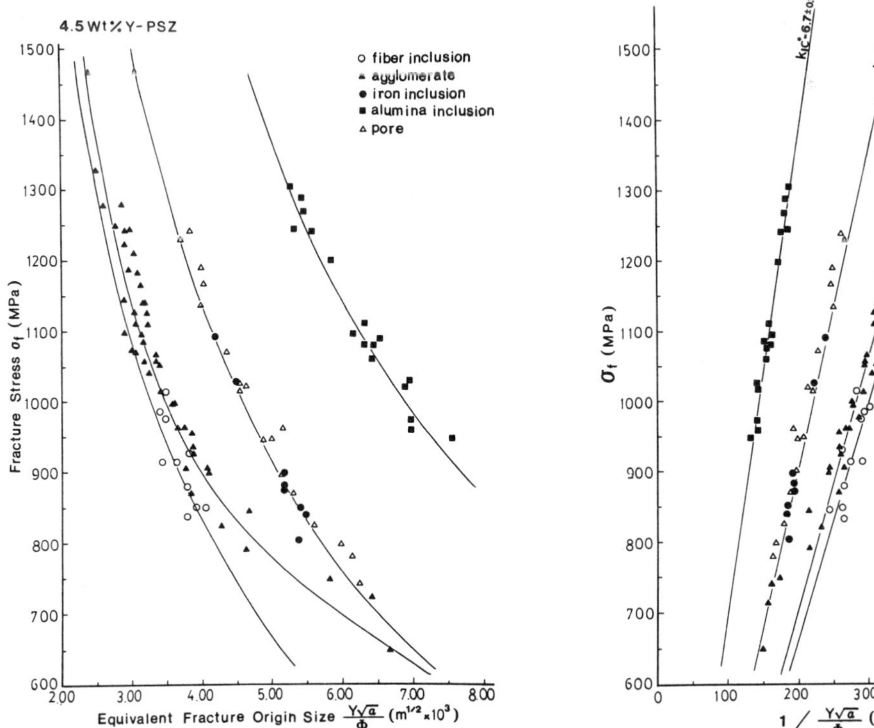

Fig. 8. $\sigma_f - Y\sqrt{a}/\phi$ plot for 4.5% Y-PSZ.

Fig. 9. $\sigma_f - 1/Y\sqrt{a}/\phi$ plot for 4.5 wt% Y-PSZ.

<u>Origin of the Relative Severity of Defects</u>

The relative severity of fracture origins is related to the residual stress fields developed around the defects. For agglomerates, the increased local density will cause differential sintering and tensile residual stress build-up in the local matrix. Incoherency relieves part of this stress (Lange, 1983) but the densified matrix and agglomerate prohibit complete stress relief and the residual tensile stress will superimpose on the applied stress causing fracture at lower stresses than predicted. The agglomerate fracture origin will be more severe than a pore. Similar reasoning holds for a fiber inclusion. By resisting the forming pressure during dry and isostatic pressing, a denser packing of powder results local to the fiber. During sintering this denser layer will sinter faster resulting in local residual stresses similar to those of agglomerate. This denser layer does not break away from the matrix and the residual tensile stress is unrelieved, i.e. a more severe condition results than for an agglomerate (as observed). Iron melts and diffuses into the ZrO_2 particles so no residual stresses will develop. Porous unsintered ZrO_2 particles inside this defect act like a pore and exhibit a similar severity (as observed). An alumina inclusion will develop a residual compressive radial stress and a residual tensile tangential stress around its periphery due to its lower thermal expansion $(8\text{-}9 \times 10^{-6}/°C$ vs. $10\text{-}11 \times 10^{-6}/°C$ for $ZrO_2)$ (The induced tangential tensile stress may promote the t→m transformation of the ZrO_2 and the 4% expansion involved will change the peripheral stress state to compression). The superposition of this stress on the applied stress results in a higher fracture stress than predicted by the crack model. An alumina inclusion thus has a severity less than a pore. The size-effect of the agglomerate fracture origin severity is explained by the magnitude of the residual stresses. When agglomerates exceed $\sim 4 \times 10^{-3}$ $m^{1/2}$ equivalent size (corresponding to a fracture stress of ~ 900 MPa) the residual stresses are further relieved by the defects, becoming pores of $\sim 7 \times 10^{-3}$ $m^{1/2}$ equivalent size. This is due to microcrack formation, the occurrence of which is predicted on exceeding a critical defect size (Green, 1982).

<u>Quantitative Expression for the Relative Severity of Defects</u>

If equation (3) is rearranged, i.e.;

$$\sigma_f = \left(\frac{K_{IC}}{\sqrt{1-v^2}} \right) \frac{1}{Y\sqrt{a}/\Phi} \qquad (4)$$

and σ_f vs $1/Y\sqrt{a}/\Phi$ is plotted for the various fracture origin types, the resultant line has slope of $K_{IC}/\sqrt{1-v^2}$. Setting $v = 0.25$ for the tetragonal ZrO_2, the "apparent" K_{IC} value $(K_{IC}{}^*)$ for each fracture origin type can be obtained. Such plots are shown in Fig. 9, and give the following $K_{IC}{}^*$ values: 3.2 ± 0.1 MPa $m^{1/2}$ for fiber inclusions, 3.4 ± 0.1 MPa $m^{1/2}$ for agglomerates, 4.4 ± 0.1 MPa $m^{1/2}$ for iron inclusions and pores and 6.7 ± 0.1 MPa $m^{1/2}$ for alumina inclusions. These values also reflect the relative severity of the fracture origin types. The microindentation technique (Niihara, 1982) was used to measure the actual K_{IC} of the ZrO_2 matrix and gave $K_{IC} = 6.9 \pm 0.3$ MPa $m^{1/2}$. Comparison with the theoretical value calculated from the part-through crack extension model for pores gave a ratio of $\simeq 1.55$. This result is close to that of a previous study (Seshadri, 1981).

The existence of a differential severity of fracture origin types make further correction of the crack model necessary. A "fracture origin severity parameter" (X) can be defined as;

$$K_{IC} = \sigma_f \cdot X \cdot \frac{Y\sqrt{a}}{\Phi} \qquad \text{(for plane stress)} \qquad (5a)$$

$$\frac{K_{IC}}{\sqrt{1-v^2}} = \sigma_f \cdot X \cdot \frac{Y\sqrt{a}}{\Phi} \qquad \text{(for plane strain)} \qquad (5b)$$

and X = 1.03 for the alumina inclusions, 1.57 for the pores and iron inclusions, 2.03 for the agglomerates and 2.16 for the fiber inclusions. These values can predict fracture stresses when the type, dimension and position of a defect is known (by NDE for example).

MODELLING OF THE RELATIVE SEVERITY OF DEFECTS

To obtain physical insight into the existence of a relative-severity-of-origins, spherical defect models were investigated and numerical values calculated.

A model of a spherical inclusion embedded in the matrix of a four-point-bend specimen is shown in Fig. 10.

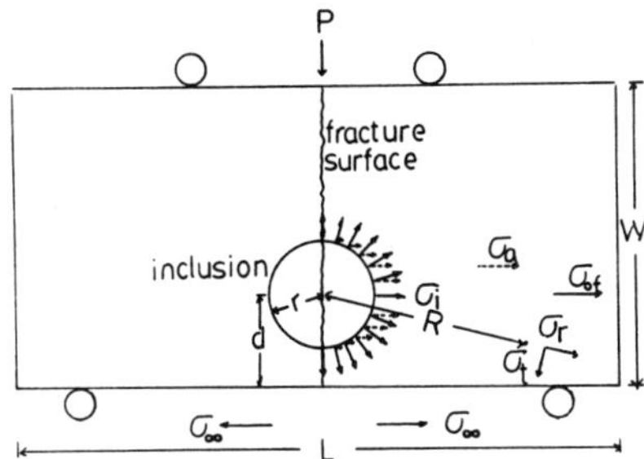

Fig. 10. Stress states of a spherical inclusion embedded in a 4-point-bend bar.

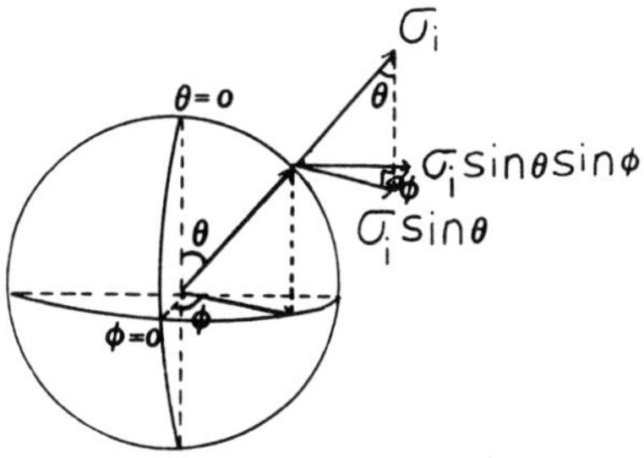

Fig. 11. Residual stress components at the inclusion-matrix interface.

This inclusion (radius r), a distance d from the tensile surface, is subjected to an applied tensile stress, σ_∞. If the inclusion thermal expansion coefficient is larger than that of the matrix (difference $\Delta\alpha$), the hydrostatic tensile stress developed inside the inclusion (σ_{ic}) upon cooling (ΔT) from sintering is (Selsing, 1961):

$$\sigma_{ic} = \frac{\Delta\alpha \cdot \Delta T}{\dfrac{1+v_m}{2E_m} + \dfrac{1-2v_i}{E_i}} \tag{6}$$

where v is the Poisson's ratio, E the elastic modulus and subscripts i and m refer to the inclusion and matrix respectively. The stress fields in the matrix, a distance R from the inclusion center, are;

$$\sigma_{rc} = \sigma_{ic}\,(r/R)^3 \qquad -\text{radial} \tag{7a}$$

$$\sigma_{tc} = -\sigma_{ic}\,(r/R)^3/2 \qquad -\text{tangential} \tag{7b}$$

where $\sigma_{rc} = \sigma_{ic}$ at the inclusion-matrix interface. The application of σ_∞ generates additional stresses inside the inclusion (σ_{ie}) due to elastic mismatch with the matrix. This is given by (Evans, 1980; Esheby, 1957);

$$\sigma_{ie} = \sigma_{o\infty}\left\{1 + \frac{2[(K_i/K_m)-1](1-2v)}{3(1-v)}\left[\frac{4G_m+3K_m}{4G_m+3K_i}\right]\right\} \tag{8}$$

where K is the bulk modulus, G the shear modulus and $\sigma_{o\infty}$ the applied stress in the inclusion. Assuming the inclusion will fracture along the cross section perpendicular to the tensile surface, the average residual stress (σ_a) applied to the inclusion-matrix interface at this section in the tensile direction is;

$$\sigma_a = \frac{1}{2\pi r^2}\int_{\phi=0}^{\pi}\int_{\theta=0}^{\pi}\sigma_{ic}\,\sin\theta\,\sin\phi\,d\theta\,d\phi \tag{9}$$

where θ and ϕ are defined in Fig. 11. At fracture initiation, σ_a is superimposed on the fracture stress at the inclusion (now σ_{of}) predicted by the crack model (equation (3)) and the flaw-strength relationship of equation (3) is expressed as;

$$\frac{K_{IC}}{\sqrt{1-v^2}} = (\sigma_{of}+\sigma_a)\cdot X \cdot \frac{Y\sqrt{\bar{a}}}{\Phi} \tag{10}$$

If one defines X = 1 for the residual-stress-free defects (iron inclusions and pores) then;

$$X = 1 - \frac{\sigma_a}{\sigma_f} \tag{11}$$

where σ_a is positive for compression and negative for tension.

A residual compressive stress is developed around an alumina inclusion and Fig. 12 shows the magnitude of

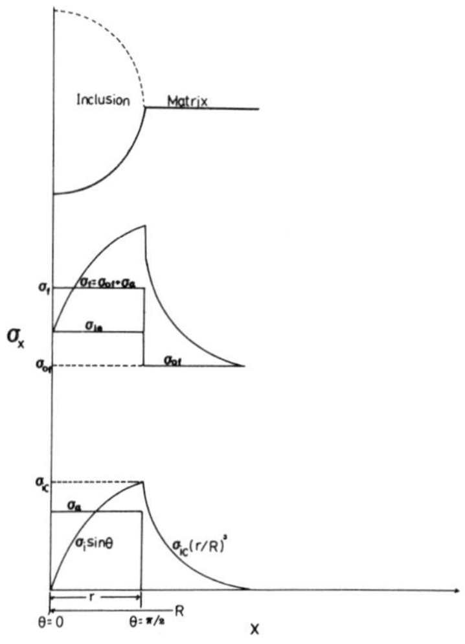

Fig. 12. Various stress components in tensile direction around an alumina inclusion in t-ZrO2.

the stress fields in tensile direction (X-direction) along a section of the inclusion-matrix interface and in matrix (indicated as a solid line) using $\Delta\alpha = 2\times10^{-6}/°C$, $\Delta T = 1000°C$, $E_m = 220\,GPa$, $E_i = 400\,GPa$ and $\nu_m = \nu_i = 0.25$, $\sigma_{ic} = 490\,MPa$ and $\sigma_a = 350\,MPa$. Now $\sigma_f = 900\,MPa$ to $1300\,MPa$ so X = 0.61 to 0.73, average 0.67. This calculated X value agrees well with the experimental value (1.03/1.55 = 0.66).

For agglomerates, the residual stress comes from the differential sintering and the relation between the sintering strain (ε) and the sintered density (ρ) is (Lange, 1983);

$$(1 - \varepsilon)^3 = \rho_0/\rho \tag{12}$$

where ρ_0 is the green density of the powder compact and ρ the fired density. So the differential strain developed between an isolated agglomerate (subscript a) and the matrix (subscript m) during sintering is;

$$\Delta\varepsilon = \varepsilon_m - \varepsilon_a = (\rho_{oa}/\rho_a)^{1/3} - (\rho_{om}/\rho_m)^{1/3} \tag{13}$$

and the hydrostatic stress in the agglomerate will be;

$$\sigma = E\Delta\varepsilon = E[(\rho_{oa}/\rho_a)^{1/3} - (\rho_{om}/\rho_m)^{1/3}] \tag{14}$$

where E is the elastic modulus of the agglomerate.

Previous studies (Pampuch, 1983; Rhodes, 1981, Reed, 1978) pointed out the effect of earlier start and finish of sintering for the agglomerates. When the developing stress intensity at the interface exceeds the fracture toughness, partial decohesion will occur and a pore-like cavity results. The residual tensile stresses in the matrix will be superimposed on the applied stress. Continued sintering leads to further matrix shrinkage and the porous cavity reduces in size. An agglomerate sintering model is shown in Fig. 13. The sintering

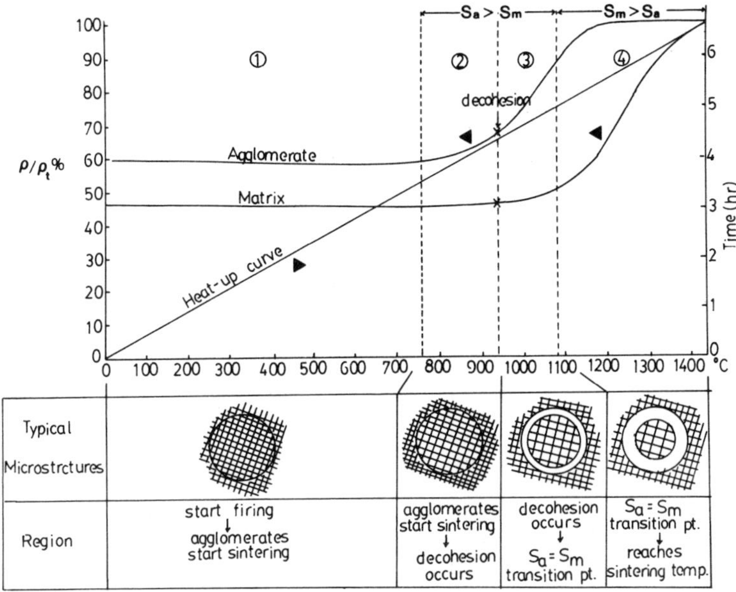

Fig. 13. An agglomerate sintering model.

behaviour of the agglomerates and matrix are expressed as a % of the theoretical sintering density (ρ_t) vs. temperature. The green density of the agglomerates is estimated as 60% ρ_t. A corresponding heat up curve is also shown in Fig. 13. Four regions appear on the diagram; (a) no sintering (b) agglomerate sintering rate (Sa) > matrix sintering rate (Sm)-before decohesion, (c) after decohesion and (d) Sm > Sa region. The corresponding microstructures are schematically shown.

Based on the above model, the residual tensile stress at Sa = Sm is several tens of GPa (e.g. if ρ_{oa} = 0.6, ρ_a = 0.9, ρ_{om} – 0.48 and ρ_m = 0.55, then σ = 18 GPa for equation (14)). This exceeds the strength of the agglomerate-matrix interface and decohesion must result. To estimate the fracture stress at the decohesion point, a critical preexisting flaw size is assumed, i.e., (0.1 nr) (r = the particle radius) (Ito, 1981). For an agglomerate of radius 10 μm and σ_f = 900 MPa (Fig. 8), the critical flaw size is ≃3 μm   If the fracture toughness is 50% of the sintered value, the fracture stress at decohesion is ≃750 MPa (equation (3)). Assuming 50% of this stress (375 MPa) remains in the matrix-cavity interface then σ_a (Equation (9)) ≃ 265 MPa. The final calculated value of X for an agglomerate is 1.29, close to the experimental value (2.03/1.55 = 1.31).

SUMMARY AND CONCLUSIONS

As a result of the work here reported;

1.   the fracture origin types and their severity in Y-PSZ have been identified and analysed using a part-through-elliptical-crack model;

2.   the relative severity of the fracture origin type is explained by the residual stress fields developed at the defect-matrix interfaces;

3.   a "fracture origin severity parameter" (X) has been defined and used to correct the crack model;

4.   modelling of the relative severity for an alumina inclusion and an agglomerate yielded satisfactory
     values of X when compared with the experimental data.

REFERENCES

Bansal, G.K. (1976). Effect of flaw shape on strength of ceramics. J. Am. Ceram. Soc., 59, 87-88.

Barratta, F.I. (1978). Stress intensity factor estimates for a peripherally cracked spherical void and a hemispherical surface pit. J. Am. Ceram. Soc., 61, 490-493.

DOD (1983). Flexural strength of high performance ceramics at ambient temperature, MIL-STD-1942 (MR).

Esheby, J.D. (1957). The determination of the elastic field of an ellipsoidal inclusion and related problem. Proc. Roy. Soc., 241A, 376-396.

Evans, A.G. and G. Tappin (1972). Effects of microstructure on the stress to propagate inherent flaws. Proc. Br. Ceram. Soc., 20, 275-297.

Evans, A.G., G.S. Kino, P.T. Khuri-Yakub and B.R. Tittmann (1977). Failure prediction in structural ceramics. Material Evaluation, 35, 85-96.

Evans, A.G., D.R. Biswas and R.M. Fulrath (1979). Some effects of cavities on the fracture of ceramics: II. spherical cavities. J. Am. Ceram. Soc., 62, 101-106.

Evans, A.G., M.E. Meyer, K.W. Fertig, B.I. Davis and H.R. Baumgartner (1980). Probabilistic models for direct initiated fracture in ceramics. J. Nond. Eval., 1, 111-122.

Green, D.J. (1982). Critical microstructure for micocracking in Al_2O_3-ZrO_2 composites. J. Am. Ceram. Soc., 65, 610-614.

Irwin, G.R. (1962). Crack-extension force for a part-through crack in a plate. J. Appl. Mech., 29, 651-654.

Ito, Y.M., M. Rosenblatt, L.Y. Cheng, F.F. Lange and A.G. Evans (1981). Cracking in particulate composites due to thermal-mechanical stress. Int. J. of Frac., 17, 483-491.

Kirchner, H.P., R.M. Gruver and W.A. Sotter (1986). Characteristics of flaws at fracture origins and fracture stress-flaws size relations in various ceramics. Mater. Sci. and Eng., 22, 147-156.

Lange, F.F. and M. Metcalfe (1983). Process-related origins: II. Agglomerate motion and cracklike internal surfaces caused by differential sintering. J. Am. Ceram. Soc., 66, 398-406.

Lange, F.F., B.I. Davisand and E. Wright (1986a). Processing-related fracture origins: IV. Elimination of voids produced by organic inclusions. J. Am. Ceram. Soc., 69, 66-69.

Lange, F.F. (1986b). Advanced processing of ceramics: Controlling flaw populations. In P.S. Nicholson (Ed.), Transactions of the Canadian University–Industry Council on Advanced Ceramics, 2nd Workshop, pp 1–29.

Mecholsky, J.J., Jr., S.W. Freiman and R.W. Rice (1976). Fracture surface analysis of ceramics. J. Mater. Sci., 11, 1310-1319.

Matsui, M., T. Soma and I. Oda (1984). Effect of microstructure on the strength of Y–TZP components. In N. Claussen (Ed.), Advances in Ceramics, Vol. 12, Science and Technology of Zirconia II. The Am. Ceramic Soc., pp. 371-381.

Masaki, T. (1986). Mechanical properties of toughened ZrO_2–Y_2O_3 ceramics. J. Am. Ceram. Soc., 69, 638-640.

Niihara, K., R. Morena and D.P.H. Hasselman (1982). Evaluation of K_{IC} of brittle solids by the indentation method with low crack-to-indent ratios. <u>J. Mater. Sci. Lett.</u>, 1, 13-16.

Pampuch, R. and K. Haberko (1983). Agglomerates in ceramic micropowders and their behaviour on cold pressing and sintering. In P. Vincenzin (Ed.), <u>Ceramic Powders</u>, Elsevier, pp. 623-634.

Parish, M. and H.K Bowen (1984). Narrow size distribution powders from commercial ceramic powder. <u>Ceramic International</u>, 10, 75-77.

Reed, J.S., T. Carbone, C. Scott and S. Lukasiewicz (1978). Some effects of aggregates and agglomerates in the fabrication of fine grained ceramics. In H. Palmour III (Ed.), <u>Materials Science Research</u>, Vol. 11, Processing of Crystalline Ceramics, Plenum, pp. 171-180.

Rhodes, W.H. (1981). Agglomerate and particle size effect on sintering Yttria-stabilized zirconia. <u>J. Am. Ceram. Soc.</u>, 64, 19-22.

Selsing, J. (1961). Internal stresses in ceramics. <u>J. Am. Ceram. Soc.</u>, 44, 419.

Seshadri, S.G. and M. Srinivasan (1981). Estimation of fracture toughness by intrinsic flaw fractography for sintered alpha silicon carbide. <u>J. Am. Ceram. Soc.</u>, 64, C-69–C-71.

Sung, J. and P.S. Nicholson (submitted to J. Am. Ceram. Soc., 1987). Strength improvement of Yttria-partially-stabilized zirconia by flaw elimination.

Taguchi, H., Y. Takahashi and H. Miyamoto (1985). Effect of milling on slip casting of partially stabilized zirconia. <u>J. Am. Ceram. Soc.</u>, 68, C-264–C-265.

Tsukuma, K. and M. Shimada (1985a). Hot isostatic pressing of Y_2O_3-partially-stabilized zirconia. <u>J. Am. Ceram. Soc.</u>, 64, 310-313.

Tsukuma, K. and K. Ueda (1985b). Strength and fracture toughness of isostatically hot-pressed composites of Al_2O_3 and Y_2O_3-partially-stabilized zirconia. <u>J. Am. Ceram. Soc.</u>, 68, C-4–C-5.

Tsukuma, K. and K. Ueda (1985c). High-temperature strength and fracture toughness of Y_2O_3-partially-stabilized ZrO_2/Al_2O_3 composites. <u>J. Am. Ceram. Soc.</u>, 68, C-56–C-58.

MECHANICAL BEHAVIOR OF CELLULAR CERAMICS

David J. Green,
Department of Materials Science and Engineering
The Pennsylvania State University
PA 16802, USA

ABSTRACT

There is a class of ceramics that can be made with very high porosities
(~ 80-90 volume percent). The microstructures of these materials are
generally based on tangled networks of fibers or on hollow cells (Green,
1984, 1986). Such materials possess attractive properties for some
applications, e.g., low weight, high temperature capability, low thermal
conductivity, permeability (unless closed cells), high surface area,
etc. A key feature in their use, however, will be their mechanical
reliability. The mechanical behavior of these materials is generally
modelled by considering the deformation of a unit cell (Green and Lange,
1982, Gibson and Ashby, 1982, Ashby, 1983). This approach allows
equations to be developed for many of the mechanical properties. This
paper reviews recent work on the mechanical behavior of open cell
alumina (Hagiwara and Green, 1986, 1987, Brezny 1988, Dam, 1988). This
work has shown that the elastic behavior (Hagiwara and Green, 1987) does
follow the predicted behavior (Gibson and Ashby, 1982) but that the
properties can be sensitive to the specific microstructure. Moreover,
the fracture behavior is complicated by variation in the strength of the
cell struts and the presence of closed faces in the cellular structure
(Brezny, 1988). Overall, the model proposed by Gibson and Ashby appears
to be a useful foundation but that quantitative characterization of the
microstructure appears to be necessary to understand the mechanical
behavior in depth.

REFERENCES

Ashby, M. F. The mechanical properties of cellular solids," _Metall._
    Trans. 14A pp. 1755-69.
Brezny, R., (1988). Fracture toughness behavior of open cell alumina,
    M. S. Thesis, The Pennsylvania State University.
Dam, C., (1988). Compressive strength behavior of open cell alumina,
    M. S. Thesis, The Pennsylvania State University.
Gibson, L. J., and M. F. Ashby (1982). The mechanics of three-
    dimensional cellular materials," _Proc. R. Soc. London, Ser. A,_ 382
    [1782] pp. 43-59.
Green, D. J., and F. F. Lange (1982). Micromechanical model for fibrous
    ceramic bodies, _J. Am. Ceram. Soc.,_ 65, pp. 138-41.

Green, D. J., (1984). Mechanical behavior of space shuttle thermal protection system tiles. In L. E. Murr (Ed.) <u>Industrial Materials Science and Engineering,</u> Marcel Dekker, New York, pp. 123-43.

Green, D. J. (1986). Mechanical behavior of lightweight ceramics. In R. C. Bradt, A. G. Evans, F. F. Lange and D. P. H. Hasselman (Eds.), <u>Fracture Mechanics of Ceramics,</u> Vol. 8, Plenum Press, New York, pp. 39-59.

Hagiwara, H., and D. J. Green (1986). Proceedings of Advanced Ceramics, II, to be published.

Hagiwara, H., and D. J. Green (1987).Elastic behavior of open cell alumina, <u>J. Am. Ceram. Soc.,</u> <u>70,</u> to be published.

RELIABILITY OF STRUCTURAL CERAMICS AT ELEVATED TEMPERATURES

D. S. Wilkinson, M. Chadwick, R. Jupp and A. G. Robertson
Department of Materials Science and Engineering
McMaster University
Hamilton, Ontario L8S 4L7

ABSTRACT

The utilisation of structural ceramic materials in critical components operating at elevated temperatures requires that their reliability over the design life be assured. The understanding of the processes which control the failure and reliability of these materials at high temperatures is currently at early stage. In this paper some of the problems associated with the use of structural ceramics under these conditions are outlined and the kind of information which is required is discussed.

KEYWORDS

Ceramics, structural reliability, failure, slow crack growth, damage, swelling, devitrification

INTRODUCTION

Developments in ceramic engineering over the last decade or so have led to increased interest in the possibility of using ceramic materials in structural applications at elevated temperatures. In particular much effort has been devoted to the development of heat engines which make extensive use of ceramics. In the short term these development programs will lead to the introduction of ceramic components into diesel engines. In the longer term ceramic gas turbine engines for automotive applications are envisioned.

One of the overriding concerns in this development is that the materials developed be reliable in service over their design life. Ceramics are, for the most part, inherently hard and brittle. Failure can occur suddenly and catastrophically from defects in the material. For this reason much of the work aimed at developing ceramic materials for advanced heat engines is aimed either at improving the processing so as to decrease the size and severity of defects, or at increasing the ability of the material to tolerate defects without failure. The interaction between these processes can be illustrated simply in terms of the Griffith equation for brittle fracture:

$$\sigma_f = K_c / \sqrt{(\pi a)} \tag{1}$$

A second key area of concern is related to slow crack growth. Flaws which are subcritical (i.e. the stress is less than that for fast fracture given by eq. (1)) can grow at room temperature due to environmental interactions until a critical flaw is developed. Therefore a great deal of work has been aimed at understanding subcritical crack growth. This process can generally be described by an equation of the form

$$v = A K^n$$

where v is the crack growth velocity and K the stress intensity factor, while A and n are environmentally-sensitive material parameters. Once A and n are known, and the initial flaw size determined, then a time to develop a critical flaw can be determined. As in the case of fast fracture, strength, lifetime under subcritical crack growth is increased by increasing toughness and decreasing flaw size, as well as by altering the microstructure in such a way as to increase the material's crack growth resistance (i.e. decreasing A).

This indicates that σ_f, the stress at which fracture occurs, depends on the fracture toughness K_c and the defect length a. Strength improvements are therefore related to increasing toughness and decreasing flaw size.

The concepts described above have successfully been applied to the development of materials exhibiting improved strength at room temperature. (See for example the paper by Sung and Nicholson in these proceedings). However, there are additional issues which arise when dealing with materials which are put in service at elevated temperatures These are the subject of this paper.

Failure at elevated temperatures generally occurs by one of three general processes. These are illustrated in Fig. 1, on a so-called fracture mechanism map for NC-132 (a MgO-doped hot-pressed silicon nitride), due to Quinn (1984). At high stresses failure occurs upon loading. This process, referred to on the diagram as "fast fracture", is the same as that which occurs at lower temperatures. Failure occurs when the stress on a suitably located and oriented flaw exceeds a critical value determined by the size of the flaw and the fracture toughness of the material. At intermediate stresses failure occurs by "slow crack growth". This process is also flaw-dependent, and generally involves a single dominant flaw in an advantageous location in the specimen. However, failure is not instantaneous. Instead the flaw grows slowly and continuously, in the form of microcracks, until the critical crack size for fast fracture is achieved. This process is similar to the process of subcritical crack growth described above with reference to fracture and room temperature. However, because of the elevated temperature, environmental assistance is not required. Thermally activated cracking from internal defects is also possible. At even lower stresses the mode of failure changes to "creep fracture". This process does not emanate from a single dominant flaw. Indeed, in Quinn's map this region is defined as that for which large artificial cracks do not lower the lifetime of the material. Grathwol (1984) has shown for example that below a threshold stress an artificial flaw produced by a string of 5 Knoop indents does not precipitate failure by slow crack growth. Instead the material fails at some other place in the specimen. While relatively little work has been done to assess the micromechanics involved here, it is generally believed that damage, in the form of cavities and microcracks, develops throughout the structure These microcracks eventually link up to create a single dominant crack which propagates through the material causing the final failure.

When studying data of this type it useful to ask how the microstructure influences the failure process For example, are the flaws which control fast fracture and slow crack growth of the same type? Does the type of flaw controlling slow crack growth change with temperature and stress? Do the flaws controlling failure by slow crack growth always exist in the starting material or can they be generated in service? Is the damage leading to creep fracture homogeneously distributed, or is it related to different types of flaw population? Answers to questions of this type are required to guide the development of new materials. Clearly it is of little value to optimize a material in terms of its resistance to fast fracture if failure in service will occur by a different process. This is especially true if the defects controlling failure are different. This also applies to non destructive evaluation (NDE). NDE can be used prior to placing components in service, in order to eliminate those which have dangerous flaws in dangerous positions. However this approach can only be successful if the type of flaw which leads to failure is known, if that type of flaw exists in the material prior to its going into service, and if it is detectable using NDE techniques.

MICROSTRUCTURAL EVOLUTION AT ELEVATED TEMPERATURES

The main difference in evaluating the structural reliability of a component going into service at elevated temperatures, as opposed to room temperature, is that the microstructure can no longer be regarded as fixed. Both thermal effects and thermally activated chemical effects can occur. Thus the microstructure evolves with time. This can have a profound effect on the nature of the failure process and on the procedure required to enable prediction of structural reliability in service. In this section, some of the processes by which microstructural evolution occurs are catalogued.

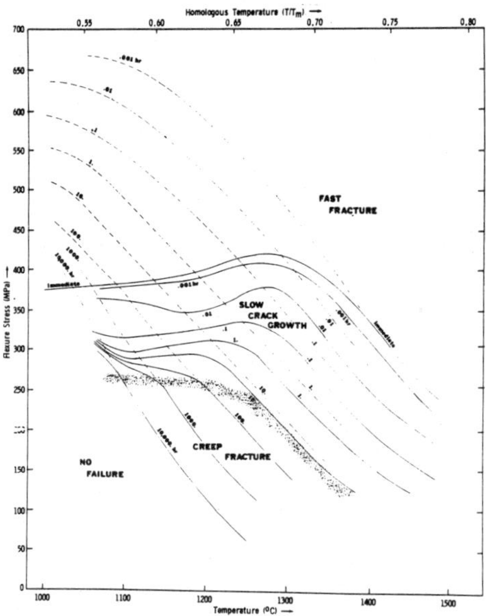

Fig. 1: A "fracture mechanism map" indicating the stress dependent times to failure for NC-132, a MgO-doped hot-pressed silicon nitride. The dotted lines indicate specimens without artificial flaws. The solid lines are for specimens containing artificial cracks. At high stresses failure is totally flaw dependent. However at low stresses the artificial flaws do not decrease the life significantly. The map is divided into three failure regimes - fast fracture, slow crack growth and creep fracture (Quinn, 1984).

Flaw healing can occur at elevated temperature. For example, a machined surface contains many topological imperfections which can act as initiation sites for failure. However, these tend to be alleviated by annealing at elevated temperatures. This is illustrated in Fig. 2 in which the surface of a sintered silicon nitride sample is shown as a function of time at 1200°C. Within the first 25 hours the machining scratches have largely disappeared. In their place a rather complex surface topography develops containing new crystalline phases and amorphous regions.

In addition to flaw healing, flaw generation is also possible, even in the absence of an applied stress. For example, in the same silicon nitride material, large pits are generated over time However, their development appears to be environmentally sensitive. The samples shown in Fig. 2 were annealed in an alumina tube furnace, and relatively few pits were formed. When the same material is annealed in a creep furnace containing a wide range of refractory materials (including Al_2O_3 and SiC) a large number of pits are generated, as shown in Fig. 3. The difference between the pit generation in the two furnaces is presumably related to interactions with the more complex chemical environment existing in the creep furnace.

All of these processes affect the residual strength of a material after exposure at elevated temperatures, whether or not this occurs under an applied load. This has been illustrated by Wiederhorn and Tighe (1983) using a "strength degradation diagram", as shown in Fig. 4, for NC-132. This diagram gives the 1200°C fast fracture strength of a material following 1200°C creep under a stress of 250 MPa for different lengths of time. It is possible in fact to identify the changes in strength with different processes such as the healing of machining flaws and the generation of pits. It should also be possible to produce such a diagram for high temperature exposure without an applied stress. In fact, if this were done for NC-132 at 1200°C, it is likely that the diagram would not change very much for times less than about 50 hours. It is only for longer times that stress driven cavitation starts to play a role in degrading the strength of the material.

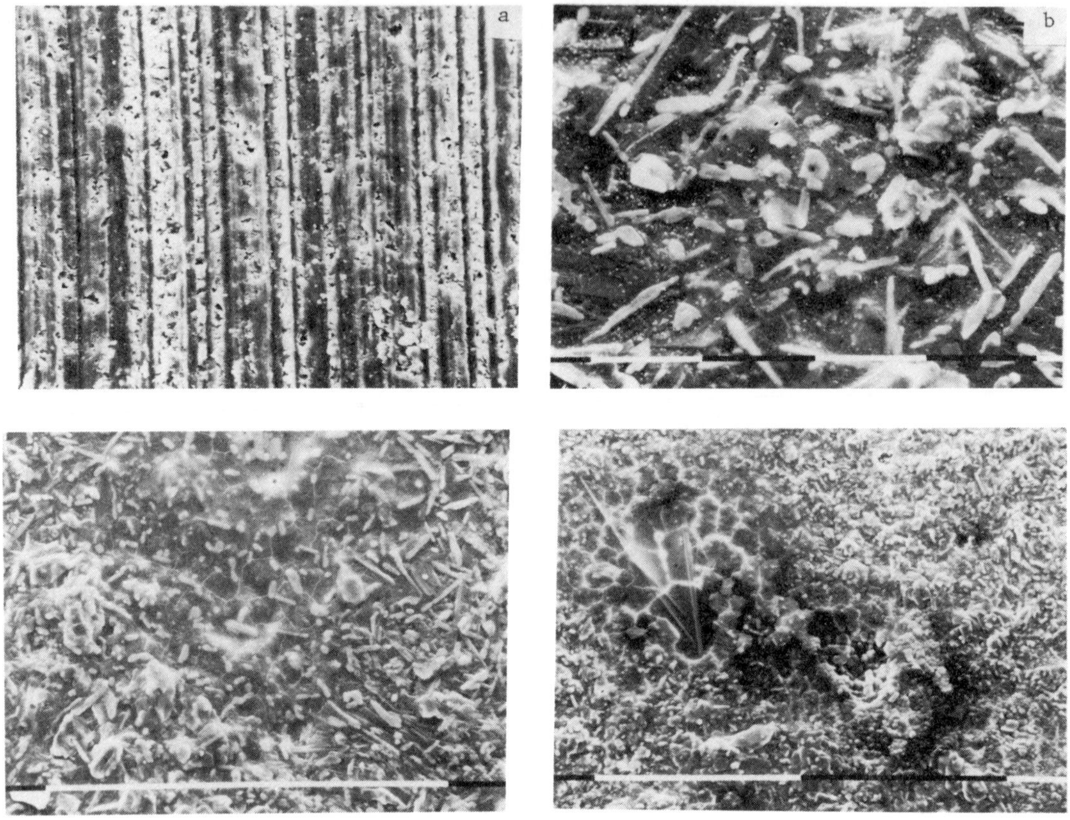

Fig. 2: SEM micrographs of a sintered silicon nitride (Kyocera SN 220) in
a) as machined form, and following annealing in air at 1200°C for
b) 25 hours, c) 50 hours and d) 200 hours.

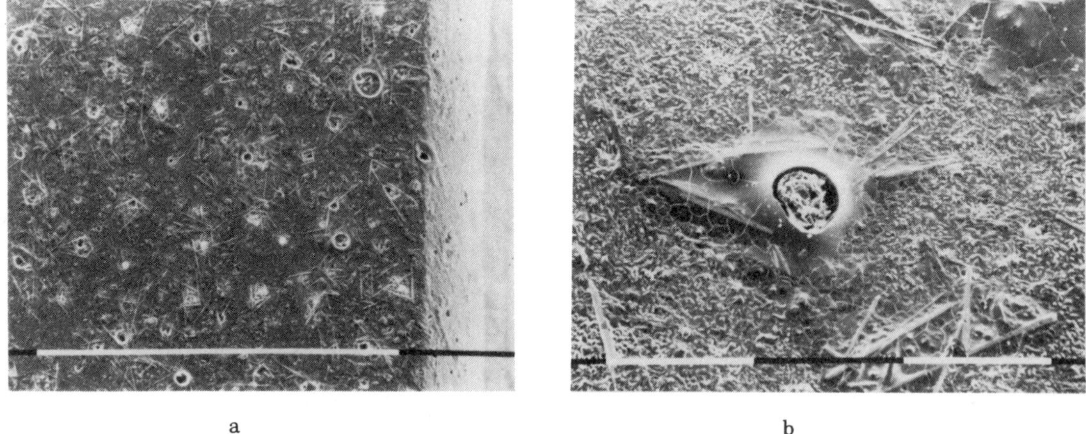

Fig. 3: a) SEM micrographs of Kyocera SN 220 after 218 hours at 1200°C. A
large number of larger pits are formed on the surface (bar = 100 µm).
b) Higher magnification of a surface pit (bar = 10 µm).

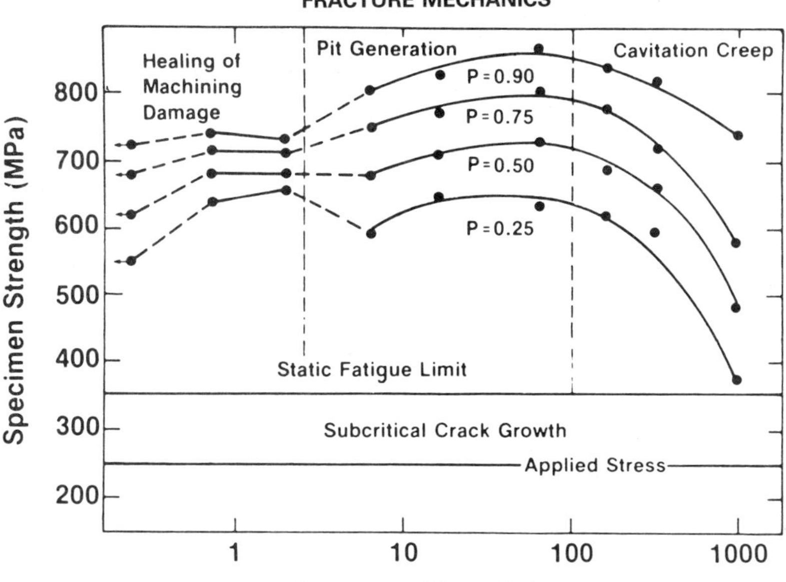

Fig. 4: A strength degradation diagram for NC-132, illustrating how the fast fracture strength of a material changes with time during high temperature exposure under load (Wiederhorn and Tighe, 1983). The ρ values represent cummulative failure probability. Thus, it is clear that the scatter in fracture strength also varies evidently at different times.

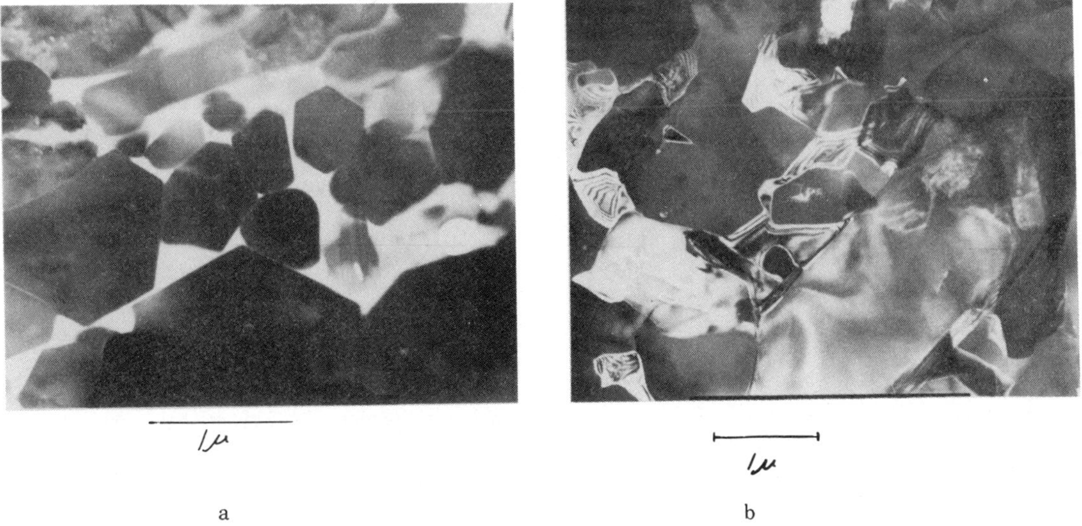

Fig. 5: Dark-field TEM micrographs of a sintered silicon nitride (Kyocera SN220) a) as-received, and b) after 100 hrs. at 1150°C and 75 MPa.

So far we have considered time dependent effects occurring primarily at the surface. However, internal changes in microstructure are also possible, and these can affect the strength and reliability of a material. For example, many types of sintered silicon nitride contain significant amounts of glass. This results from the processing of the material. The material we have been testing, Kyocera SN 220, contains Al_2O_3 and Y_2O_3 as sintering enhancers. These enable the formation of a liquid during firing, which on cooling forms a glass. At temperatures of interest for heat engine applications this glass is unstable and tends to devitrify. Figure 5 shows dark-field TEM micrographs of this material, before and after annealing for 100 hours at 1150°C. The glass has largely crystallized around the β-Si_3N_4 grains. Since the glass is primarily responsible for the creep properties of this material, devitrification has a large impact on the strength of the material. A continuous increase in creep resistance is observed with increased annealing (Jupp et al., 1987). A similar effect has also been observed in vitreous-bonded alumina, in which case also the glass crystallizes during annealing (Wiederhorn and Fuller, 1985).

An additional problem is associated with this effect. It is related to the effect of continuous devitrification on the characterization on the creep properties. Since creep resistance increases as devitrification occurs, a steady state is not achieved and the interpretation of creep results is ambiguous. This has been addressed in a recent model (Wilkinson, 1987). The main result is that so long as $\varepsilon/(\tau \varepsilon_0)$ is much less than one, then the effect of devitrification on measured creep rates is negligible. Here ε is the strain over which creep rates are measured, τ is the characteristic time associated with devitrification, and ε_0 is the initial strain rate. In many cases however, this term is not small. This is because both devitrification and creep require diffusion over distances of the order of the grain size. Therefore the time required for devitrification is comparable to that for measurable creep to occur. It can also be shown that if the rate of devitrification is stress-dependent and creep tests are performed in bending, then this will contribute to the redistribution of stress across the sample. If, for example, the rate of devitrification is enhanced in compression with respect to tension (because of the volume contraction associated with crystallization) then the strength of the bar will increase faster on the compressive side and net redistribution of stress to the compressive side will occur. Such a redistribution of stress is often observed during bend testing of vitreous bonded ceramics (Cohrt et al., 1981, 1984).

An additional form of internal damage which occurs in the absence of applied stress is swelling. When gas is evolved during pressureless sintering a phenomenon known as bloating occurs, in which the body desinters towards the end of the firing cycle (Haroun and El-Masry, 1981). Dense ceramics, particularly if hot pressed, may contain gas trapped under pressure in the residual porosity. Under certain circumstances this can lead to swelling (Howlett and Brook, 1984; Solomon and Hsu, 1980a, 1980b). This phenomenon has been extensively studied in hot-pressed Al_2O_3 (Bennison and Harmer, 1983; Robertson and Wilkinson, 1985, 1986; Robertson, Li and Wilkinson, 1987). Typically both grain growth and void growth occur simultaneously. At moderate temperatures (i.e. between about 1250°C and 1450°C) the voids are situated at grain boundary triple junctions and are uniformly distributed throughout the material. At higher temperatures (i.e. greater than about 1600°C) Bennison and Harmer see a more complex swelling process involving abnormal grain growth. However, this temperature is too high to be of practical interest. Long time exposure at 1250°C is possible however, and could lead to a degradation in both room temperature strength and creep resistance at elevated temperatures.

It is interesting to note that a similar phenomenon is found in core samples extracted from polar glaciers (Gow and Williamson, 1975). These samples are composed of ice which had densified slowly from snow under pressure. As the ice densifies air is trapped in the residual porosity. However, below a depth of 700 in, the pressure is sufficient to drive the dissociation of the trapped oxygen and nitrogen to form either ice clathrate or hydrate, and the pores disappear. When samples of this ice are removed from the glacier and stored for some time at atmospheric pressure, pores reappear in the material. This process is similar to that which we have postulated occurs in Al_2O_3 (Robertson and Williamson, 1985, 1986) whereby CO/CO_2 gas trapped in residual pores is thought to dissociate and to dissolve in the Al_2O_3 during hot pressing, and subsequently reappear during atmospheric annealing at 1350°C and above.

EFFECT OF ANNEALING ON LOW TEMPERATURE FRACTURE

We now turn our attention to the effect of exposure at elevated temperatures on structural reliability at room temperature. Clearly any change in bulk microstructure may have an effect on strength and toughness. One example of this is illustrated in Fig. 6. This shows the room temperature bend strength of a commercial $Al_2O_3/SiC_{(w)}$ composite following annealing in air for 300 hours at a range of temperatures. There is significant loss of strength produced by this exposure (Wilkinson et al., 1987). As indicated in Fig. 7, the failure origins in this material consist of SiC-free regions, which presumably result from an inhomogeneous dispersion of the SiC whiskers in the alumina powder during initial processing. The exact cause of the degradation in properties has yet to be determined. However there is a range of possibilities. First of all, grain growth in the SiC-free regions occurs during annealing, which can lower the strength of these regions. Furthermore, it has been established (Mah et al., 1984) that SiC whiskers degrade on exposure to air at elevated temperatures due to the formation of a glassy surface layer. This can also lead, by reaction with the alumina, to the formation of mullite at the whisker/matrix interface.

HIGH TEMPERATURE RELIABILITY

Since the microstructure of structural ceramics evolves with time at elevated temperatures, the lifetime of a component in service can not be predicted on the basis of short time testing. The answers to the questions posed at the end of the Introduction are therefore important. To date relatively little information is available to do this. It is therefore clear that for the present the design of systems using structural ceramics at elevated temperatures will be restricted to a temperature range in which both the rate of microstructural evolution and the rate of creep, are negligibly small. This is a very conservative approach however, and must be rectified in the long term by developing a greater understanding of material behaviour.

For example, there is much to be learned about the effect of devitrification on the creep strength of silicon nitride ceramics. In this way a time-dependent model for creep can be developed. In addition, greater understanding of the effect of stress on devitrification is required. Since the rate of creep can be very sensitive to the volume fraction of glass in the material, small effects due to stress can lead to extensive stress redistribution across a complex structure. Moreover, certain ceramic components will be exposed to large temperature gradients in service. Since the rate of devitrification is temperature dependent this may also lead to stress redistribution.

Given the need for ceramic materials to last in service for times between 10,000 and 100,000 hours, we need to know much more about the damage processes which lead to failure over this time period. Quinn's fracture mechanism map (Fig. 1) suggests a transition from slow crack growth to "creep fracture" in times less than 10,000 hours. However the transition may not be as clear as this diagram suggests. Recent results in our laboratory on creep damage in hot-pressed alumina suggests that a gradual transition may occur as the time to failure and/or the temperature is increased. This transition involves a decreasing sensitivity to defects on the part of the material. Thus at short failure times, slow crack growth and failure from a single dominant flaw occurs (Fig. 8a). At longer failure times, multiple microcracking is found (Fig. 8b) These emanate from a variety of flaw types including porous regions, large grain size regions and inclusions. As shown previously by Dalgleish et al. (1985), this leads to coalescence by intense localized damage and consequently to failure by a process involving the shear linkage of nearby microcracks. At longer failure times still (i.e. lower stresses or higher temperatures) the material becomes very insensitive to defects. Instead, cavities develop generally at triple junctions in the material (Fig. 8c). These are very resistant to coalescence and the material is effectively superplastic.

On the basis of this evidence it is clear that the mechanism fields drawn on fracture maps produced to date must be seen as tentative and cannot be used to define a region in stress-temperature space in which extrapolation of lifetime data is justifiable. Indeed, very little data is available which can be used to guide the design of components for long time service at elevated temperatures, or the development of materials for these components. Most testing has involved relatively short failure times. Extrapolation of these data to conditions required in service is not now possible, given the current uncertainty about damage processes and failure modes. It is also clear that the stress and temperature at which transitions between the various failure modes occur will be very sensitive to material, and to material "quality". Even for material of the same nominal composition, variations in starting powder and processing, which lead to variations in microstructure defect population, can dramatically affect the failure process. We have found this in our own work on hot-pressed Al_2O_3. We have tested material made by two different manufacturers. These have very

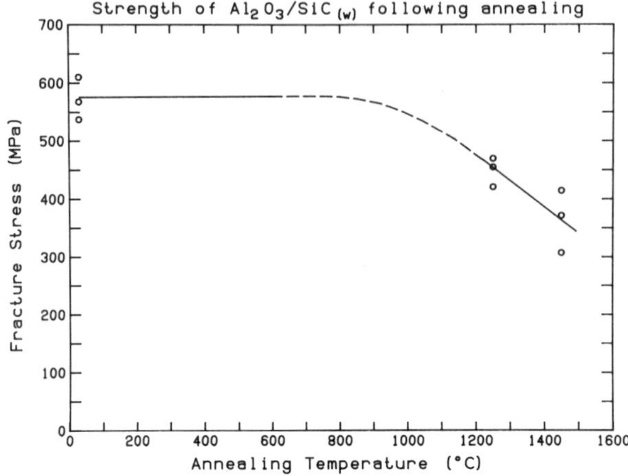

Fig. 6: The strength of a commercial $Al_2O_3/SiC_{(w)}$ composite in the as-received condition and after exposure in air for 300 hours at two different temperatures. There is a clear degradation in strength.

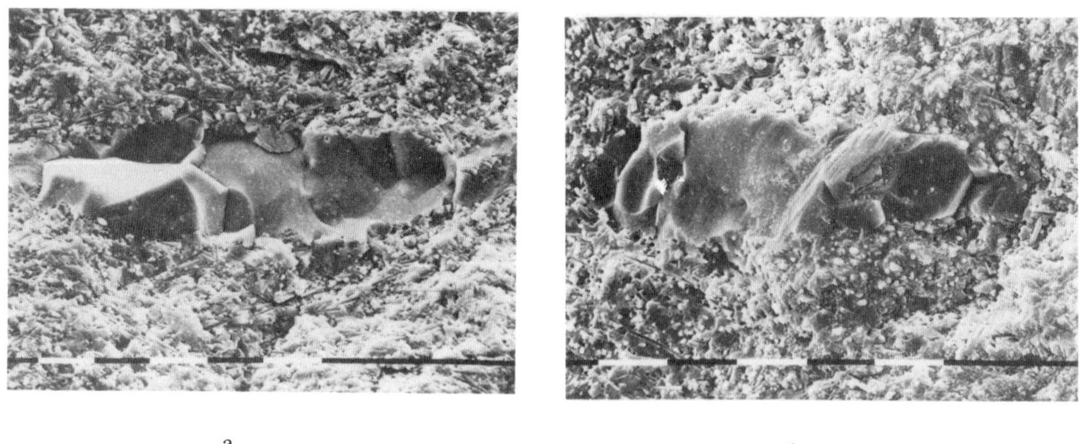

a   b

Fig. 7: The fracture can be traced to SiC-free regions, both in a) the as-received material (bar = 10 μm), and b) annealed material, in this case for 300 hours at 1450°C (bar = 10 μm).

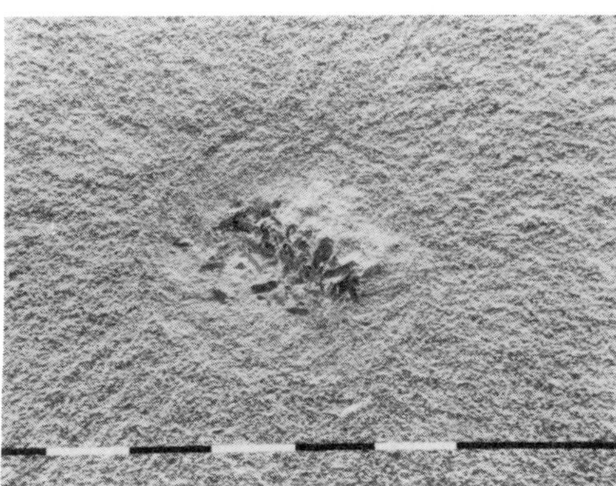

Fig. 8a: At high stress (short failure time) failure is by slow crack growth from a single dominant flaw. In this case the flaw is a region of large grains. The right figure clearly shows the extent of the slow crack growth region, about 400µm in diameter, which preceded fast fracture (bar = 1mm, left; bar = 100µm, right)

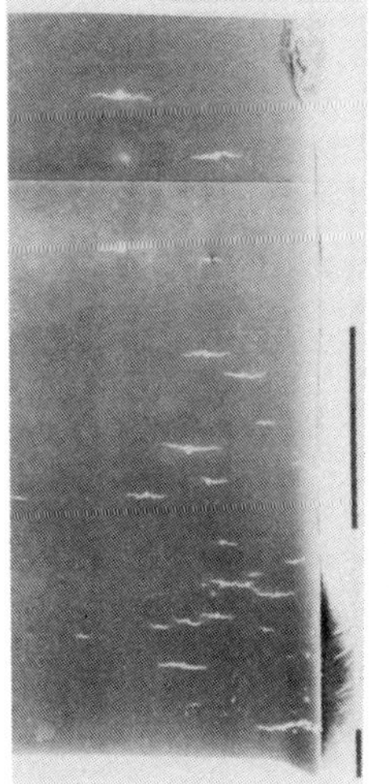
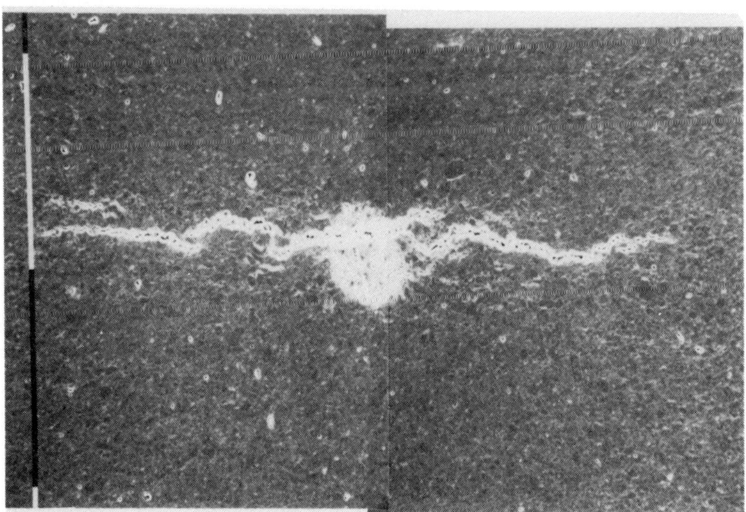

Fig. 8b: At lower stresses or higher temperatures, the hot-pressed Al_2O_3 fails by a process involving multiple microcracking. These pictures are taken from the tensile surface of a bend bar creep test interrupted before failure. There is evidence of microcrack linkage in the left figure. The bottom micrograph on the right shows a typical crack, emanating in this case from a porous region and extending about 150µm in each direction (bar = 1mm, left; bar = 100µm, right)

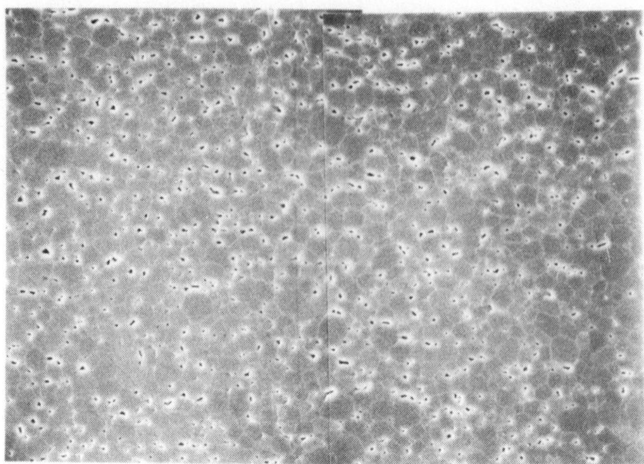

Fig. 8c: At higher temperatures and lower stresses, damage becomes very
           uniform and the material is flaw-tolerant. This figure illustrates the
           kind of internal damage found by polishing behind the tensile
           surface which is on the right hand edge of the micrograph. It consists
           of cavities, no longer than a grain facet, with little evidence of cavity
           linkage.

different defect populations. As a result, one is much more prone to failure by slow crack growth and
multiple microcracking. The other, which contains relatively few defects, exhibits uniform cavitation
behaviour to a lower temperature.

We are forced to return therefore to the questions posed at the end of the Introduction. One would like if at
all possible to put materials into service in a regime in which failure occurs by uniform cavitation, and in
which defect tolerance is high. The development of suitable materials is dependent on a better understanding of long time failure processes. The types of defect morphologies which are most likely to generate
microcracks must be identified, and ways found to improve processing techniques to minimize the
occurrence of such defects.

REFERENCES

Bennison, S. J. and Harmer, M. P. (1983). in Advances in Ceramics, **6** (Am. Ceram. Soc., Columbus, OH).

Cohrt, H., Grathwol, G., Thümmler, F. (1981). Res. Mech. Lett., **1**, 159.

Cohrt, H., Grathwol, G., Thümmler, F. (1984). Res. Mech., **10**, 55.

Dalgleish, B. J., Slamovich, E. B. and Evans, A. G. (1985). J. Amer. Ceram. Soc., **68**, 575.

Gow, A.J. and Williamson, T. (1985). J. Geophys. Res., **80**, 5101.

Grathwol, G. (1984). Proc. 2nd Intl. Conf. on Creep and Fracture of Engineering Structures (B. Wilshire and
           D.R. J. Own, eds., Pineridge, Swansea), 565.

Haroun, N.A. and G. Masry, M.A.A. (1981). Ceram. Intl., **7**, 38.

Howlett, S.P. and Brook, S.J. (1984). Proc. 2nd Intl. Durability and Fatigue Conf., 320.

Jupp, R., Chadwick, M. and Wilkinson, D.S. (1987), to be published.

Mah, T. et al. (1984). J. Mater. Sci., **19**, 1191.

Quinn, G. D. (1984). Ceram. Eng. Sci. Proc., **5**, 596-602.

Robertson, A. G. and Wilkinson, D. S. (1985). Proc. Intl. Conf. on Fracture Mechanics,

Robertson, A. G. and Wilkinson, D. S. (1986). J. de Physique, Colloque C1, suppl. au no. 2, Tome 47, C1.661

Robertson, A. G., Li Shujie and Wilkinson, D. S. (1987), to be published.

Solomon, A.A. and Hsu, F., (1980a). J. Amer. Ceram. Soc., **63**, 467.

Solomon, A.A. and Hsu, F., (1980b). Materials Science Research, 13, (Plenum, N.Y.; G.C. Kuczynski, ed.), 485.

Wiederhorn, S. M. and Tighe, N. J. (1983). J. Am. Ceram. Soc., **66**, 884-889.

Wiederhorn, S.M. and Fuller Jr., E.R. (1985). Mater. Sci. Engg., **71**, 169.

Wilkinson, D. S. (1987), to be published.

Wilkinson, D. S. and Robertson, A. G. (1987), to be published.

POTENTIAL DROP TECHNIQUE FOR FOLLOWING HIGH-TEMPERATURE
CRACK GROWTH IN CERAMICS

Tom B. Troczynski and Patrick S. Nicholson

Ceramic Engineering Research Group
Department of Materials Science and Engineering
McMaster University
Hamilton, Ontario, Canada

ABSTRACT

An electrical potential drop technique has been developed for the determination of curved crack front advance in conductive ceramics. The solutions obtained are applied to chevron-notched four-point bend specimens. Fracture of ionically conducting zirconium oxide ceramics is studied in the range 1000° to 1300°C. A driving force of the order of 1 J/m^2 is sufficient to initiate fracture of the ceramic at high temperatures.

INTRODUCTION

The electrical potential drop (PD) technique has been widely adopted for crack growth and fatigue studies in metals. An analytical solution was found and verified experimentally (Gilbey and Pearson, 1966) It was shown by Clark and Knott (1975), Knott (1980) and Druce and Booth (1980) that the technique is highly sensitive and variation of the potential probe position close to the crack has relatively little effect In this work we propose utilization of the PD technique for following high-temperature crack growth in conducting ceramics (for example, stabilised zirconia or silicon carbide). There are some advantages to the application of the method to ceramics as compared with metals. The resistivity of zirconia at 1000 to 1300°C is in the 10 ohm·cm range, thus a potential drop of ~10 mV is experienced if ~1 mA current is passed through the sample The resistivity of metals is in the μohm·cm range, resulting in a potential drop of 1 μV on the passage of 1 A. Such large current densities (typical sample cross-section is <1 cm^2) can result in excessive heating of the specimen. It is much more difficult to initiate stable cracks (or fatigue cracks) in ceramics than in metals One technique to assure stable fracture initiation in a ceramic is to introduce a chevron (instead of a straight-through) initial crack front. To use the PD technique with such a notch, the formula for the potential drop for a crack front of arbitrary shape (in particular, chevron notches) has been developed and applied to the high-temperature fracture study of stabilised zirconia cermics.

POTENTIAL DROP ACROSS A CRACK FRONT OF ARBITRARY SHAPE

Solution of the Laplace equation for the electrical potential distribution around a crack in a plane specimen (Gilbey and Pearson, 1966) gives the dimensionless crack length x as:

$$x = \frac{2}{\pi} \cos^{-1}\left[\frac{1 - P_1^2}{P_2 - P_1^2 \cdot P_3} \right] \tag{1}$$

where:

$$P_1 = \left\{ \exp\left[\frac{U(x)}{2k} \right] - 1 \right\} / \left\{ \exp \frac{U(x)}{2k} + 1 \right\} \tag{2}$$

$$P_2 = \mathrm{sech}^2\left[\frac{\pi y}{2W} \right] \tag{3}$$

$$P_3 = \mathrm{sech}^2\left[\frac{\pi d}{2W} \right] \tag{4}$$

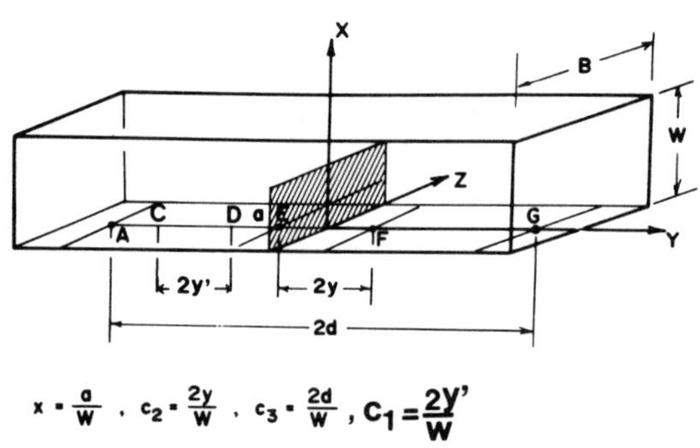

$$x = \frac{a}{W} \ , \ c_2 = \frac{2y}{W} \ , \ c_3 = \frac{2d}{W} \ , \ C_1 = \frac{2y'}{W}$$

Fig. 1 Geometry and cross section of straight through notched specimen.

The parameters of these equations are illustrated via the geometry of the straight-through-notch (ST) specimen, Fig. 1. $U(x)$ is the electrical potential drop between surface points E, F (distance = 2y). A constant current (dc) is applied between the surface points A and G (distance apart = 2d). The proportionality constant, k, is found by measurement of $U(x_0)$, for a known initial crack length, $x = x_0$. In particular, for $x_0 = 0$ and $d > 1W$ (the normal experimental situation for bend bars), it follows from eq. (1) that

$$k = \frac{U(0)}{2y} \cdot \frac{W}{\pi} \tag{5}$$

Equations (2) and (5) suggest that the potential drop versus crack growth data should be stored in dimensionless form, i.e.

$$Q(x) = U(0)/U(x) = R(0)/R(x) \tag{6}$$

where R(0), R(x) are the electrical resistances of the reference (unnotched) and ST notched samples respectively. R(x) is always measured across the crack, between lines $Y = \pm y$, whilst current is applied between lines $Y = \pm d$, Fig. 1. R(0) is measured either as R(x) if the initial notch has not been cut, or between any pair of lines at C,D (distance 2y' apart) whilst the current is applied between lines $Y = 0$ and $Y = -d$, Fig. 1.

If the notch cannot be approximated by the straight-through geometry, solution (1) is invalid and the potential distribution around the crack must be described by the three-dimensional equation. To avoid the complicated boundary conditions involved in the three-dimensional solution, a semi-empirical simplified approach was taken and subsequently verified by experimental calibration.

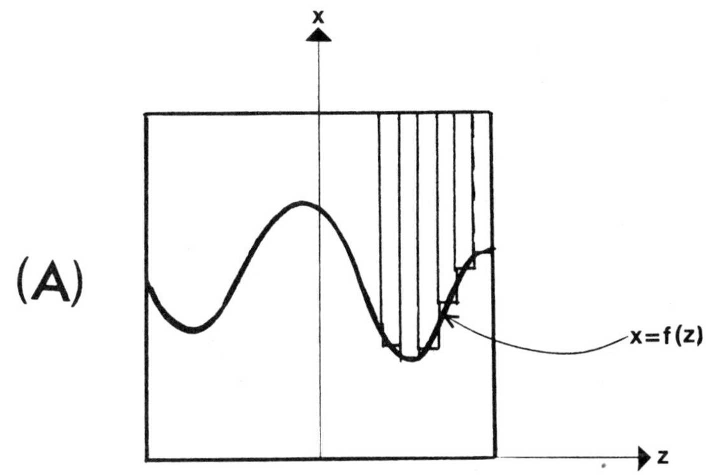

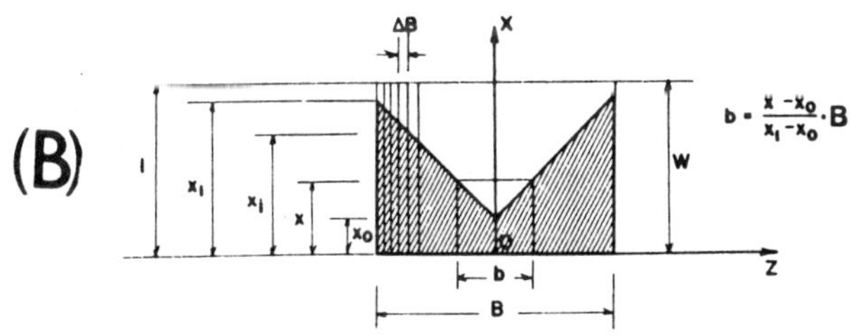

Fig. 2 (A) Cross section of the specimen with arbitrary crack front shape.
(B) Cross section of chevron notched specimen.

The sample with arbitrary crack front shape $x = f(z)$, Fig. 2A, is divided into a series of slices of approximately ST configuration. The system is treated as a set of resistors connected in parallel. The procedure is simplified for the symmetric shape of the chevron notch. As a result of the analysis of Troczynski and Nicholson (1985), a relative potential drop $Q_{CN} = U(0)/U(x)$ for the CN specimen can be expressed as a function of specimen geometry, i.e.,

$$Q_{CN}(x) = \frac{2 \cdot (1 - x)(x - x_0) \cdot Q_{ST}(x) + I(x)/N}{(x_1 - x_0) \cdot (2 - x - x_1) + (x - x_0) \cdot (x_1 - x_0)} \tag{6}$$

where the geometric parameters are shown in Fig. 2B. $Q_{ST}(x)$ is the relative potential drop for a ST specimen of the same geometry and notch x. $I(x)$ results from the slicing procedure and $N(= 500)$ is the number of slices. The validity of eq. 6 was verified by numerous experimental calibrations (Troczynski, 1987a). Fig. 3 shows an example of the results for simultaneous determination of the relative crack length $x = a/W$ obtained by the PD technique (x_{PD}) and that obtained by determination of the compliance change of a partially fractured graphite speciment (x_{CA}). Both values are in satisfactory agreement and were further verified by optical measurements (Troczynski, 1987a).

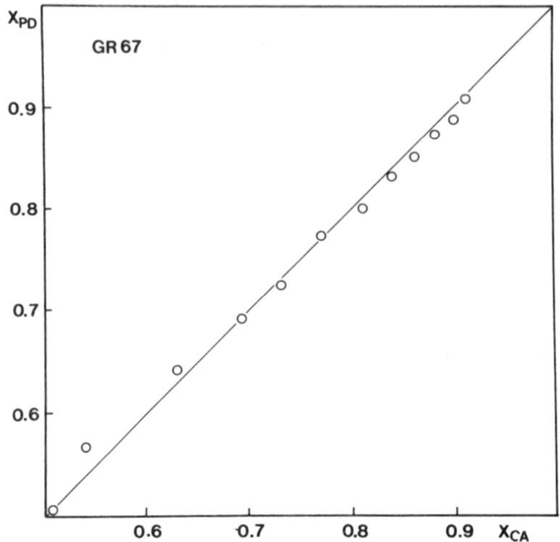

Fig. 3 Crack length determined by potential drop (x_{PD}) vs. that determined by compliance analysis (x_{CA}).

THE SENSITIVITY OF THE PD TECHNIQUE

The sensitivity of the potential drop technique is defined as the increment of the relative potential drop against the increment of the crack length, $\Delta Q/\Delta x$. $Q(x)$ is analysed against x in a series of 3-dimensional plots, the third axis being the dimensionless interprobe distance $C_1 = 2y'/W$, $C_2 = 2y/W$ and $C_3 = 2d/W$ (Fig. 4). Typical experimental conditions were used to construct these figures, i.e. fixed $x_0 = 0.4$ and $x_1 = 1.0$, variable C_1 (from 0.1 to 1.1 in 0.1 steps), C_2 (from 0.25 to 0.75 in 0.05 steps) and C_3 (from 2.0 to 12.0 in 1.0 steps). x varied from 0.4 to 0.95 in 0.05 steps. A common axis system was used. The corner Z is common and x is plotted to the left and C_i (i = 1,2,3) to the right of Z. Considering Fig. 4A (i.e. plots E1, E2, E3), the sensitivity improves markedly as C_1 increases and less markedly as C_2 decreases. The sensitivity of the PD technique to detect crack initiation in a chevron-notched specimen is low. This is indicated by the levelling of the Q vs x curves at $x \simeq x_0$. Figure 4B (plots E4, E5, E6) shows that an increase of C_2 (at constant C_1) decreases the sensitivity slightly. The variation of the voltage probe distance, C_2, by 100% (from 0.25 to 0.50) causes a decrease of the output signal Q of only ~4%. In other words, a large error in the C_2 measurement (resulting, for example, from a large area of contact between the probes and the material), causes a small error in the crack length (x) calculation. The variation of Q with the current probe distance, C_3, is shown in Fig. 4C (plots E7, E8). Q is constant as long as $C_3 > 3$ (a slight bending of the Q vs C_3 lines is evident for $C_3 < 3$). Thus it is concluded that precise measurement of the distance between the current probes is also noncrucial. Consequently, despite the difficulty of making good contact between the metallic probes and the ceramic surfaces, the PD technique is an effective way of monitoring the high-temperature fracture of conducting ceramics.

HIGH TEMPERATURE FRACTURE STUDIES OF ZIRCONIUM OXIDE CERAMICS

Chevron-notched four-point-bend bars of partially stabilised (by 8 wt% Y_2O_3, PSZ) and fully stabilised (by 12 wt% Y_2O_3, FSZ) zirconium oxide were prepared for elevated temperature fracture tests (Troczynski, 1987a). Fig. 5 shows the bottom, B, top, T, and two sides, S, of the specimen. Current (1) and voltage (2) probes were pushed into grooves (4) in the bottom of the specimen and fastened to the top through Al_2O_3 plates (3). Electrical contact was established by applying conducting high-temperature platinum based enamel over all mechanically supported contacts (for example, contact (5) between the temporary specimen probes and permanent Pt leads).

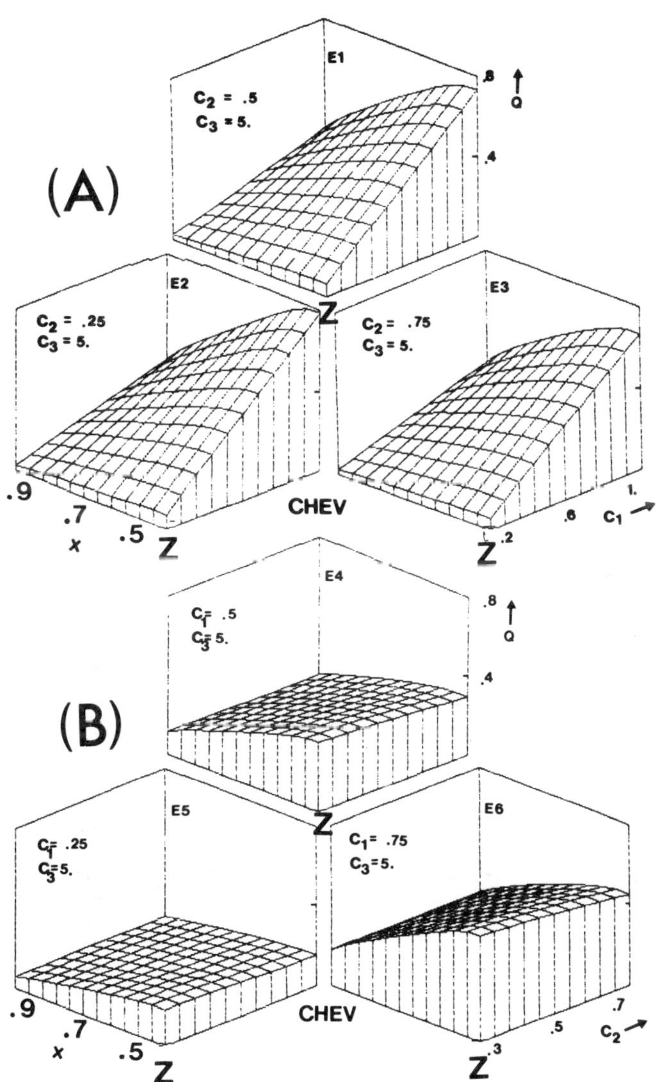

Fig. 4   Variation of the relative potential drop Q against the crack length x and (A) the reference voltage probes distance C_1; (B) the working voltage probes distance C_2; (C) the current probes distance C_3.

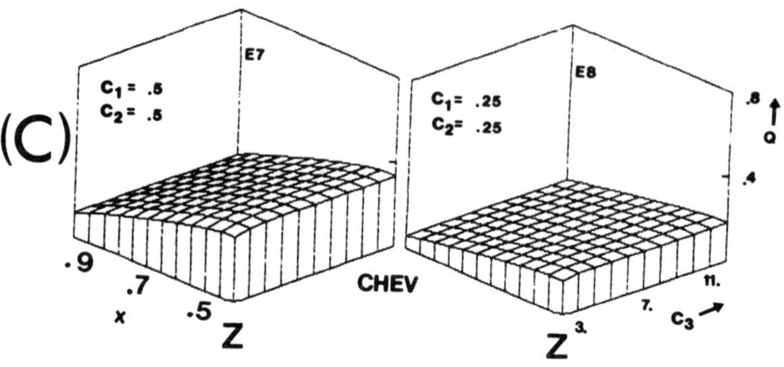

Fig. 4   continued.

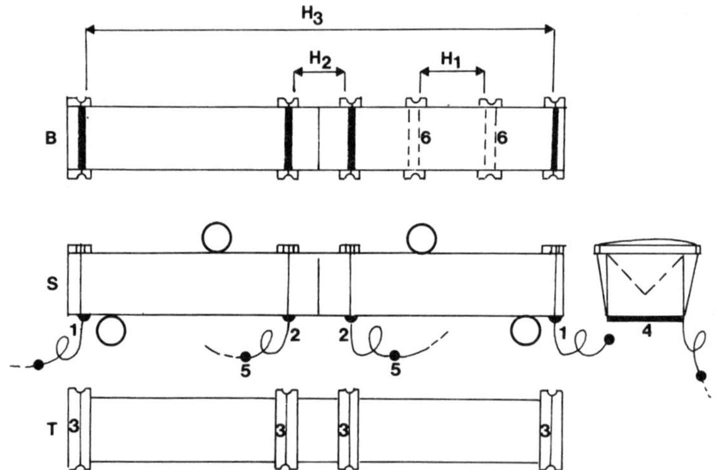

Fig. 5 Attachment of the potential and current probes to the specimen's surface

The samples were cycled through multiple load-fracture-unload (LFU) tests. Troczynski and Nicholson (1987b) showed that the work performed in the consecutive cycles tends to the resistance-to-fracture initiation value, in the limit of complete fracture. A typical experimental set up and the output of the LFU/PD experiment are shown in Fig. 6. On the load vs. potential drop graph, fracture initiates at point S but, due to the slow initial crack advance, the first increase in the potential drop is recorded at point I. Unloading starts at point T and in the last stage of unloading the potential drop decreases due to partial closure and electrical contact of the fractured surfaces. The values of load and potential drop increment (i.e. crack length increment) are sufficient to calculate the crack driving force (Troczynski 1987a). Typical results for PSZ at 1300°C are shown in Fig. 7A (load vs. potential drop) and 7B (resistance-to-fracture R_e, compiled for four specimens). According to the previous discussion and the data of Fig. 7B, fracture can be initiated in PSZ at 1300°C by driving forces as low as 2 J/m^2 (this should not be confused with the energy dissipated nonelastically into viscous deformation of the specimen; this energy, R_n, can reach 1 kJ/m^2 (Troczynski, 1987a)). The vertical unloading lines in Fig. 7A indicate a viscous type of material deformation, i.e. the elastic recovery is insufficient to partially close the crack upon unloading. These effects originate in the glassy grain-boundary phases present in the zirconia.

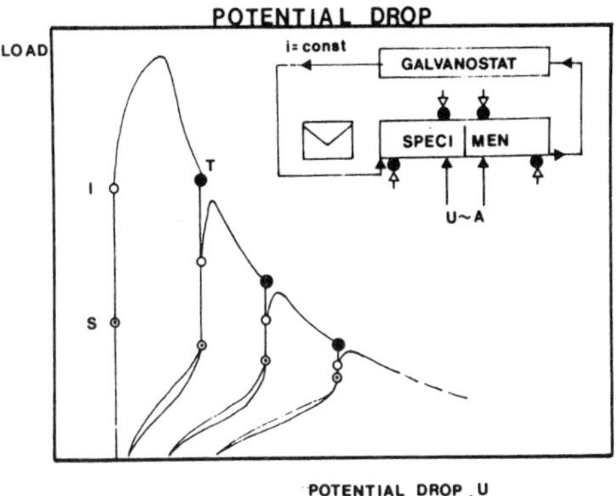

Fig. 6 Schematic load against potential drop record for load-fracture-unload experiment.

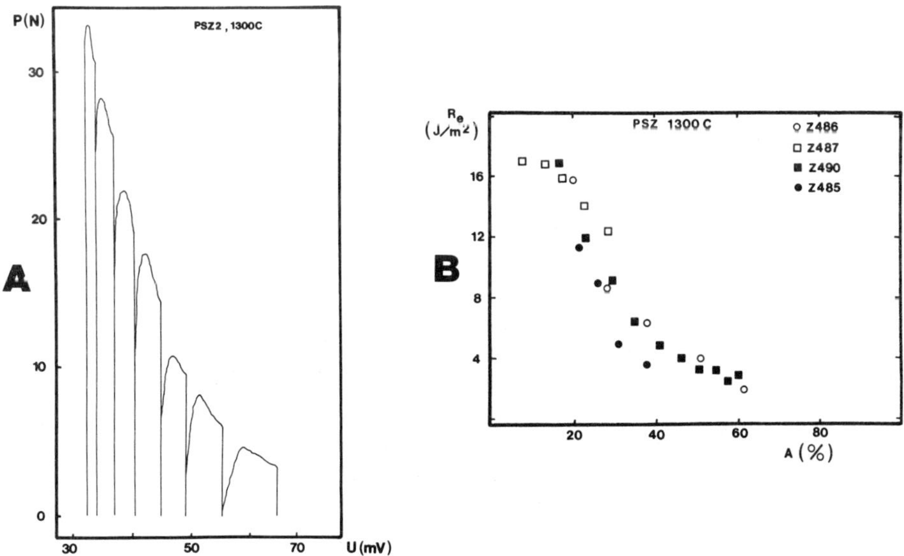

Fig. 7   Fracture test results for PSZ at 1300°C; (A) load vs. potential drop; (B) resistance to fracture vs. fracture area.

The elastic crack closure phenomenon can be clearly distinguished on the load vs. potential drop graph for FSZ at 1000°C (Fig. 8A) and 1300°C (Fig. 8B). Apparently, higher amounts of yttria in these alloys result in a more rigid grain-boundary glassy phase. The resulting values of resistance-to-fracture initiation are 0.5 and 0.1 J/m^2 at 1000°C and 1300°C respectively (Fig. 9). These considerably lower values as compared to those for PSZ seem to indicate a dispersion-type toughening mechanism must be operating in the PSZ via the stable tetragonal inclusions in the cubic matrix.

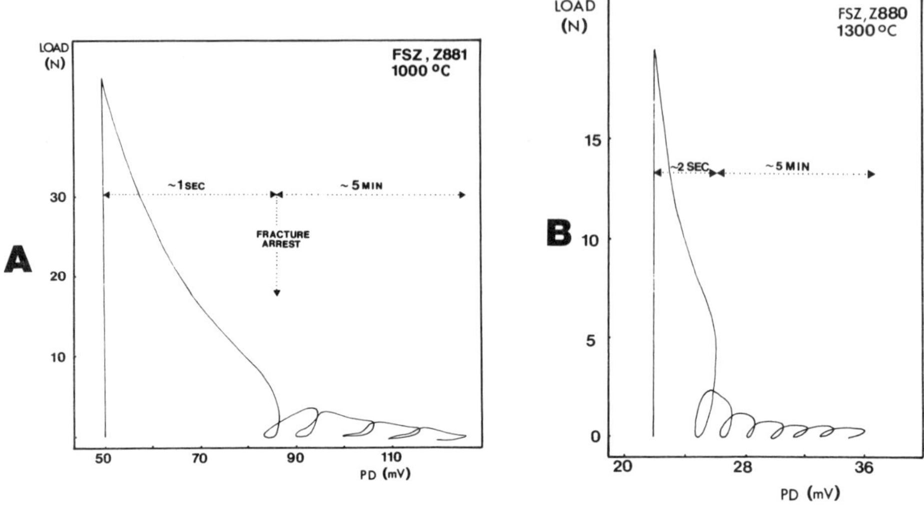

Fig. 8 Load vs potential drop for FSZ at (A) 1000°C; (B) 1300°C.

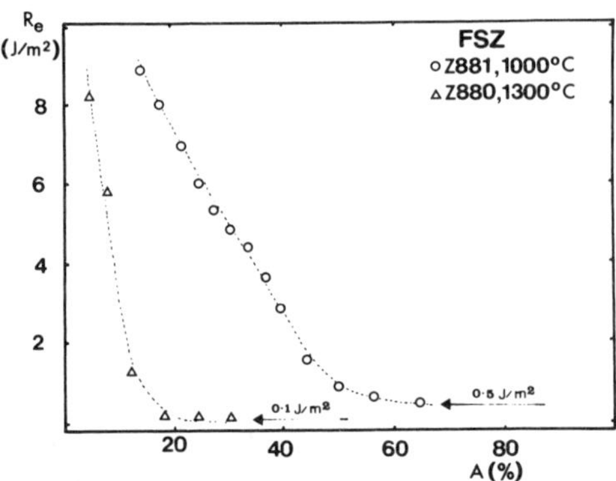

Fig. 9 Resistance to fracture vs. fracture area for FSZ at 1000°C and 1300°C.

The fracture surface of PSZ at 1300°C (Fig. 10A) clearly indicates liquid-phase-assisted grain separation. Fig. 10B shows the plane normal to the crack front, with the crack tip arrested after the last unloading cycle. It seems that crack extension is assisted by accumulated crack-tip cavitation. This type of cavitation is clearly visible on the fracture surface of FSZ at 1000°C, Fig. 11A. Due to the high cavity density, it is impossible to locate the arrested crack tip at 1300°C, Fig. 11B.

The PD method of high-temperature resistance-to-fracture determination is unique as it provides data unavailable by other techniques. It allows determination of crack extension into the bulk of the specimen at minimum force. In this respect, the generated values provide design data for long-term, low-stress, high temperature applications of ceramic materials.

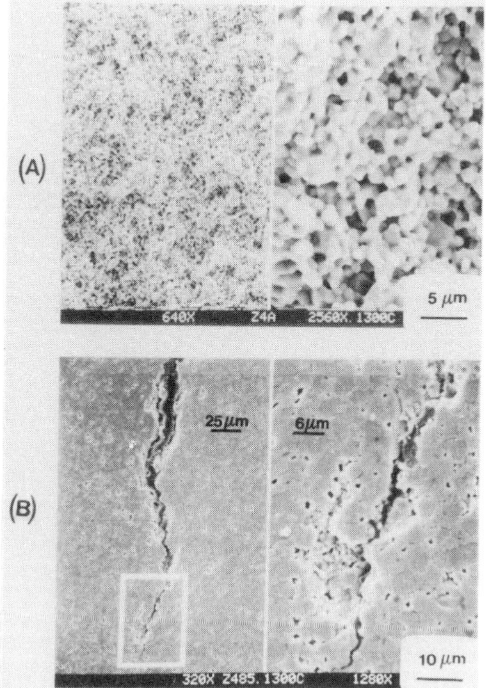

Fig. 10 Fractography of PSZ at 1300°C;
(A) fracture surface; (B) arrested crack.

Fig. 11 Fractography of FSZ; (A) fracture surface
at 1000°C; (B) arrested crack at 1300°C.

CONCLUSIONS

The potential drop technique for following the movement of a crack front of arbitrary shape has been developed numerically and verified by experimental calibration. The sensitivity of the technique was analyzed for the chevron notch geometry. It was shown that the PD method is relatively immune to errors arising from determination of the potential and current probe location. Therefore, the technique is an effective way to monitor the high-temperature fracture of electrically-conducting ceramics. The technique was used to determine the resistance-to-fracture of stabilised zirconias, broken in a cyclic load-fracture-unload sequence. It was found that the resistance to fracture initiation at 1300°C in monolithic PSZ is ~2 J/m^2 while in FSZ it is a fraction of 1 J/m^2. These results clearly indicate the ease of high-temperature slow fracture evolution in monolithic zirconia ceramics which contain glassy grain-boundary phases.

REFERENCES

Clark, G. and J.F. Knott (1975). Measurement of Fatigue Cracks in Notched Specimens by Means of Theoretical Electrical Potential Calibrations. J. Mech. Phys. Sol. 23, 265-276.

Druce, S.G. and G.S. Booth (1980). The Effect of Errors in the Geometric and Electrical Measurements on Crack Length Monitoring by the Potential Drop Technique. In J. Beevers (Ed.), The Measurement of Crack Length and Shape During Fracture and Fatigue, EMAS, U.K.

Gilbey, D.M. and S. Pearson (1966). Measurement of the Length of a Central or Edge Crack in a Sheet of Metal by an Electrical Resistance Method. RAE Tech. Rep. No. 66402, Farnborough, U.K.

Knott, J.F. (1980). The Use of Analogue and Mapping Techniques with Particular Reference to Detection of Short Cracks. In C.J. Beevers (Ed.), <u>The Measurement of Crack Length and Shape During Fracture and Fatigue</u>, EMAS, U.K.

Troczynski, T.B. and P.S. Nicholson (1985). Analysis of the Electrical Potential Drop Technique for Crack Length Measurement in a Chevron-Notched Specimen. <u>Am. Cer. Soc. Bull.</u>, <u>64</u> 1272-1275.

Troczynski, T.B. and P.S. Nicholson (1986). Application of the Potential Drop Technique to the Fracture Mechanics of Ceramics. In R.C. Bradt, A.G. Evans, D.P.H. Hasselman and F.F. Lange (Eds.), <u>Fracture Mechanics of Ceramics</u>, Vol. 8, Plenum Press, New York, pp. 199-211.

Troczynski, T.B. (1987a). Energy Analysis of Brittle Fracture and its Application to Zirconium Oxide Ceramics, Ph.D. Thesis, McMaster University

Troczynski, T.B. and P.S. Nicholson (1987b). Effect of Subcritical Crack Growth on Fracture Toughness and Work-of-Fracture Tests Using Chevron-Notched Specimens. <u>J. Am. Cer. Soc.</u>, <u>70</u>, 78-85.

SESSION 3:   FATIGUE AND FRACTURE OF WELDMENTS

CHAIRMAN: R. COOTE
             NOVA,
             Calgary

INVESTIGATION OF PLASTIC INSTABILITY CRITERIA FOR
FRACTURE OF PIPELINE GIRTH WELDS CONTAINING DEFECTS

M.J. Worswick and R.J. Pick

Department of Mechanical Engineering, University of Waterloo
Waterloo, Ontario, N2L 3G1

ABSTRACT

Plastic zone development and crack mouth opening displacement in circumferentially surface-flawed line pipe subjected to pure bending have been predicted using three-dimensional finite elements. The analysis considered both material (elastic-plastic) and geometric nonlinearities.

Predicted failure moments were shown to compare favourably with a number of analytical plastic instability models. The results from this investigation have been used to assess the degree of conservatism inherent in the newest version of the Canadian Gas Pipeline Standard CSA Z184.

KEYWORDS

Pipeline girth weld; defect; elastic-plastic fracture; plastic instabilty.

INTRODUCTION

In recent years, various gas pipeline codes (eg. British Standard BS 4515:1984 and American Petroleum Institute API 1104) have adopted fitness-for-purpose approaches for the assessment of defects in pipeline girth welds. In Canada, Nova, An Alberta Corporation, and the American Gas Association have sponsored a series of full scale fracture tests on line pipe with intentionally flawed girth welds at the Welding Institute of Canada (Glover, Coote and Pick, 1981; Glover and Coote, 1983) and the University of Waterloo (Pick, Glover and Coote, 1980). The results of these tests have been used to support the development of a Canadian defect assessment procedure which is summarized in a new appendix to Canadian Gas Pipeline Standard CSA Z184 (Appendix K).

The method of brittle fracture assessment in Appendix K is similar to that in British Standard BS 4515:1984, which is based on the crack tip opening displacement (CTOD) approach. In addition, a plastic instability analysis, limiting the extent of yielding in the pipe, is stipulated. This second requirement was introduced to prevent plastic collapse or ligament instability in line pipe made from high toughness, low work hardening materials.

In this investigation, a three-dimensional elastic-plastic finite element analysis has been used to examine circumferentially surface-flawed line pipe subjected to pure bending. The plastic zone development and crack (mouth) opening displacement (COD) have been predicted as a function of applied moment and were used to examine the degree of conservatism inherent in the plastic instability criterion of CSA Z184, Appendix K. The analysis described herein complements previous experimental investigations into flawed line pipe behaviour, providing greater detail in describing the growth of plasticity and COD.

FINITE ELEMENT MODEL

Of the various methods of stress analysis, the finite element method is the preferred technique for modelling the large deformations and plasticity occurring in cracked line pipe subjected to high bending loads. This method permits inclusion of the effects of local weld geometry and properties and consideration of various types of loading. For this investigation, the authors have chosen to use the commercial code ABAQUS, developed by Hibbitt, Karlsson and Sorensen Inc. In principal this code has all the necessary features to facilitate modelling of this problem. However since the modelling of plasticity and large deformations was accomplished by stepping incrementally through the loading history, it was necessary to establish reasonable increment sizes and other parameters to ensure that the analysis converged to an accurate solution. A number of studies were made with respect to element type, boundary conditions and applied loading technique in an effort to minimize computing time. These studies made use of various features available in ABAQUS and user-written subroutines. For brevity only the major developments in this area will be described.

Model Geometry

Figure 1 shows the general arrangement of the line pipe that was analysed. A pipe of 914 mm outside diameter and 11.1 mm wall thickness was chosen since this pipe size has been used extensively in full scale fracture testing (Glover, Coote and Pick, 1981; Glover and Coote, 1983; Pick, Glover and Coote, 1980). An interior circumferential defect was located at the centre of the pipe and the pipe was loaded by application of a bending moment oriented such that the crack experienced the maximum tensile bending stress. Table 1 summarizes the three crack sizes considered.

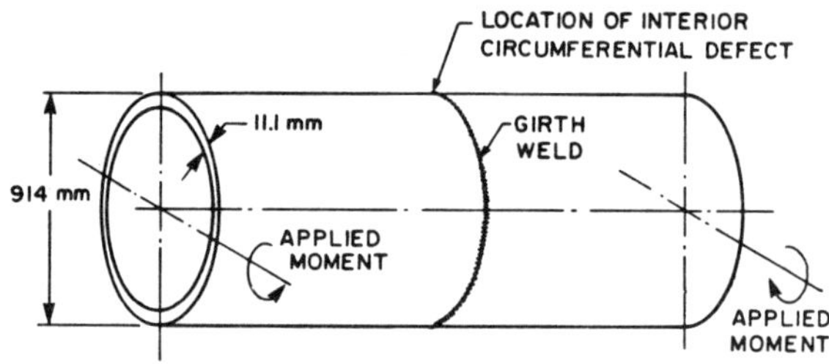

Fig. 1. General pipe arrangement.

Table 1: Defect sizes considered

Defect	Defect Length (L) (mm)	Defect Depth (a) (mm)	Depth to Wall Thickness Ratio (a/t)	Length to Circumference Ratio (L/c)
Short Deep	111.1	7.4	2/3	3.87 %
Short Shallow	111.1	3.7	1/3	3.87 %
Long Shallow	287.1	3.7	1/3	10 %

Due to symmetry only one quarter of the pipe shown in Fig. 1 need be modelled. This reduced section is shown in Fig. 2. Appropriate symmetry conditions were applied along the edges of the reduced pipe section. That is, the x-direction displacements were constrained to be zero in the symmetry plane defined by $x = 0$ and the y-direction displacements were set to zero in the other symmetry plane, $y = 457$ mm (Fig. 2), except along the crack face.

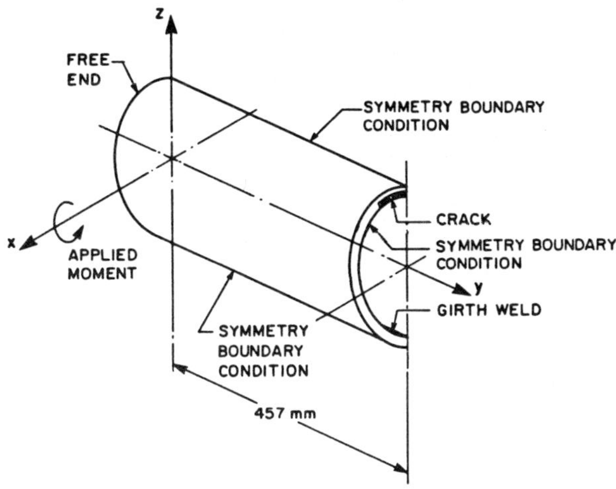

Fig. 2. Section of analysis.

Material Properties

Material properties were determined from tensile tests of a typical Grade X449 pipeline steel by Glover (1984). The elastic response was described by Young's Modulus, $E = 213.1$ MPa, and Poisson's Ratio, $\nu = 0.3$. The initial yield strength, σ_{ys}, was 475 MPa, and a linear hardening model was assumed up to ultimate strength, $\sigma_{ut} = 573$ MPa, at 36% axial strain (Fig. 3). Yielding was assumed to be governed by the Von Mises yield criterion.

Finite Element Mesh

The finite element mesh used to model the long deep defect has been plotted in Figs. 4 and 5. Note that the section plotted in Fig. 5 fits into the cut-out in Fig. 4. Twenty node, reduced integration isoparametric elements were used in this analysis.

In order to accurately model the behaviour of a crack, it was necessary to use numerous finite elements in the vicinity of the crack due to the severe stress gradients. Figure 6 shows a section through the pipe wall at the centre of the crack in the plane $x = 0$. In modelling cracks, it is common to move the midside nodes of the elements adjacent to the crack to the quarter point to achieve a stress singularity at the crack tip (Barsoum, 1976,1977). This modification permits accurate modelling of the crack tip singularity in an elastic analysis but not necessarily when yielding of a strain hardening material occurs. Therefore the mesh was not modified in this manner. The section of the mesh near the crack, shown in Fig. 6, was designed to focus enough elements about the crack tip to allow accurate calculation of a COD value. It is doubtful that there were a sufficient number of elements to allow accurate prediction of other fracture mechanics quantities such as CTOD or J-Integral. In any case, there was no provision in the version of ABAQUS available at the time (Version 4.5) for the calculation of a J-Integral magnitude when large displacements occur.

The crack was considered to run circumferentially on the inner surface of the pipe as though in a girth weld. No attempt was made to model weld reinforcement or weld metal properties however this will be the topic of a future study. The two dimensional studies of Worswick and Pick (1985) suggest that weld metal overmatch and weld reinforcement may delay increases in COD with increasing load. Confirmation of this result in three dimensions would be of interest. During the analyses, the growth of the plastic zone and COD were monitored as the moment (end rotation) on the pipe was increased. The COD was defined as being twice the axial displacement of the node at the center of the crack on the interior surface of the pipe (Fig. 7).

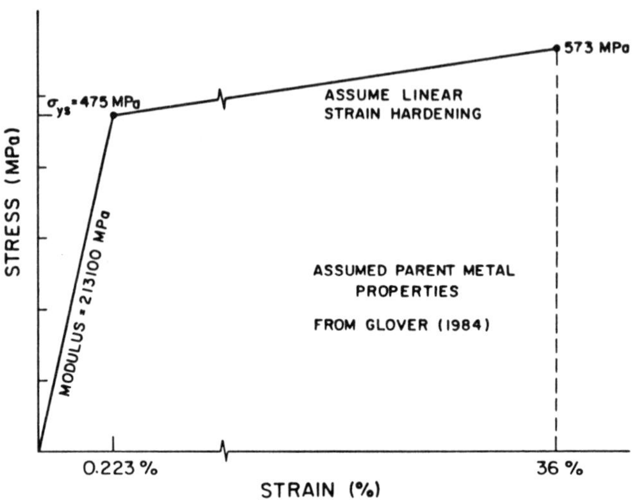

Fig. 3. Uniaxial stress-strain curve — CSA X449

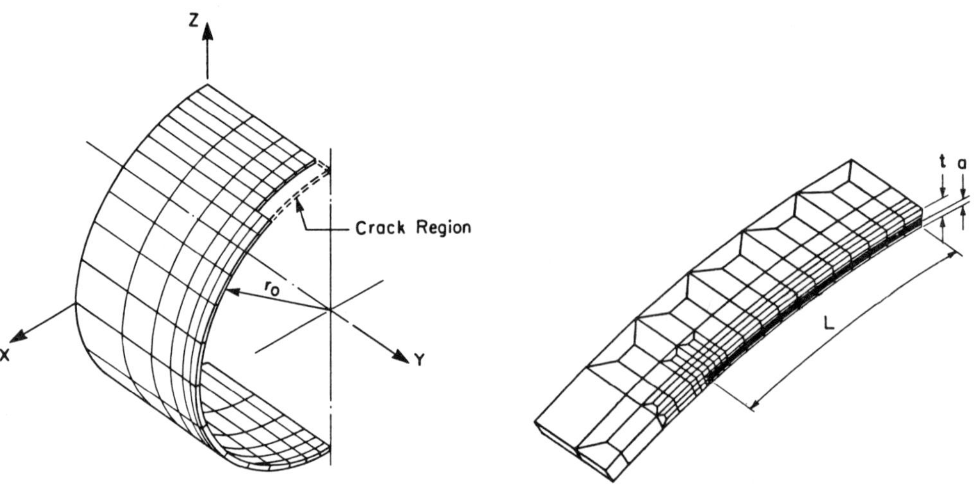

Fig. 4. Finite element mesh - nominal section.   Fig. 5. Finite element mesh - crack region.

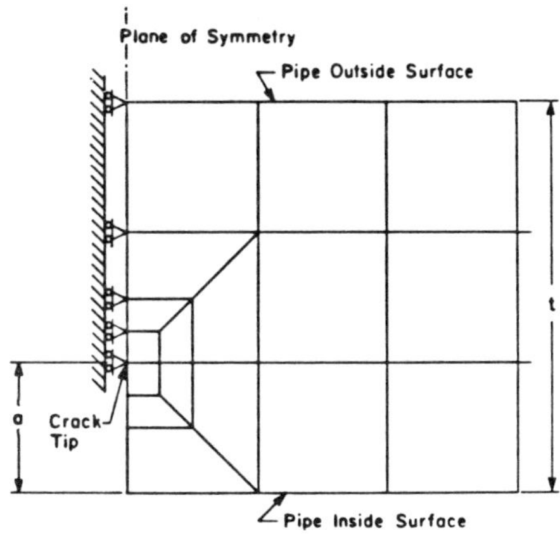

Fig. 6. Section through mesh at defect.

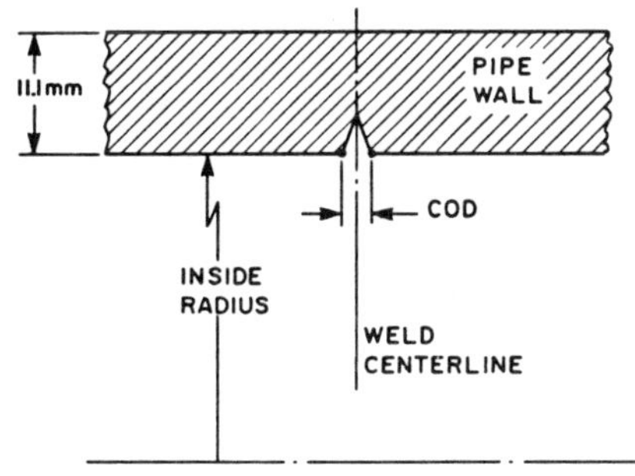

Fig. 7. COD measurement.

Load Application

The pipe section was to be loaded by a bending moment applied at the free end of the pipe away from the crack (Fig. 2). The application of a series of loads to the elements at the end of the pipe, to simulate an applied moment, can often cause local stress concentrations. While one could analyse a pipe long enough for the effect of these local pertubations to be negligible at the section of interest, modelling such a long pipe was not feasible because of the size of the analysis. Therefore the moment was applied to a shorter length of pipe by attaching all of the nodes at the free end of the pipe to a plane and then causing the plane to rotate about the pipe neutral axis. The nodes were allowed to move within the plane and therefore ovalling of the pipe was modelled correctly. In plane pipe this corresponds to the application of a pure bending moment. This is also true in a cracked pipe of sufficient length. For this analysis a half length of 457.0 mm was chosen as a compromise between accuracy and computing cost. The applied bending moment for a given amount of end rotation was calculated by summing the effect of the reactive loads developed at the nodes on the end of the pipe.

RESULTS

General Load-Displacement Behaviour

Figure 8 shows the load-displacement behaviour of the pipe for the three crack sizes. Plotted is the end-rotation, ϕ versus the applied moment, $M_{applied}$. Also plotted is the pipe ovality, D_x/D_z, in which D_x and D_z are the pipe diameters measured along the x-axis and z-axis, respectively. There was little difference between the curves for each analysis demonstrating that cracks of the sizes considered exerted little influence on the gross load-displacement behaviour of the pipe.

Growth of The Plastic Zone

Figure 9 shows the extent of the plastic zone predicted on the outer surface of the pipe at applied moments of 2.9 and 3.9 MN-m from each analysis. Figure 10 shows the plastic zone on the inner surface of the pipe for the same analyses and loads. The curves drawn represent the demarcation between material subjected to elastic stress levels and material at stresses above yield. In each figure, the crack geometries considered have been indicated.

First consider the extent of the plastic zone on the outer surface at $M_{applied} = 2.9$ MN-m (Fig. 9, curves 1–3). At this load level, plastic flow was confined to the ligament region above the crack. In the case of the shallow

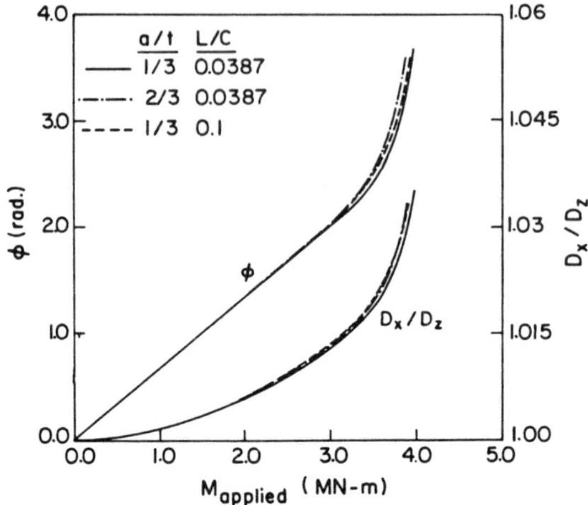

Fig. 8. Pipe end rotation and ovality versus applied moment.

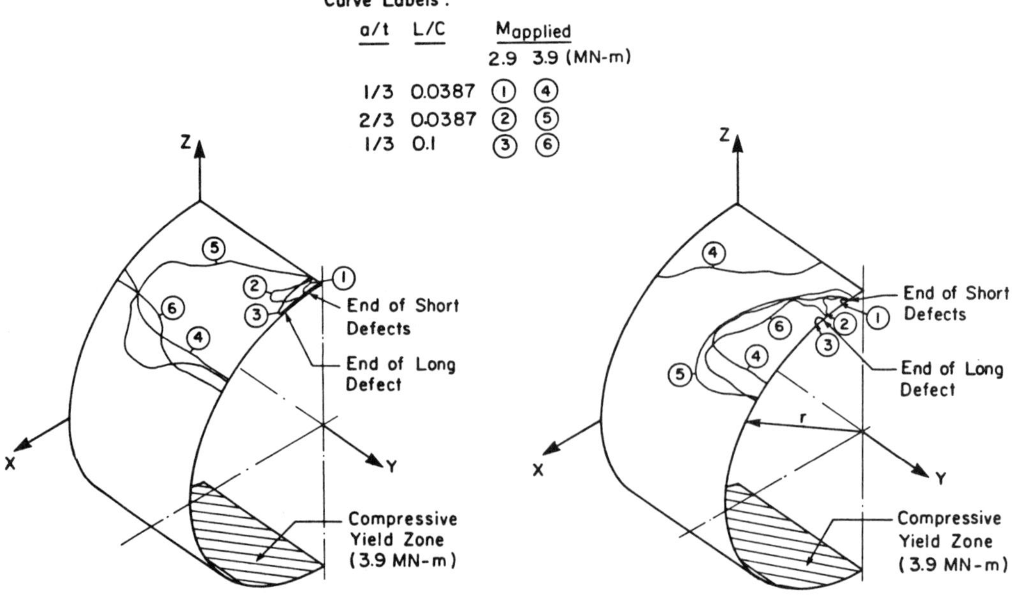

Fig. 9. Plastic zone on outer surface of pipe.   Fig. 10. Plastic zone on inner surface of pipe.

defects, circumferential growth of the plastic zone had not extended past the end of the crack. In the deeper crack case, the plastic zone had extended circumferentially past the end of the defect and axially away from the plane of the defect. Note that the nominal bending stresses were below yield and that the compressive region of the pipe remained elastic at $M_{applied} = 2.9$ MN-m.

On the inside face of the pipe at $M_{applied} = 2.9$ MN-m (Fig. 10, curves 1–3), plastic flow was confined to the immediate region surrounding the end of the crack (Note that the crack mouth opens onto the inner surface of the pipe). The plastic zone size was larger for the deep crack compared to the two shallow crack analyses.

With increased load, the plastic zone extended both circumferentially and axially and compressive yield occurred on the bottom of the pipe as seen in Figs. 9 and 10 for $M_{applied} = 3.9$ MN-m (curves 4–6 in both figures). It is important to note that the extent of the tensile yield zone in the circumferential direction did not differ greatly between the three analyses. However some difference was seen in the axial growth of the plastic zones. In general, there was a tendency for deformation to localize in the pipe wall near the crack, through formation of a shear band originating at the crack tip and extending at roughly 45° to the pipe outer surface (Fig. 11). The effect of this localization was to relieve the stresses in the tensile region away from the crack, inhibiting the axial growth of the plastic zone as seen in Figs. 9 and 10. Note that this effect on the axial plastic zone development increased with crack size and was enhanced on the inside face of the pipe by rotation of the crack face. It was anticipated, however, that the tensile plastic zone at the end of the pipe away from the crack would eventually conform to that of an uncracked pipe if a greater pipe length were considered.

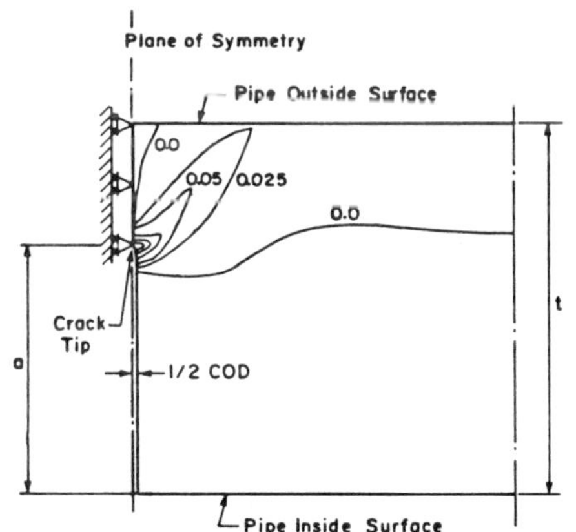

Fig. 11. Plastic strain contours near defect.

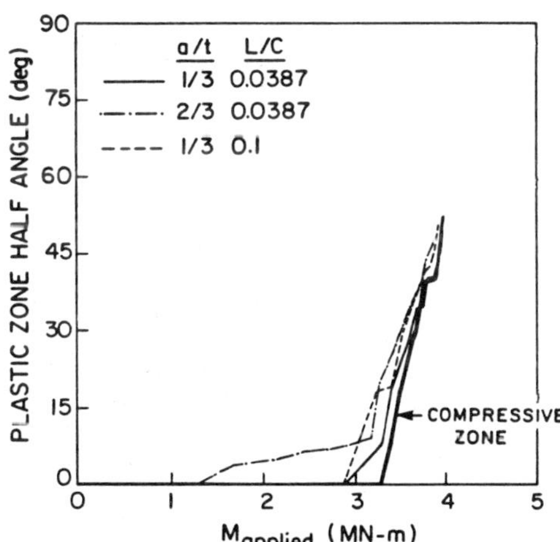

Fig. 12. Circumferential growth of plastic zone.

Figure 12 shows the growth of the plastic zone around the circumference of the pipe. The plastic zone half angles plotted were based on interpolation of the nodal plastic strains on the outer surface of the pipe. The tensile angle was measured from the positive z-axis to the tensile plastic zone boundary in the plane of the crack and the compressive angle was measured from the negative z-axis to the compressive plastic zone boundary, also in the plane of the crack. The three lighter curves plotted show the tensile yield angle development from each analysis. Only one curve was plotted for the compressive yield angle (heavy line) since it exhibited little variation between the three analyses. Some ambiguity exists in this definition of tensile plastic zone angle since yielding initially intersected the outer surface of the pipe away from the plane of the crack (Fig. 11) and only gradually grew to envelope the ligament with increased load. In the deeper crack case, the plastic zone near the crack grew more rapidly to envelope the ligament than in the shallow crack cases. This behaviour was reflected in Fig. 12 in the growth of the tensile plastic zone half angle at lower loads for the deeper defect (1.3 MN-m versus 2.9 MN-m). Note however that this difference in growth of the tensile plastic zone occurred at load levels in the elastic range and therefore does not concern the discussion to follow (The applied moment at which yield would first occur was calculated as 3.33 MN-m for a plane pipe and 3.08, 3.21 and 3.02 MN-m for pipes with short deep, short shallow and long shallow cracks, respectively, based on classical beam theory without consideration of the stress concentration of the crack).

Significant growth of the tensile plastic zone began at an applied moment of approximately 3.0 MN-m (Fig. 12). Once the plastic zone had grown past the end of the crack, a rapid increase in the plastic zone size occurred with increased applied moment. The circumferential growth of the plastic zone on the compressive side of the pipe was similar for all defects and was also similar to the growth of the plastic zone on the tensile side of the pipe once the zone had grown past the ligament (6.97° and 18° for the short and long defects, respectively). That is, once past the crack, the plastic zone grew in a similar fashion in each pipe. The analyses were terminated at an applied moment of approximately 3.9 MN-m. At this load level, the plastic zone was growing rapidly towards the neutral axis of the pipe which represents the first step in the formation of a plastic hinge and plastic collapse.

By assuming a value for the material flow stress, the moment required for the formation of a plastic hinge (plastic collapse) was calculated using classical theory. Table 2 shows the results of this type of calculation for the various crack sizes considered and a number of flow stress definitions that have been proposed in the technical literature.

Table 2: Predicted plastic collapse moment

Defect	Predicted Plastic Collapse Moment (MN-m) Using flow stress defined as:			
	σ_{ys}	$\sigma_{ys} + 69$ MPa	$(\sigma_{ys} + \sigma_{ut})/2$	σ_{ut}
Plane Pipe	4.29	4.92	4.74	5.18
Short Deep	4.12	4.72	4.55	4.97
Short Shallow	4.21	4.82	4.65	5.08
Long Shallow	4.07	4.67	4.49	4.91

Two points should be noted from this table. First, cracks of the size considered in this investigation exert only a minor effect on the calculated moment for plastic collapse. Secondly, the choice of flow stress greatly influences the value of the plastic collapse moment.

The curves of Fig. 12 rose rapidly with increasing applied load as the plastic zone grew towards the neutral axis. However as the plastic zone becomes near to the neutral axis, the slope of the curves in Fig. 12 will decrease and the curves will become asymptotic to the angle corresponding to the neutral axis location (approximately 90°). That is, in the absence of ligament instability, the plastic zone will grow to approximately the neutral axis and further loading will be carried by strain hardening of the pipe material until the applied moment approaches the plastic collapse moments listed in Table 2 and failure by plastic collapse occurs. The exact value of this moment depends upon the amount of average strain hardening as quantified by the flow stress.

Growth of COD With Applied Moment

Figure 13 shows the predicted growth of COD with applied moment from each analysis. It can be seen that COD increased rapidly as the plastic zone extended beyond the ligament (approximately 3.0 MN-m). While growth of the plastic zone after this point appeared similar for all cracks (Fig. 12), COD grew more rapidly for the short deep crack and the long shallow crack. The analyses were continued to an applied moment of approximately 3.9 MN-m. At this load level, COD increased rapidly with small increases in applied moment, suggesting that failure by tearing or shear of the ligament would be imminent. As an estimate of ligament instability the asymptotes of the curves in Fig. 13 have been estimated and tabulated in Table 3 as the predicted failure moment from the finite element model (FEM model).

The plastic collapse moment, calculated using a flow stress equal to $\sigma_{ys} + 69$ MPa, has also been entered in Table 3 (PC model). It is important to note that the finite element prediction of the failure moment of the pipe by ligament instability was well below the classical definition of plastic collapse or the formation of a plastic hinge. While the plastic zone predicted by the finite element analysis near ligament instability was quite large, a portion of the pipe remained elastic and those parts of the pipe that had experienced plastic flow (except near the defect) were at relatively low strain levels so that little strain hardening had taken place. Thus the moment at ligament instability would be low compared to the plastic collapse moment since the flow stress had only been reached locally near the

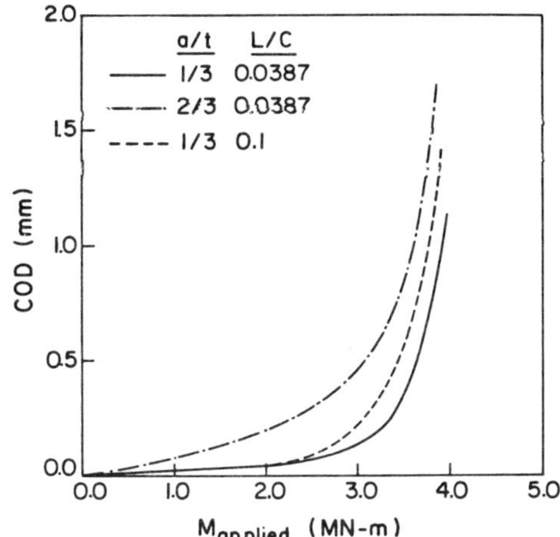

Fig. 13. COD versus applied moment.

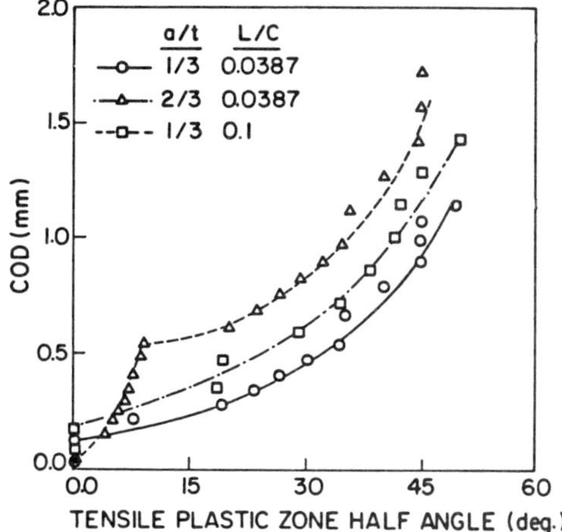

Fig. 14. COD versus plastic zone size.

Table 3: Predicted pipe failure moments

Defect	Failure Moment (MN-m)			
	FEM model	PC model	LIM model	CSA model
Short Deep	4.0	4.72	4.08	3.16
Short Shallow	4.2	4.82	4.25	3.33
Long Shallow	4.1	4.67	3.36	3.06

crack. Note that use of a lower assumed value for the material flow stress in the plastic collapse calculations (Table 2) would lead to better agreement with the finite element predictions of failure moment.

DISCUSSION

Growth of plasticity in the pipe began with yielding in the ligament over the crack. The early growth of plasticity in the ligament appeared to be dependent on the crack depth, with deeper cracks leading to earlier plasticity. Once the plastic zone grew to the end of the crack, the pipes behaved in a similar fashion with rapid growth of the plastic zone with increased moment. It is important to note that the rapid increase of COD coincided with the rapid growth of the plastic zone as seen in Fig. 14 which plots COD as a function of plastic zone size. That is, large COD values did not occur until after the plastic zone had grown to beyond the end of the crack. This behaviour was most noticeable in the shallow crack analyses.

A rapid increase in growth of COD with applied moment would suggest that the ligament was becoming unstable, inevitably leading to tearing of the ligament and fracture of the pipe. The results indicate that the onset of ligament instabilty (in pipe materials of sufficient toughness) would not occur until after the plastic zone grew beyond the crack at which point the crack loses the elastic restraint of the surrounding material. Ligament instability becomes imminent, only controlled by the strain hardening of the ligament material. In general, it is not possible to determine the size of plastic zone that would lead to ligament instability since the instability process appears to be governed more by the asymptotic load-COD behaviour than by plastic zone size (Erdogan and Delale, 1981; Ezzat and Erdogan, 1982). It is also difficult to determine an exact value for the COD to cause ligament instability. However the point at which the plastic zone grows beyond the ends of the defect may be considered a conservative estimate of failure for the crack sizes examined. Unfortunately use of this estimate suggests, unrealistically, that the moment to cause ligament instability would be higher for long cracks compared to that for short cracks of equal depth. Therefore a better estimate of the point at which ligament instability occurs might be when the plastic zone reaches a certain size.

Ligament Instability Model

In a recent publication, Worswick, Coote, Pick and Burns (1986) recommended the use of a ligament instability model for the prediction of the failure moment of cracked line pipe. This model was based on the assumption that failure would occur when the stress in the ligament reached the flow stress and the corresponding moment was empirically corrected for crack length. Using this criterion and a flow stress defined as $\sigma_{ys} + 69$ MPa, the ligament instability moment was calculated and entered in Table 3 (LIM model).

The effect of the empirical correction for crack length was to decrease the predicted failure moment for longer cracks. It can be seen from Table 3 that the agreement was quite good for the short cracks, however the predicted failure moment from the finite element analysis of the long defect was considerably larger than that from the ligament instability model. The reason for this discrepancy was not clear however it should not necessarily be construed as an error in the ligament instability model since it is in agreement with a considerable body of experimental data. The finite element model may have been inaccurate due to the large size of the elements at the crack tip leading to excessive stiffness and incorrect modelling of crack tip blunting.

CSA Z184 Code Calculation

Coote, Glover, Pick and Burns (1986) have recommended a defect acceptance standard which has been adopted in CSA Z184. A plastic collapse/fracture section of this standard has been based on the assumption that failure will occur when the tensile plastic zone extends 35.4° around the circumference (17.7° half angle). This angle was arrived at by assuming a linear bending stress distribution in the pipe with the outer fibre having reached 1.05 times the yield stress. The predicted failure moments for the defects considered in this investigation have been calculated using this model and entered in Table 3 (CSA model).

Table 3 indicates that the assumed failure criterion of the CSA defect acceptance standards was consistently conservative for the sizes of cracks studied. Furthermore examination of Fig. 12 indicates that the selection of 17.7° as the plastic zone half angle at ligament instability is reasonable.

<u>Mode of Loading</u>

The present research has examined flawed line pipe subjected to pure bending. The effects on ligament instability of significant superimposed axial loads or internal pressure loading of the crack face require further investigation. There exists some concern over the conservatism of failure criteria based on plastic zone size when applied to combined axial and bending load cases since, in the limit of pure axial tension, the entire pipe section will become fully plastic within a very small range of applied load. To alleviate this concern, the approach taken in CSA Z184 has been to decrease allowable stress used in Engineering Critical Assessment of defects should significant axial loads be present.

CONCLUSIONS

It has been shown that a finite element model can effectively model the behaviour of a circumferentially cracked thin walled pipe, including the effects of plasticity and large deformations.

The results indicate that if sufficient fracture toughness exists in the pipe and weld metal, failure will occur by instability of the ligament above the crack. For the crack sizes considered in this investigation, the instability will occur after growth of the plastic zone beyond the length of the crack and it appears that the instability can be thought of as being brought on by a loss of support of the ligament.

The use of COD as a design parameter is impractical since it is difficult to relate COD to applied load for a specific pipe and crack geometry. In addition, determination of the exact COD value at the point of instability would be a major task both experimentally and numerically. Therefore for design purposes it is more practical to relate the ligament instability to the plastic zone size and loss of restraint as has been done in CSA Z184.

The relationship between plastic zone size and ligament instability will be a function of both the crack geometry and the weld material and geometry. For all crack sizes analysed, the CSA Z184 estimate of a critical plastic zone size of 35.4° was reasonable.

ACKNOWLEDGEMENT

Funding for this work was provided by Nova, An Alberta Corporation and the University of Waterloo Faculty of Engineering. The authors express their appreciation to Dr. R.I. Coote of Nova, An Alberta Corporation and Dr. A.G. Glover of the Welding Institute of Canada for assistance with this project.

REFERENCES

Barsoum, R.S. (1976). On the Use of Isoparametric Finite Elements in Linear Fracture Mechanics, *Int. J. Num. Meth. Eng.*, **10**, 1976, 25–37.

Barsoum, R.S. (1977). Triangular and Quarter-Point Elements as Elastic and Perfectly-Plastic Crack Tip Elements, *Int. J. Num. Meth. Eng.*, **11**, 1977, 85–98.

Erdogan, F. and Delale, F. (1981). Ductile fracture of Pipes and Cylindrical Containers With a Circumferential Flaw, *J. Press. Vess. Tech.*, **103**, 1981, 160–168.

Ezzat, H. and Erdogan, F. (1982). Elastic-Plastic Fracture of Cylindrical Shells Containing a Part-Through Circumferential Crack, *J. Press. Vess. Tech.*, **104**, 1982, 323–330.

Coote, R.I., Glover, A.G., Pick, R.J. and Burns, D.J. (1986). Alternative Girth Weld Acceptance Standards in the Canadian Gas Pipeline Code, In *Proc. 3rd Int. Conf. on Welding and Performance of Pipelines*, U. K. Welding Institute, November, 1986.

Glover, A.G. (1984). Welding Institute of Canada Contract RC-182, October, 1984.

Glover, A.G. and Coote, R.I. (1983). Full Scale Fracture Tests of Pipeline Girth Welds, In *Proc. of 4th Nat. Congress on Pressure Vessels and Piping Technology*, June, 1983, Portland, ASME.

Glover, A.G., Coote, R.I. and Pick, R.J. (1981). Engineering Critical Assessment of Pipeline Girth Welds, In *Proc.*

Int. Conf. on Fitness for Purpose Validation of Welded Constructions, U. K. Welding Inst., London, England, November, 1981.

Pick, R.J., Glover, A.G. and Coote, R.I. (1980). Full Scale Testing of Large Diameter Pipelines, In *Proc. Int. Conf. Pipelines and Energy Plant Piping*, Calgary, November, 1980, Pergamon Press.

Worswick, M.J., Coote, R.I., Pick, R.J. and Burns, D.J. (1986). Investigation of Plastic Collapse Criteria for Defects in Line Pipe Girth Welds During Bending, In *Proc. 3rd Int. Conf. on Welding and Performance of Pipelines*, U. K. Welding Inst., November, 1986.

Worswick, M.J. and Pick, R.J. (1985). Influence of Weld Misalignment and Weld Metal Overmatch on the Prediction of Fracture in Pipeline Girth Welds, *Int. J. Press. Vess. & Piping*, **21** (1985) 209–234.

WELD METAL CHARPY – CTOD CORRELATIONS FOR COLD MARINE APPLICATION

J.E.M. Braid
Physical Metallurgy Research Laboratoires, CANMET
568 Booth St., Ottawa, Canada K1A 0G1

ABSTRACT

Empirical correlations between CTOD and 35 J Charpy transition temperature have
been made for weld metals at –10°C and –30°C test temperatures. Results at –10°C
agree with current U.K. Department of Energy requirements for weld metals in
offshore structures. The correlation analyses show that for a given 35 J Charpy
temperature, submerged arc welds give higher CTOD values than shielded-metal-arc or
flux-cored arc welds. However, for the purpose of design codes and standards it is
recommended that correlations based on shielded-metal-arc weld and flux-cored arc
weld data be used since achievement of specified toughness levels is necessary in
out-of-position welds.

The 35 J Charpy temperature is shifted to lower temperatures as the CTOD test
temperature is decreased from –10°C to –30°C. The shift is approximately 14°C for
CTOD values above 0.1 mm and approximately 21°C for CTOD values below 0.1 mm. A 35
J Charpy requirement could be established for a given level of CTOD, calculated
using an engineering critical assessment, from the correlations. For a CTOD level
of 0.1 mm at –30°C the 35 J temperature would be –55°C.

KEYWORDS

Charpy test, Transition temperatures, Crack tip opening displacement, Correlations,
Weld metals

INTRODUCTION

Welded joints for critical applications in Canadian offshore and Arctic structures
and vessels will have to meet toughness requirements for service at temperatures as
low as –30°C for East Coast regions and –50°C for Arctic regions. Although the
fabrication of such structures will rely on past experience, codes and standards
developed for regions such as the North Sea, existing toughness criteria will find
limited use owing to the much lower service temperatures. Existing Arctic
structures provide some experience. However, this is limited due to the small
number of exploration structures and vessels operating in these regions, all of
which have only experienced a small proportion of their design lives.

Besides a possible loss of life, the danger posed to the highly sensitive ecology of Canada's northern areas by a serious brittle fracture will probably result in requirements for a level of resistance to the initiation of brittle fracture, particularly in the weld and heat-affected-zone(HAZ), as well as an ability to arrest brittle fractures, primarily in the base plate. Fracture mechanics tests are used to establish fracture initiation resistance for these applications. In particular the crack tip opening displacement (CTOD) parameter is used extensively in areas such as the North Sea. The current U.K. Department of Energy Guidance Notes toughness requirements for weld metals are based on CTOD test data (Pisarski and Harrison, 1985).   There are several other codes and standards which use fracture mechanics as the basis for material requirements such as the American Association of State Highway and Transporation Officials (AASHTO) materials requirements for steel bridges (Barsom, 1975), as well as the requirements for thick-walled nuclear pressure vessels. However, the actual toughness values specified in these standards are given in terms of Charpy impact energies. The use of Charpy tests by consumable manufacturers, fabricators and standards organizations is related to its simplicity, lower cost and extensive prior exper- ience in the fabrication of welded structures. Hence, correlations between frac- ture mechanics parameters, such as CTOD, and Charpy tests become necessary to utilize the fracture mechanics based toughness requirements.

There are several problems which need to be considered when attempting to correlate fracture toughness, as measured by CTOD or K_{Ic}, and Charpy V-notch values:

(a)   fracture toughness tests measure crack initiation whereas
      Charpy tests include both initiation and propagation charac-
      teristics;
(b)   the notch acuity varies from a sharp fatigue crack for the
      fracture mechanics specimens to round notches for the Charpy
      tests;
(c)   the loading rate varies from slow for CTOD and K_{Ic} to impact
      rates for Charpy;
(d)   the difference in size of the test specimens can be sub-
      stantial. For weld metals, which are not homogeneous, the
      position of the specimens becomes critical, particularly for
      the smaller Charpy testpiece.

Consideration of the above points and, in particular, the non-homogeneity of weld metals results in fracture toughness – Charpy correlations being empirical only.

One of the earliest correlations between K_{Ic} and upper shelf CVN is that due to Barsom and Rolfe (1970) and Rolfe and Novak (1970):

$$(K_{Ic}/\sigma_{ys})^2 = (5\ CVN/\sigma_{ys})-0.25$$

$$(1)$$

where K_{Ic} is in ksi$\sqrt{in}$, σ_y is the yield stress at room temperature in ksi and CVN is the upper shelf Charpy energy in ft-lb. The use of this equation is restricted to steels with yield strengths greater than 690 MPa (100 ksi). For steels with yield strengths lower than this, K_{Ic} values can undergo a substantial increase within a given temperature range whereas Charpy impact values in the same temper- ature range remain at a constant value. Correlations in the transition temperature range need to consider notch acuity and loading rate. Such an approach was used in

developing the Charpy requirements for the AASHTO bridge steels (Barsom, 1975).
The temperature shift between dynamic and slow bend test values is given as:

$$T_{Shift} = 215 - 1.5\sigma_y \tag{2}$$

for 36 ksi < $\sigma_y \leq$ 140 ksi (248 MPa – 965 MPa) and

$$T_{Shift} = 0 \tag{3}$$

for $\sigma_y >$ 140 ksi (965 MPa). The correlation between K_{Ic} and CVN is then given as:

$$K_{Ic}^2/E = 5 \text{ (CVN)} \tag{4}$$

where E is Youngs modulus in ksi, K_{Ic} is given in ksi$\sqrt{in}$ and CVN is given in
ft-lbs. Estimates of K_{Ic} using such approaches are found to be reasonable for
steels with yield strengths above 500 MPa while predictions become very
conservative for lower yield strength steels (270 MPa to 400 MPa).

Several attempts have been made to correlate Charpy data with CTOD results in the
literature (Dolby 1981, Kanazawa and co-workers, 1985, Chaudhuri and co-workers,
1986). Many are restricted to one type of steel or one combination of welding
consumables making the correlation unsuitable for general use. For weld metals the
correlations first proposed by Dolby (1981), later incorporated into the U.K. Dept.
of Energy (DoE) Guidance Notes (Pisarski and Harrison, 1985), represent a
reasonably large selection of welding processes and consumables used for offshore
structure fabrication. However, this correlation is suitable for minimum design
temperatures of –10°C only. Hence, there is a need to establish a similar
correlation for minimum design temperatures of –30°C and –50°C. Towards this goal,
this paper will investigate the possibility of establishing a correlation between
Charpy V and CTOD test data for weld metals for a minimum design temperature of
–30°C.

TEST DATA CONSIDERATIONS

Following the arguments of Dolby (1981) two points require consideration when
establishing Charpy – CTOD correlations for weld metals.

1. The micromechanism of fracture should be the same in both
   specimen types.
2. Both test specimens should sample the same region of the weld
   deposit.

In considering the first point, unstable failure by cleavage is of most concern and
is the common failure mode of these types of weld metals at temperatures of –30°C
to –50°C. For Charpy tests a high level of crystallinity due to cleavage failure
occurs on and near the lower shelf region of the energy–temperature transition
curve, typically below 50 J. In existing codes and standards Charpy energy levels
used to define the temperature transition often range between 20 J and 40 J.
The 27 J level has been related to the NDT temperature measured by the drop weight
test. Using a Charpy energy level corresponding to the nominal yield strength of
the base material, in MPa, divided by 10 to define the transition temperature
allows for an increase in the energy requirement for increasing base material yield
strength. This has been used in several codes including the DoE Guidance Notes,

partially to compensate for a decrease in upper shelf energy with increase in yield strength. The base material in the present study has a nominal yield of 350 MPa. Hence, a 35 J level is used to define the Charpy transition level. For CTOD tests, failure initiation by cleavage was confirmed by observing fracture surfaces in a scanning electron microscope.

There are several reasons for having both test specimens sampling the same region of the weld metal. For design purposes CTOD test specimens use the full plate or joint thickness, which can be considerably larger than the 10 mm x 10 mm Charpy specimen. The high triaxiality of the stresses at the crack tip in the centre one-third of the CTOD specimen normally results in the initiation of brittle failure in this region in a zone of low toughness. For weld metals the low toughness zone is primarily an as-deposited microstructure (Tweed and Knott, 1983). Hence proportions of as-deposited and reheated microstructure, as well as the type of as-deposited microstructures, can result in wide variations in measured toughness. Thus the Charpy test specimen should be located within the centre one-third of the CTOD testpiece. A further consideration, however, is the location of the root region of the welded joint. The root region of ferritic welded joints can suffer an embrittlement effect due to strain ageing damage during welding (Cochrane and co-workers, 1976, Berkhout and van den Brink, 1978, McRobie and Knott, 1985). This can result in variations in CTOD values for single "V" and double "V" joints. Test results (Dawson and Judson, 1982) have shown that single "V" joints had higher CTOD values than double "V" joints but root region Charpy values for single "V" welds were lower than those for double "V" welds. Single "V" CTOD values were higher because the root region was located near the specimen surface, a low constraint (plane stress) region. The single "V" joint often results in larger distortions and hence higher root region strain than double "V" joints for plates of the same thickness. The result is a larger amount of dynamic strain ageing damage in the root region of the single "V" weld compared to the double "V" weld. For the data used in the present study, Charpy specimens were either located in the weld mid--thickness or in the region of highest strain ageing damage consistent with the centre one-third location of the CTOD crack front.

EXPERIMENTAL

The Charpy – CTOD data used in this study are taken from a series of welds used to evaluate the suitability of current welding consumables for applications in Canadian offshore and Arctic regions. Twenty-four shielded-metal-arc welds (SMAW) in 20 mm (Braid and Gianetto, 1986a) and 40 mm (Braid and Gianetto, 1985) thicknesses, 20 flux-cored arc welds (FCAW) also in 20 mm and 40 mm thicknesses (Bala and Santyr, 1986) and 12 submerged-arc welds (SAW), 20 mm thick (Braid and Gianetto, 1986b) were used.

The welds were fully restrained using jigs and consisted of a single "V" weld preparation. All welds were backgouged to remove the root passes and filled. They were all X-ray examined and found to be free of defects. The welds underwent a detailed evaluation including quantitative metallography, fractography, chemical analysis and analysis of the composition and size distribution of non-metallic inclusions. These evaluations are reported elsewhere (Braid and Gianetto, 1985, 1986a, 1986b, Bala and Santyr, 1986). All the welds had yield strengths in the range 465 MPa to 660 MPa. The SMA welds were of C-Mn or C-Mn-(1 Ni, 2½ Ni or 3½ Ni) compositions corresponding to AWS classifications E70XX or E80XX and of basic formulation. The FCA welds were made using basic, rutile or metal-cored wires of AWS E71TX-X or E81TX-X classifications. The compositions of the submerged-arc welds were of the C-Mn, C-Mn-Ni, C-Mn-Ti-B or C-Mn-Ti-B-Mo types.

Nine through–thickness, BX2B, centre–notched CTOD specimens were machined and
tested for each of the 20 mm thick welds and for the 40 mm thick FCA welds. For
the 40 mm SMA welds, the CTOD specimens were also the centre–notched, through
thickness, BX2B type with six specimens tested in the as–welded condition and
another six tested in the post–weld heat treated (PWHT) condition for each weld.

A minimum of 18 Charpy specimens were machined from the mid–thickness location of
each of the 20 mm thick welds to obtain full temperature transition curves. The
notches were located along the weld centreline in the same plane and direction as
the fatigue crack of the CTOD specimens. This location is above the backgouged
root and represents the area of the weld with the most remaining strain ageing
damage as well as the region of highest constraint in the CTOD tests. For the 40
mm thick welds, both subsurface and root region Charpy specimens were machined and
tested to give full temperature transition curves for both regions. The root
region specimens were located above the backgouged area in weld metal with the most
remaining strain age damage. This region is also within the centre one–third of
the CTOD specimen. Hence, the root region Charpy results for the 40 mm welds were
used for correlation purposes.

For each of the 20 mm thick welds for SMAW and SAW consumables five of the nine
CTOD specimens were tested at −30°C and the remaining four were tested at −10°C.
For each of the 40 mm SMA welds three CTOD specimens were tested at −30°C and the
remaining three at −10°C. For the FCA welds three specimens were tested at −30°C
with the remainder tested at temperatures selected from between −50°C to +23°C
(mostly −50°C and −10°C).

REGRESSION ANALYSIS AND RESULTS

The CTOD results are plotted against the 35 J Charpy transition temperature in
Figs. 1 to 3 for CTOD tests at −30°C and Figs. 4 to 6 for CTOD tests at −10°C.

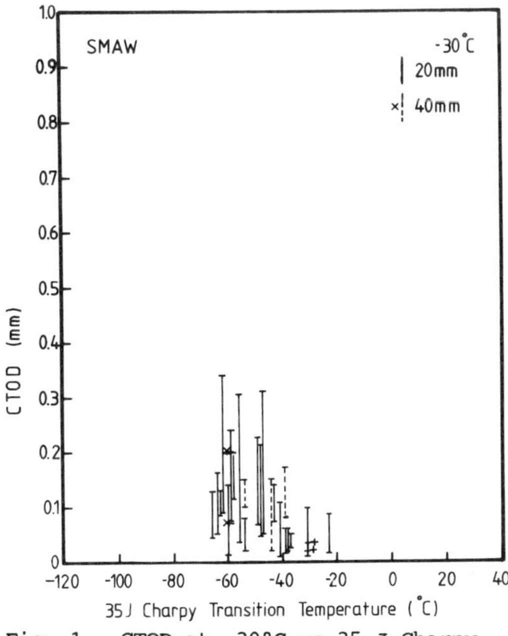

Fig. 1. CTOD at −30°C vs 35 J Charpy temperature for shielded metal arc welds (SMAW).

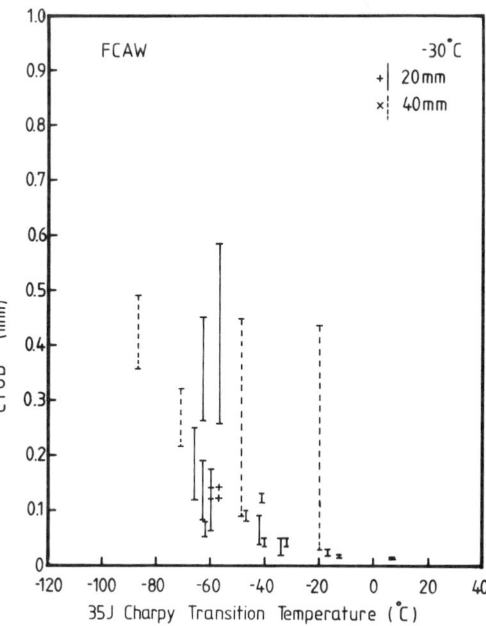

Fig. 2. CTOD at −30°C vs 35 J Charpy temperature for flux–cored arc welds (FCAW).

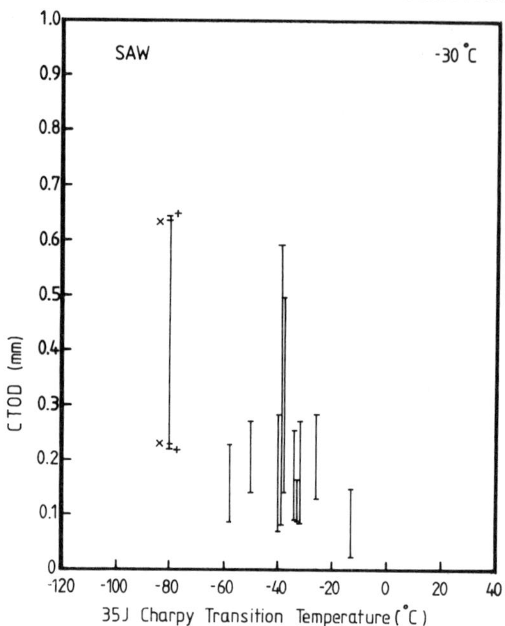

Fig. 3.   CTOD at −30°C vs 35 J Charpy temperature for submerged arc welds (SAW).

Fig. 4.   CTOD at −10°C vs 35 J Charpy temperature for SMA welds.

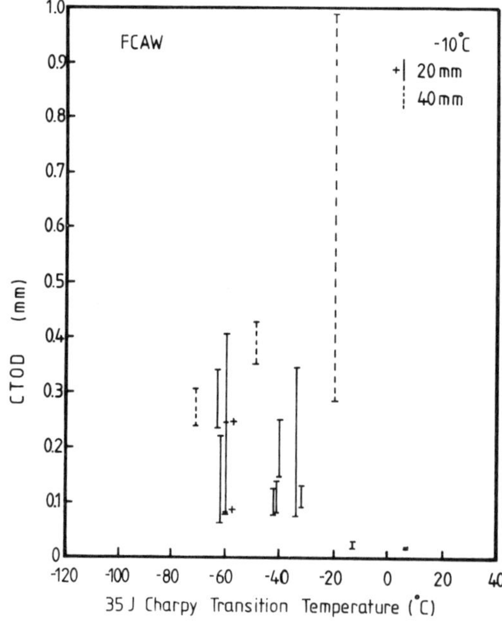

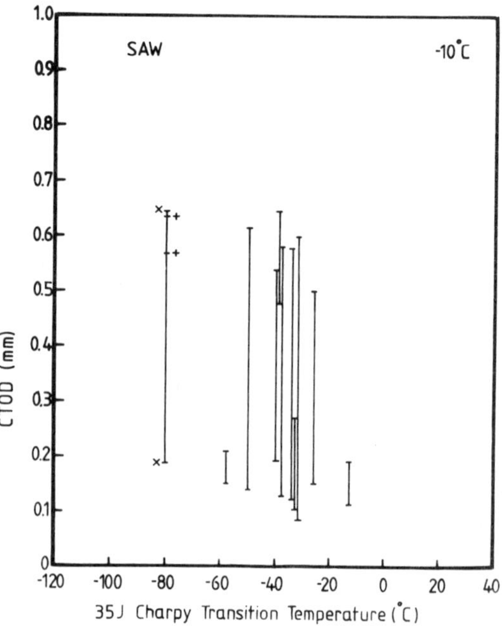

Fig. 5.   CTOD at −10°C vs 35 J Charpy temperature for FCA welds.

Fig. 6.   CTOD at −10°C vs 35 J Charpy temperature for SA welds.

In establishing correlations between the CTOD results and the 35 J Charpy temperature two approaches were considered. One approach is to establish a transition temperature for the CTOD tests for a given level, commonly 0.1 mm or 0.2 mm, and correlate this with the 35 J Charpy temperature. The other approach is to correlate CTOD data obtained at a fixed temperature with the 35 J Charpy temperature. This (second) approach was adopted for several reasons. As shown by the data in Figs. 1 to 6 considerable scatter exists for most of the welds. In their study on flux-cored arc welding consumables Bala and Santyr (1986) attempted to establish CTOD - temperature transition curves. For those welds where trends could be established the CTOD values were either generally low (less than 0.2 mm) or the variations at higher test temperatures indicated that an insufficient number of tests existed for an accurate determination. The scatter in CTOD values at -30°C and -10°C for the SMA and SA welds is such that seven of the 38 welds have minimum values at -10°C lower than at -30°C and 22 of the 38 welds have minimum values at the two test temperatures within 0.05 mm of each other. The scatter in results was considered to be too large to make a determination of 0.1 mm or 0.2 mm CTOD transition temperatures with any reliable accuracy. Also, the existence of the correlation between CTOD at -10°C and CVN given by Dolby (1981), which forms part of the correlation used to establish the U.K. DoE requirements (Pisarski and Harrison, 1985), provides a means for comparison.

SMAW and FCAW Deposits

A least squares regression analysis was performed on the various sets of data for the SMAW and FCAW deposits for both linear-linear and log-log equations of the form:

$$\delta = A_o + A_1 T_T \tag{5}$$

or

$$\log \delta = A_o + A_1 \log (T_T) \tag{6}$$

where δ is the value of CTOD (mm), T_T is the 35 J Charpy temperature (°C) and A_o and A_1 are constants. There was, statistically, no significant difference between

the sets of data for the SMAW and FCAW deposits, similar to that observed by Dolby (1981), for both -10°C and -30°C test temperatures. These data sets were therefore combined and least squares regression analysis were performed using equations (5) and (6). The coefficient of correlation for the least squares analysis, r^2, showed that equation (6) (r^2 = .407 for -30°C and .284 for -10°C) provided a better fit than equation (5) (r^2 = .285 for -30°C and .067 for -10°C). The low values of r^2 reflect the large scatter in CTOD values. The results, using equation (6), are shown in Fig. 7 for -10°C test temperature and Fig. 8 for -30°C test temperature. The data points shown in the two figures represent the average values of CTOD for each weld. Also shown in Figs. 7 and 8 are the results of regression analysis using the minimum values of CTOD only.

SAW Deposits

A least squares regression analysis using equations (5) and (6) was performed on the SAW data sets for -10°C and -30°C. The results are shown in Fig. 9 for -10°C test temperature and Fig. 10 for -30°C test temperature. Also shown in these two figures are the results of regression analysis using the minimum values of CTOD only. The differences in r^2 between equations (5) and (6) are small and, in general, equation (5) (r^2 = 0.389 for -30°C average values, 0.706 for -30°C minimum

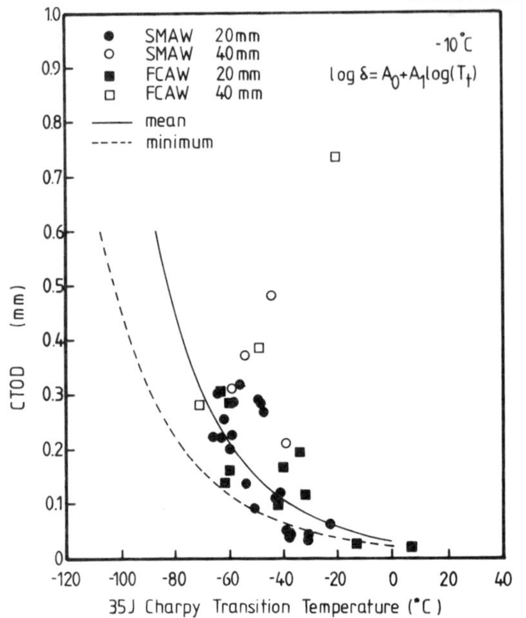

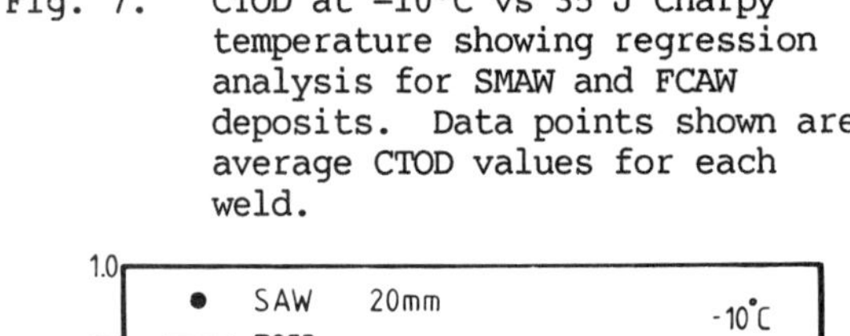

Fig. 7.   CTOD at −10°C vs 35 J Charpy temperature showing regression analysis for SMAW and FCAW deposits. Data points shown are average CTOD values for each weld.

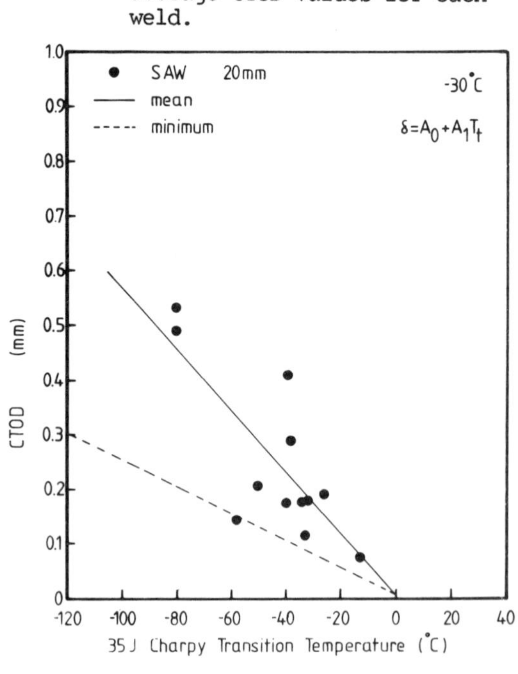

Fig. 8.   CTOD at −30°C vs 35 J Charpy temperature showing regression analysis for SMAW and FCAW deposits. Data points shown are average CTOD values for each weld.

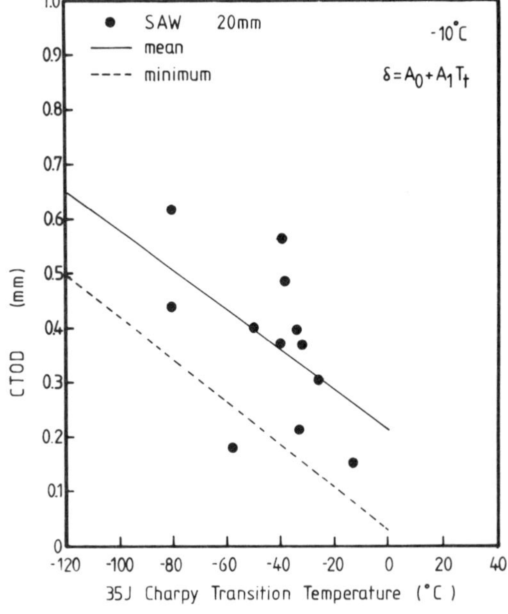

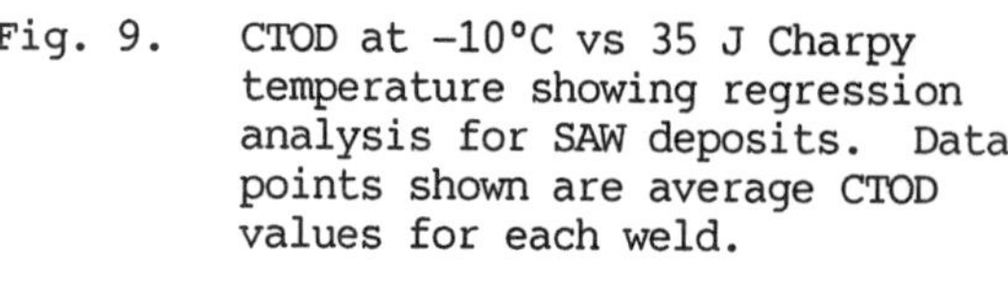

Fig. 9.   CTOD at −10°C vs 35 J Charpy temperature showing regression analysis for SAW deposits. Data points shown are average CTOD values for each weld.

Fig. 10.   CTOD at −30°C vs 35 J Charpy temperature showing regression analysis for SAW deposits. Data points shown are average CTOD values for each weld.

values, 0.129 for -10°C values, 0.262 for -10°C minimum values) provides a better fit than equation (6) (r^2 = .367 for -30°C values, .583 for -30°C minimum values, .116 for -10°C values and .315 for -10°C minimum values).

Comparison of Figs. 9 and 10 with Figs. 7 and 8 show a distinct difference in the data (and regression lines). The SAW data is shifted to higher CTOD values compared with the SMAW and FCAW values.

DISCUSSION

The SMAW and FCAW results indicate no effect of thickness between 20 mm and 40 mm in agreement with the observations of Dolby (1981). One would normally expect thicker plates to have lower CTOD values due to higher levels of constraint. It has been shown that dynamic strain aging in the weld root region can control toughness by causing a decrease in toughness in this region (Cochrane and co-workers, 1976, Berkhout and van den Brink, 1978, McRobie and Knott, 1985). A degradation of root-region toughness, compared to the subsurface region, was measured for the 40 mm SMAW (Braid and Gianetto, 1985) and 40 mm FCAW (Bala and Santyr, 1986) deposits and attributed to dynamic strain aging.

The similarity in behaviour of the SMAW and FCAW deposits allowing both sets of data to be considered together can be accounted for by the relative similarity in the weld deposits including yield strength, the amount of as-deposited microstructure sampled and dynamic strain ageing damage. The SMA and basic FCA welds also have similar weld cleanliness with oxygen levels between 290 and 430 ppm. FCA welds using rutile or metal-cored wires had higher oxygen contents of 400 to 1200 ppm with resulting lower toughness, although some had good toughness as a result of improved microstructures with high proportions of acicular ferrite.

The difference in behaviour between the SMAW and FCAW deposits and the SAW deposits does not appear to be due to differences in yield strength as suggested by Dolby (1981), but rather the microstructure of the deposits. The yield strengths of the SA welds ranged from 500 MPa to 599 MPa which is a narrower range than that for the SMAW and FCAW deposits (465 MPa to 660 MPa). In terms of microstructure, however, the SAW deposits consist of a high proportion of as-deposited microstructure (Braid and Gianetto, 1986b) compared to the SMAW and FCAW deposits (Braid and Gianetto, 1985, 1986a, Bala and Santyr, 1986). Although as-deposited structures generally have poorer toughness than reheated structures (Stout and co-workers, 1969, Dawson and Judson, 1982, Tweed and Knott, 1983, McRobie and Knott, 1985) the SAW deposits have significantly lower oxygen and nitrogen levels (<280 ppm O and <100 ppm N) than the SMAW and FCAW deposits. Gianetto and Braid (1985) have qualitatively rationalized differences in toughness measured by Charpy testing and CTOD between an SMA weld and an SA weld using the model of Ritchie and co-workers (1986). Both welds had similar as-deposited microstructures (approx. 64% AF), however, the SA weld had 85% as-deposited structure compared with 46% for the SMA weld. Tensile and Charpy properties were similar as was the chemical composition of the deposits except for the oxygen levels. The CTOD levels were significantly higher for the SA weld. The SA weld had 260 ppm O while the SMA weld had 440 ppm O. An analysis of the inclusions of the two welds showed the SMA weld had a significantly higher number of inclusions $\geq$ 1 μm dia. than the SA weld. It has been shown by several workers (Tweed and Knott, 1983, McRobie and Knott, 1985, Braid and Gianetto, 1985, 1986a, 1986b, Bala and Santyr, 1986) that cleavage fracture initiation in ferritic weld metals primarily occurs at non-metallic inclusions $\geq$ 1 μm dia. Considering inclusions of size $\geq$ 1 μm as cleavage initiation sites, the greater number of these inclusions in the SMA weld (higher O level weld) results in a smaller distance, "l", from the crack tip to an inclusion compared with the SA weld in CTOD tests as shown in Fig. 11. Fracture occurs when the stress in front of the crack tip, σ_{yy}, becomes greater than the fracture stress, σ_F, over the distance, "l". For a rounded notch, as in a Charpy specimen, fracture occurs when the maximum value of σ_{yy} exceeds σ_F at the plastic–elastic interface (which is much greater than "l"). Hence, the much lower oxygen levels of the SA welds should result in a lower number of large, $\geq$ 1 μm dia., inclusions and hence higher CTOD toughness levels for similar Charpy performance compared with the SMA and FCA welds.

The results of the regression analyses shown in Figs. 7 to 10 indicate that the 35 J Charpy temperature must decrease for increasing CTOD values for a given CTOD test temperature. The higher CTOD levels attained in some of the welds, particularly at the −10°C test temperature, were δ_u values exhibiting some stable ductile tearing. However, final failure was by cleavage. The only δ_m values were recorded for some of the SAW deposits due to the low oxygen levels and for some of the PWHT 40 mm SMAW deposits (Braid and Gianetto, 1985). The regression analysis did not include PWHT test results. Previous work on effects of PWHT on weld metal (Braid and Gianetto, 1985) has indicated that CTOD levels can increase as a result of PWHT. However, there was no corresponding improvement in Charpy properties and in one case CTOD levels deteriorated as a result of the formation of large Fe_3C particles. Correlations for PWHT condition were not made due to an insufficient data base.

Comparison of the SAW results (Figs. 9 and 10) with the SMAW and FCAW results (Figs. 7 and 8) show that CTOD levels for a given 35 J Charpy temperature are higher for SA welding compared with SMA or FCA welding. Alternatively, for a given CTOD level the Charpy requirement is less for SAW than for SMAW or FCAW. Often, critical joints in offshore and Arctic structures have to be welded out of position and the benefits of improved CTOD performance using SAW cannot be utilized. Hence, toughness requirements need to reflect those obtainable by out-of-position techniques such as SMAW and FCAW. For a given CTOD level of, say, 0.2 mm at −30°C the 35 J Charpy temperature is considerably reduced from −35°C for SAW to −72°C for SMAW and FCAW. These lower levels can be difficult to achieve, however, there are some consumables available which are suitable.

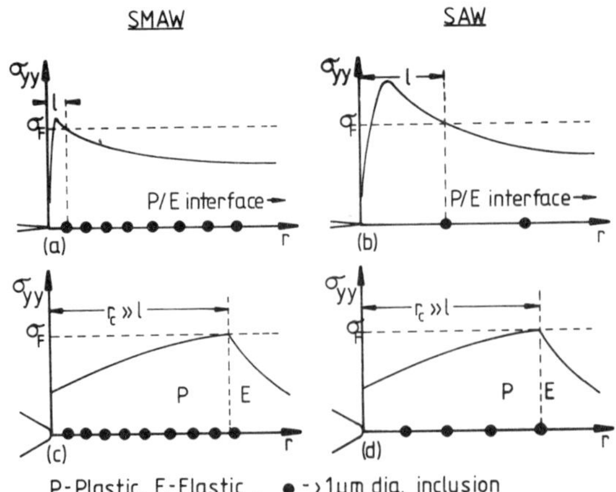

Fig. 11.   Stress distribution as a function of distance, r, from sharp crack tip (a and b) and rounded notch (c and d). For sharp cracks, fracture occurs when the stress, σ_{yy}, exceeds the critical fracture stress, σ_F, over the distance "l". For rounded notches, fracture occurs when the stress, σ_{yy}, reaches σ_F.

TABLE 1.   Shifts in 35 J Charpy temperature with change in CTOD test
           temperatures from −10°C to −30°C, for different CTOD levels

CTOD (mm)	Temperature Shift (C°)
0.5	21
0.1	17
0.15	14
0.2	13
0.3	12
0.4	12

The summary of results for SMAW and FCAW shown in Fig. 12 indicates the extent of
the shift in the 35 J temperature with a decrease in CTOD test temperature.   The
shift is tabulated for given CTOD levels in Table 1.   The current U.K. DoE
Guidelines (Pisarski and Harrison, 1985) use a shift of 0.7°C in Charpy test
temperature for every 1°C that the design temperature differs from −10°C.   This
shift is limited to the range −20°C to +10°C.   Using this shift to −30°C would give
20 x 0.7 = 14°C lower which is the same size of temperature shift indicated by the
regression analysis for CTOD above 0.1 mm (Table 1).   Using this temperature shift
for the Charpy test temperatures for weld metal in high stress regions the DoE
requirement would be 35 J at −34°C for welds ≤20 mm thick and −54°C for welds >20
mm thick up to 100 mm thick.   Using the required levels of CTOD calculated by
Pisarski and Harrison (1985) these requirements are shown in Fig. 12 and indicate
that some failures would be expected.   This results from the larger temperature
shift for CTOD values below 0.1 mm as shown by the regression analyses as well as
the original position of the DoE requirements at −10°C compared with the current
data (Fig. 12).

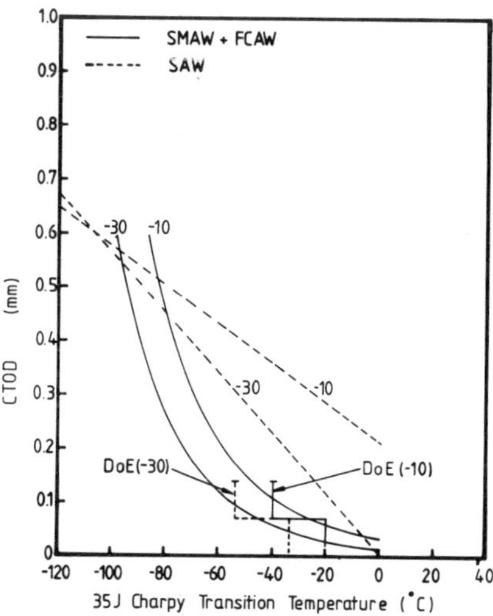

Fig. 12.   Summary of CTOD − 35 J Charpy temperature regression results showing
             shift in 35 J Charpy temperature due to a decrease in CTOD test
             temperature from −10°C to −30°C.

CONCLUSIONS

Empirical correlations between CTOD and 35 J Charpy transition temperature have
been made for weld metals at −10°C and −30°C test temperatures. The 35 J Charpy
temperature decreases for increasing CTOD levels at both −10°C and −30°C CTOD test
temperatures. The results at −10°C agree with those of Dolby and the current U.K.
Department of Energy weld metal toughness requirements.

The results of the correlation analysis show that for a given 35 J Charpy
temperature, submerged−arc welds give higher CTOD values than shielded−metal−arc or
flux−cored arc welds. This is thought to be due to the lower oxygen levels of the
SAW deposits and the resulting lower number of large (> 1 μm dia.) non−metallic
inclusions which act as cleavage crack initiation sites.

For the purposes of design codes and standards it is recommended that the correlations based on the SMAW and FCAW data be used since achievement of specified toughness levels, based on CTOD, is necessary in out−of−position welds for critical
applications.

The 35 J Charpy temperature is shifted to lower temperatures as the CTOD test
temperature is decreased from −10°C to −30°C. The shift is approximately 14°C for
CTOD values above 0.1 mm which corresponds to the 0.7°C per 1°C change in design
temperature rule in the U.K. DoE requirements. For CTOD values below 0.1mm the
shift is approximately 20°C.

RECOMMENDATIONS

The level of CTOD required for a joint with a specific risk and consequence of
failure should be evaluated using an engineering critical assessment concurrent
with the level of quality control suitable for the joint. Once established, the
correlations previously described could then be used to establish minimum charpy
toughness levels. Such an approach would help to minimize overly conservative and
unrealistic toughness specifications. The engineering critical assessment should
consider large scale structural behaviour and, in particular, the possible
beneficial effects of weld metal over matching as discussed by Denys (1987).

ACKNOWLEDGEMENTS

The author would like to thank Dr. J.T. McGrath, Dr. W.R. Tyson, Dr. R. Thomson,
J.A. Gianetto and B.A. Graville for many useful discussions and valuable help
during the course of this work.

REFERENCES

Bala, S.R. and S. Santyr, (1986). "Evaluation of Flux−Cored Arc Welding
Consumables for Offshore and Arctic Structure Fabrication", Report prepared by
AMCA Int'l for PMRL−CANMET, D.S.S. File No. 15SQ−23440−4−9028.

Barsom, J.M., (1975). Eng. Fract. Mech., 7,(3), pp 605−618.

Barsom, J.M. and S.T. Rolfe, (1970). In ASTM STP466, American Society for Testing and Materials, Philadelphia, pp 281-302.

Berkhout, C.F. and S.H. van den Brink, (1978). Weld. Metal. Fabric., 46,(5), pp 347-351.

Braid, J.E.M. and Gianetto, J.A., (1985). Proc. Int. Conf. Welding for Challenging Environments, Ed. by Welding Institute of Canada, Pergamon Press, Toronto, pp 145-155.

Braid, J.E.M. and Gianetto, J.A., (1986a). "Toughness of Weld Metals for Fabrication of Cold Marine Structures and Vessels, Part 1: Shielded Metal Arc Welding", Division Report PMRL 86-99(TR), CANMET, Ottawa,.

Braid, J.E.M. and Gianetto, J.A., (1986b). "Toughness of Weld Metals for Fabrication of Cold Marine Structures and Vessels, Part 2: Submerged Arc Welding", Division Report PMRL 86-100(TR), CANMET, Ottawa.

Chaudhuri, S.K., Ojha, S.N. and Ramaswamy, V., (1986). Int. J. Pres. Ves. and Piping, 22,. pp 23-30.

Cochrane, R.C., Terry, P. and Garland, J.G., (1976). Weld. Metal. Fabric. 44, (4) and.(6),. pp 316-320 and 439-448.

Dawson, G.W. and Judson, P., (1982). Proc. 2nd Int. Conf. Offshore Welded Structures, Paper 2, Welding Institute, Abington, Cambridge.

Denys, R.M., (1987). "Difference between small and large scale testing of weldments", Proc. AWS Annual Meeting, Chicago.

Dolby, R.E., (1981). Metal Const., 13,(1). pp 43-51.

Gianetto, J.A. and Braid, J.E.M., (1985). "An Evaluation of Weld Metal Toughness for Low Temperature Offshore Structure Applications", Division Report, PMRL 85-71(TR), CANMET, Ottawa.

Kanazawa, T., Watanabe, I. and Suzuki, M., (1985). IIW Doc. No. X-1085-85, International Institute of Welding.

McRobie, D.E. and Knott, J.F., (1985). Metal Sci. Technol., Vol. 1, pp 357-365.

Pisarski, H.G. and Harrison, J.D., (1985). Proc. Int. Conf. Welding for Challenging Environments, Weld. Inst. of Canada, Toronto, Canada, October. pp 131-144.

Ritchie, R.O., Francis, B. and Server, W.L., (1976). Met. Trans., Vol. 7A. p 831.

Rolfe, S.T. and Novak, S.R., (1970). In ASTM STP463, American Society for Testing and Materials, Philadelphia. pp 124-159.

Stout, R.D., McLaughlin, P.F. and Strunk, S.S., (1969). Weld. J., Vol. 48. pp 155s-160s.

Tweed, J.H. and Knott, J.F., (1983). Met. Sci., Vol. 17, pp 45-54.

A CANADIAN FACILITY FOR WIDE PLATE TESTING OF WELDMENTS

A. G. Glover, NOVA, An Alberta Corporation
(formerly of the Welding Institute of Canada)

R. J. Pick, Department of Mechanical Engineering
University of Waterloo

INTRODUCTION

The long term development of Canadian oil and gas resources in frontier regions will require the capability of assessing fracture at low temperatures and under severe operating conditions. These developments will also require development of major structures, support vessels and ancillary equipment all of which will be subjected to potentially severe operating conditions. Other countries have faced similar problems but not under such low temperatures or severe conditions. To address some of the concerns about such equipment the Welding Institute of Canada (WIC) and the University of Waterloo have jointly designed, constructed and commissioned a Wide Plate Test Facility (WPTF) to study the fracture characteristics of materials and welded components. The cost of the facility, owned and operated by the Welding Institute of Canada, was funded in a shared cost program by CANMET of the Department of Energy Mines and Resources and WIC.

This present paper describes the philosophy that led to the specifications of the facility and how the facility was designed, constructed, commissioned and assessed.

BACKGROUND

In the past the design of large scale structures used conventional design criteria based on the following principles:

(a) The structure was designed to keep stress levels below the elastic limit (yield) stress in order to prevent permanent deformation and large-scale deflection.

(b) The structure was designed so that the applied load was below the ultimate load to prevent mechanical instability such as buckling or necking.

(c) Local yielding near discontinuities was allowed with elastic constraint being relied upon to maintain the structural integrity of a member.

This design procedure, based on average stresses, is acceptable for many engineering structures, and works whether the design is based on yield or ultimate stress provided that material with adequate uniaxial strength is utilized. These criteria inherently assume that the fracture strength is equal to or greater than the ultimate strength (Figure 1). Unfortunately none of these design criteria address the problem of brittle fracture directly. Even when the above criteria are satisfied failures in large scale structures have occured, many of which involve the unstable propagation of a crack through the structure. In many cases these failures have always been produced by applied stresses less than the yield and design stresses (Figure 2) and have been termed brittle fracture. From a material point of view brittle fracture occurs by rapid propagation of a crack after little or no plastic deformation. From a structural engineering point of view fracture occurs before the design deformation is reached. In either case the failure is seldom preceded by load shedding deformations and is unexpected and often catastrophic.

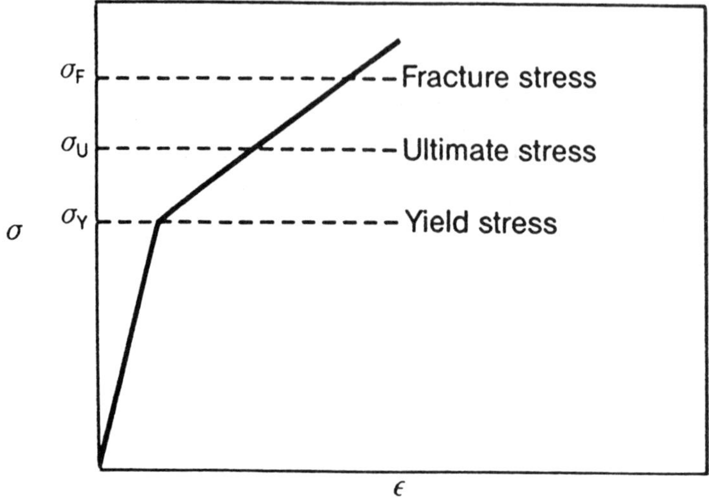

Figure 1: Fracture occurs after extensive plastic deformation, $\sigma_f > \sigma_u$ ductile material.

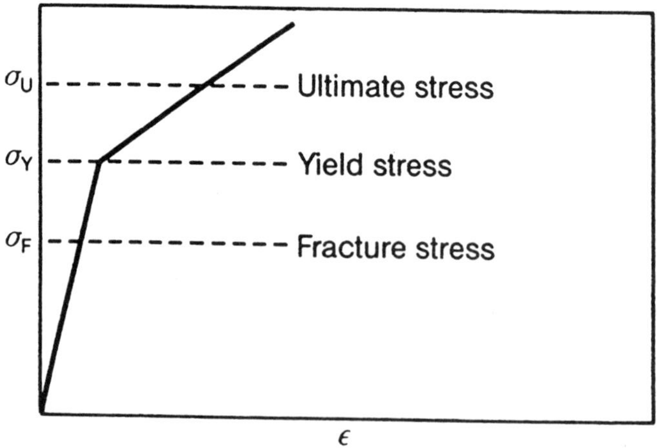

Figure 2: Fracture occurs before elastic limit is reached, $\sigma_f < \sigma_y$, brittle material.

The above criteria do not recognize the significance of an initial defect or stress concentration in the behaviour of the structure. If a component is to fracture in a brittle manner it must contain a stress concentrator because it is necessary to confine the initial event to a localized region.

Initially, therefore, notch brittleness was recognized as a problem of design significance and this led to a variety of tests being developed, most of which were notched bars tested at various temperatures and loaded in tension and/or bending. A number of major failures occuring in the 1940's emphasized that brittle fracture was a serious problem and led to a new series of standardized tests. The most widely used test becoming the Charpy V-notch impact test which measured the materials ductile to brittle temperature transition behaviour [1]. While the Charpy test has become an accepted standard there are difficulties in its interpretation because of strain rate and thickness effects and because it measures both initiation and progation energy. These concerns led to the development of the Drop Weight Test [2] and Crack Arrest Tests [3] making use of specimens containing natural sharp tipped cracks. The combination of data from Drop Weight Tests and Crack Arrest Tests led to the development of the Fracture Analysis Diagram [4]. More recently the need to precisely define material behaviour has led to the development of Fracture Mechanics Tests [5] based on both elastic and elastic-plastic behaviour.

As a result of these developments many test techniques are now available to assess the behaviour of the parent material and welded joints in environments that may promote brittle fracture. It is therefore important to show that these tests provide the essential data for applying realistic design methods. To interpret test results with confidence a fundamental understanding of the fracture philosophies are required. For example, which parameters are involved in initiation and propagation of fracture and will the fracture test give this information? In order to relate design criterion chosen to prevent brittle fracture to various test techniques it is appropriate to group the fracture tests into three categories:

(1) Large Scale type tests:
   In these tests the extreme service conditions are reproduced on a full size specimen to provide a direct measure of the brittle fracture risks in service.

(2) Intermediate Scale tests:
   These tests measure the risk of fracture by providing sufficient fundamental information for analysis of the behaviour of the full sized component, i.e. fracture mechanics tests.

(3) Small Scale tests:
   These are used to investigate the capacity of the material and as quality control tests. They can be used, after correlation with full scale tests, to predict the possibility of fracture in the full structure, i.e. quality control tests.

In order to provide a full understanding of fracture behaviour and so allow estimation of full size structural behaviour the test needs to take account of:

(a) thickness

(b) effects of welding embrittlement, residual stress

(c) stress, strain magnitude and distribution

(d) overall size effects

(e) crack size and location

(f) strain rate

(g) absolute and relative strengths of the weld and parent material regions

(h) temperature

and, for propagation only, the stored energy of the structure.

For many engineering components intermediate and small scale testing have proven to be effective in defining and preventing brittle fracture. However in large scale welded structures where the cost of construction and of failure is high it is often prudent to undertake large scale testing to assess the size and thickness effects and because it is difficult to effectively scale welds and welding parameters.

The complexities of large scale structures such as offshore rigs and the consequences of a failure often mean that in order to check the safety of a structure it is necessary to test a structural element in which the strain concentrations are similar or larger than any detectable defect. This led to the consideration of a Wide Plate Test Facility as an appropriate apparatus to study brittle fracture mechanisms. In studies using the wide plate configuration the test specimens are of realistic thickness; do contain a defect; have sufficient width to negate edge effects on the defect behaviour; and are large enough to study the effects of residual strain fields resulting from welding.

PROGRAM

Discussions within the regulatory bodies had shown a concern for adequate material and welding technology for Canadian offshore development and marine use. Some of these concerns related to what type of design philosophy was required, however in all cases a lack of data hindered the decision making process. As a result a 50/50 program was developed between CANMET and the Welding Institute of Canada the objective of which was to:

(a) design, construct and commission a wide plate test facility.

(b) establish a national facility available for Canadian industry and government, owned and operated by the Welding Institute of Canada.

(c) study the fracture characteristics of welded components.

The design of the Wide Plate Test Facility was carried out by Professor Pick at the University of Waterloo. Prior to detailed design of the facility it was necessary to decide which fracture test philosophy would be used. In designing to prevent fracture there are two design philosophies that can be use: one based on crack initiation resistance and the other based upon crack propagation resistance. In the former it is assumed that a defect is present and fracture can be controlled through the prevention of initiation by a material specification and a knowledge of the stress state, material toughness and defect description. The latter it assumes that initiation will occur and prevention of fracture is controlled by the material having sufficient toughness to arrest the crack. Very early in the preliminary design it became apparent that because of size limitations and cost the test facility could not be designed to study crack propagation resistance of large cracks. The design of the basic facility was therefore based on a rigid machine intended for studying crack initiation and/or short crack propagation and arrest.

DESIGN OF WIDE PLATE TEST FACILITY

The Wide Plate Test Facility was designed with a capacity of 4000 metric tons. This allows the generation of tensile stresses of 414 MPa in the maximum size of plate (1220 mm width by 75 mm thickness). The maximum length of test plate is 1830 mm which enables it to be shortened for re-use in additional tests.

The design of the test facility provides a rigid machine with a limited stroke and a compact assembly. Figure 3 shows the assembled facility. Four hydraulic capsules operate against end lugs which were electroslag welded to the end plates. The plate specimen is placed in the assembly and also electroslag welded to the end plates. The lugs, end plates, specimen and the hydraulic capsules contain the load path. The capsules have been placed as close as possible to the specimen to minimize the bending in the end plates and lugs and provide a compact arrangement. Of concern in the design is the distribution of the end load from the lugs into the test plate to ensure an even stress distribution at the test section. Strain gauge measurements during commissioning indicated that the positioning of the end lugs provided almost constant stress across the test section of the plate specimen.

The hydraulic capsules have a stroke of approximately 35 mm which is adequate for elastic, average stress levels in the specimen. A maximum hydraulic pressure of 207 Mpa (30,000 psi) is used to allow the use of commercial piping components while minimizing the size of the hydraulic capsules. The hydraulic capsules (Figure 4) were designed and constructed at the University of Waterloo using an O-ring seal with steel backing rings and brass

Figure 3: Assembled Wide Plate Test Facility, showing location of hydraulic capsules and end lugs.

piston guides. Extensive development of various high pressure seals at the University has led to some confidence in this design. The cylinder was constructed by boring an annealled SAE 4340 rod, 406 mm in diameter, to provide a blind cylinder. The ram is annealled Atlas SPS rod, 254 mm in diameter. Pressure is supplied to the cylinder through a hole in the ram to minimize stress concentrations. Once extended from the cylinder, the ram can be retracted through the use of threaded cap screws attached to a collar on the ram. Various lengths of test plate can be accomodated through the use of cylindrical shims added to the end of the ram. The hydraulic capsules are arranged in banks on either side of the specimen within a sub assembly. This sub assembly can be removed intact allowing easy access to the specimen.

The assembly is mounted on edge on a base frame and surrounded by a portal frame. This framework is used to support the assembly and restrain the specimen pieces after fracture. Between the specimen and the framework are a series of pipes (Figure 5). After fracture the kinetic energy of the two halves of the specimen is absorbed by plastic deformation of these pipes. Commissioning tests have shown this to be an inexpensive but effect restraint system up to at least 60% of the full load capacity of the facility. The shock absorbing pipes are the only components requiring replacement between tests.

Hydraulic Control System

To provide even loading of the plate specimen it is necessary to ensure that each of the four hydraulic capsules generates the same force. Ideally this loading could be guaranteed if each hydraulic capsule was operated from a servo-controlled pressure supply activated by a load cell. However, servo controlled hydraulic components for pressures of 207 MPa are not readily available. This, the cost and overall control and reliability of such a complex system led the authors to consider a much simpler arrangement. Pressure is supplied from a single low volume variable pressure hydraulic supply driven by pneumatic pressure of .7 MPa. Control of the air pressure, throttling of the hydraulic supply and the large volume of the hydraulic capsules provides a very fine control on the pressure supply with simple manual operation. The same pressure is supplied to each hydraulic capsule and provided that each capsule has the same seal friction the same end force should be generated in each capsule. Because of the large end forces involved and the special design of the ram seal, losses due to friction at the seals are minimum and similar in each capsule. Monitoring of the stress levels in each ram is used to ensure that the rams are providing equal loading during each test.

Figure 4: Hydraulic capsules showing pressure connection through the rams.

Figure 5: Details of support frame, end portals and restraint system.

<u>Cooling Equipment</u>

The plate specimen can be cooled locally with an insulated cold box surrounding the plate specimen (Figure 6). Circulation of nitrogen gas from boiling liquid nitrogen allows cooling of 75 mm thick plate to -100° C .

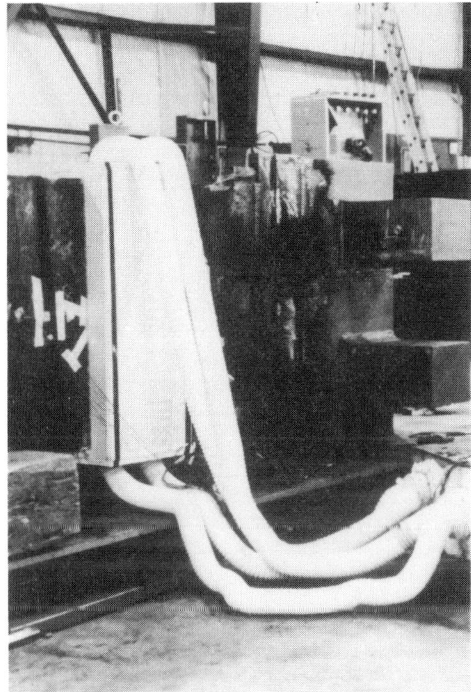

Figure 6: Local cooling system surrounding test plate.

Trials have shown that precooling the plate with dry ice assists in developing good low temperature profiles. In addition insulation of the test plate surface adjacent to the local cooling box is necessary. A final consideration is the temperature of the end lugs during testing. At very low test temperatures it is anticipated that the lug regions may have to be warmed. This could be achieved by simply circulating water or hot air around that region.

<u>Instrumentation</u>

During testing each specimen is fully instrumented with an array of strain gauges and thermocouples to ensure that stress and temperature distributions are controlled. In addition clip gauge, LVDT and potential drop transducers can be added to provide data concerning the behaviour of defects in the plate specimen. Data is recorded on a 64 channel data acquisition system with appropriate computer control. Analysis of the data provides stain fields, temperature gradients, crack opening, initiation and propagation characteristics and crack velocity.

CALIBRATION AND COMMISSIONING

The capacity of the four capsules is 4000 metric tons at a pressure of 207 MPa (30,000 Psi). The maximum plate width is 1220 mm (48 inches) and hence the limiting fracture stress for different plate thicknesses is given in Figure 7. The maximum stress that can be generated prior to fracture is 320 MPa for 100 mm plate, 425 MPa for 75 mm plate, 650 MPa for 50 mm plate and greater than 1000 MPa for 24 mm plate. The facility is primarily designed to test plates from 12 to 75 mm in thickness.

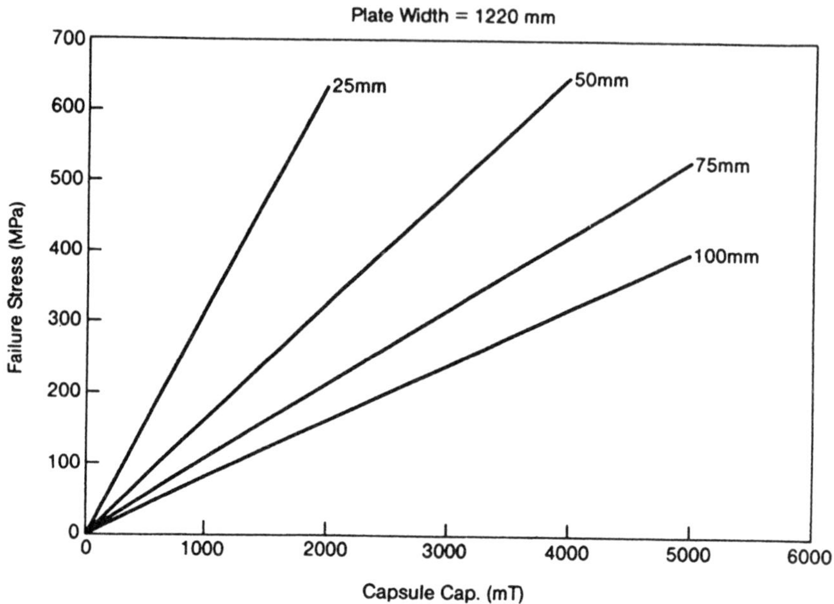

Figure 7: Relationship between fracture stress and plate thickness.

Initial calibration trials were carried out on a plain plate 1000 by 1000 by 25 mm thick. The measured yield stress of the plate was 340 MPa and the loading was carried out at room temperature. Although the plate was instrumented the main objective of the trial was to assess the performance of the capsule seals, the uniformity of loading and the strain distribution around the end lugs. No problems were encountered with the seals during these preliminary trials, however initial load readings were in error because of the time required to overcome frictional effects at low pressure. Subsequently at higher pressures the tests gave load readings which agreed with calculations to within 5%. These trials also indicated that there was negligable bending at the lugs, but some uneven loading on the test plate. This problem was traced to the alignment of the test plate within the end plates. Although care had been taken in alignment prior to welding no alignment frame had been used. The effect of this misalignment is shown in Figure 8, where a consistent difference in strain readings from one side of the plate to the other is apparent. The theoretical strain at the applied load for Figure 8 is 632 microstrain, which is in good agreement with the average results. The results also reflect a lateral buckle that occurred when the relatively thin plate was one-sided welded into the end lugs.

To achieve better alignment a heavy duty alignment frame was designed and built by the University of Waterloo. Subsequent welding of test plates into the end lugs was carried out in the alignment frame by using electroslag welding, which maintained the axial and vertical linearity and gave much more even loading of the plate.

After these initial trials a full scale fracture test was carried out to assess the loading configuation, the local cooling and the restraint systems. A test was conducted at - 30° C on material with a measured COD of 0.30 mm at the test temperature. Details of the test arrangement are shown in Figure 9 and the instrumentation in Figure 10. Loading proceeded normally and uniform loading of the test plate was obtained. Failure occurred as the stress approached the yield strain of the plate and the restraint system arrested the plate halves (Figure 11). The facility was operating at about 30% capacity when failure of the test plate occurred. Analysis of the test showed that the predicted tolerable strain was 0.0768% and the actual failure strain was 0.159%. Subsequent fracture tests have been carried out on 60 mm thick weldments at 60% capacity of the facility and the equipment and the restraint system was shown to be workable.

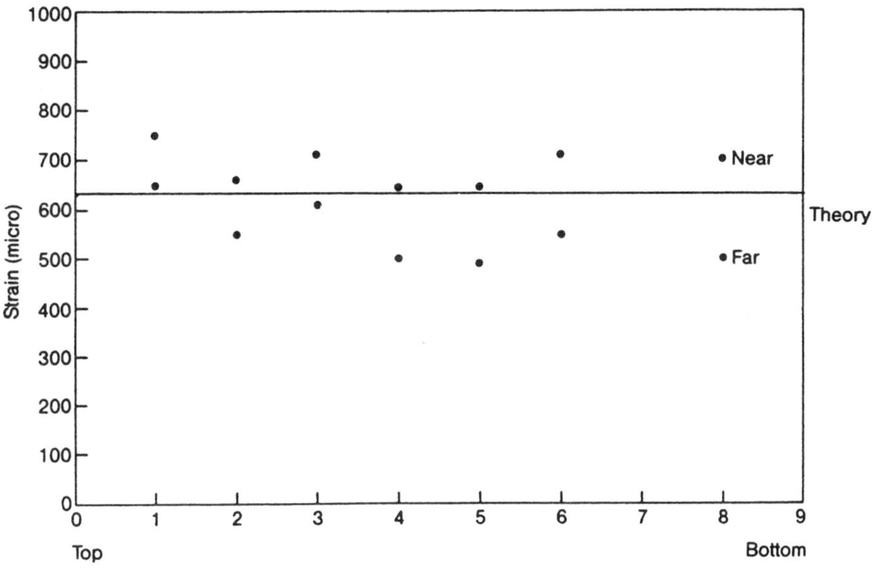

Figure 8: Strain output across 25 mm plate for initial trial showing effect of misalignment.

Figure 9: Final installation of clip gauges prior to full scale test at -30°C.

FRACTURE MECHANICS

Figure 10: Instrumentation on test plate prior to placement of the cooling system.

Figure 11: Final configuration of the restraint system after fracture test.

FUTURE PROGRAMS

The Wide Plate Test Facility is now fully operational and areas of interest that are currently being studied include safety assessments of marine vessels and Canadian offshore technology. These studies will include weld metal correlation between CTOD and CVN to obtain indexing of small scale tests. Programs will also be developed to assess the heat affected zone behaviour of presently developed high toughness steels, and to study the effect of short crack arrest.

ACKNOWLEDGEMENTS

The authors would like to acknowledge the support of CANMET, Department of Energy, Mines and Resources and the Welding Institute of Canada in developing the facility. Rememberance is acknowledged of the late Arie DeGroot who provided much of the detailed design of the facility. Thanks is given Jeremy Price, Dirk Varo, Neil Sweetman, Azael Aquirne, Paul Koornneeff who assisted in the construction, assembly and commissioning of the facility.

REFERENCES

1. Handbook for Welding Design, Vol. 1 2nd Ed. The Welding Institute, 1967.

2. Naval Research Laboratory, Washington (Metallurgy Division) Reports NRL 5920, March 1963, NRL 6300, November 1963.

3. The Robertson Crack Arrest Test, British Welding Journal, August 1968.

4. Pellini, W.S. and Puzak, P.P., "Fracture Analysis Diagram Procedures for the Fracture - Safe Engineering Design of Steel Structures", US Naval Research Laboratories, Washington No 6030, 1963.

5. ASTM Standards, Designation E399-70T "Experimental Techniques in Fracture Mechanics", BS5762:1979 "Methods for Crack Opening Displacement (COD) Testing.

6. Wells, A.A., British Welding Journal, 8 No. 5, 1961.

7. Soete, W. and Renys, R., "Welding Plate Testing as a Basis for Plastic Fracture Mechanics, 11W-X-919-79.

8. Pisarski, H.G. and Harrison, J.D., "Fracture Toughness Consideration for Offshore Structures" Metal Construction, December 1986.

THE INFLUENCE OF THICKNESS ON THE FATIGUE ENDURANCE OF
NON-LOAD CARRYING SPOT WELDED JOINTS

F.G. Hamel and J. Masounave

Industrial Materials Research Institute
National Research Council Canada
75 De Mortagne Blvd.
Boucherville (Quebec) Canada J4B 6Y4

ABSTRACT

Non-load carrying spot welded joints of 301L stainless steel sheets were tested
in fatigue. Different combinations of 1.5, 3 and 4.5 mm thick sheets were
evaluated. The initiation of the fatigue crack was detected by a compliance
method. The obtained S-N curves revealed that the presence of a non-load
carrying connection considerably reduces the fatigue endurance limit as
compared to plain specimens. It is also shown that the fatigue limit decreases
when the thickness of the non-load carrying sheet is increased. However, the
thickness of the load carrying sheet has little influence on the fatigue
limit. The comparison with other results also showed that the alloy has a
small influence on the high-cycle regime.

These results indicate that for design purpose in fatigue, the load carried by
a spot welded connection is not the only factor to consider. More
specifically, in the case of a low-load carrying joint the stress in each sheet
and the thickness of the sheet attached to it are the controlling factors in
high-cycle fatigue.

KEYWORDS

Fatigue; spot welds; thickness; endurance limit; non-load carrying; stainless
steel; resistance welding.

INTRODUCTION

Spot welded joints are extensively used for the assembly of sheet metal
structure especially in the transportation industry. For thin sheets, the
diameter and spacing of the welds can be designed such that multi spot
assemblies can carry a static load comparable to the static strength of the
sheets.

However, the situation is different for the fatigue resistance especially in
the high-cycle regime. Spot welded joints are severe stress concentrators and
can even be considered as crack-like defects. The fatigue crack initiation
period is almost non-existant and most of the fatigue life is in the crack

growth regime (Davidson, 1983a; Pollard, 1974; Overbeeke, 1974). The situation
is more critical for high strength alloys (stainless steels and HSLA steels)
because their fatigue properties are not improved as much as their strength.
In fact, the fatigue crack growth properties of HSLA steels are similar to mild
steel and are inferior for austenitic stainless steel. It follows that the
high-cycle fatigue endurance of spot welded joints of high strength alloys is
of greater concern than for plain carbon steel because structures made of
sheets of high strength alloys are usually more stressed. Moreover, the quality control tests used in the industry (peel, tensile shear and cross tension)
which are adequate to characterize the static strength of a spot welded joint
show a poor correlation with the high-cycle fatigue endurance (Davidson,
1983a).

In consideration of these facts, the fatigue behaviour or resistance spot
welded joints has been extensively studied by many investigators (Cooper, 1985,
1986; Davidson, 1983b, 1983c, 1984; Iwasaki, 1983; Orts, 1981; Overbeeke, 1984;
Pollard, 1984; Pook 1975; Rivett, 1983; Wang, 1985). Review papers have also
been published by Davidson (1983a) and Defourny (1985). Most of these studies
have been performed on single or multi spot joints in the single lap shear
configuration. In this configuration, 100% of the stress in the sheets is
caused by the load transmitted through the spot weld and these joints can be
designated fully load carrying.

Fully load carrying joints are not representative of many practical situations. The other extreme is the non-load carrying joint in which a short
sheet is spot welded on a longer sheet (Fig. 1). Only the longer sheet is
stressed and the smaller one acts as a stiffener producing a stress concentration area in the long stressed sheet. Very few studies (Defourny, 1981, 1985;

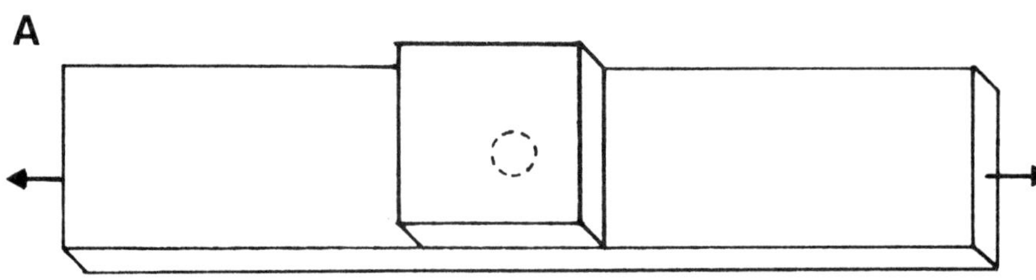

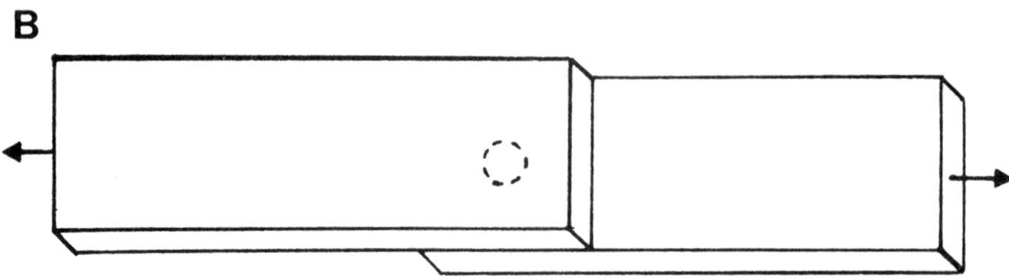

Fig. 1 Schematic illustration of a) non-load carrying spot welded joint
and b) fully load carrying spot welded joint.

Sperle, 1984) on the non-load carrying configuration were reported in the literature.

Defourny (1981, 1985) obtained load vs number of cycle curves of non-load carrying samples of 1 mm thick sheets. They compared these results to 100% load carrying configurations including the single lap shear joint. They concluded that the fatigue resistance for a non-load carrying spot corresponds roughly to the performance of the plain base material. They outlined the importance of the notch effect in the load carrying configurations.

Sperle (1984) performed similar tests on 1.2 mm thick non-load carrying samples. Fatigue crack initiation in his specimens occured in the base metal for single spot weld. In the case of samples with 3 spot welds in a row, the fatigue strength was found to be slightly lower than for plain specimens. He concluded that "spot welded attachments can, if necessary, be placed on strength members without being detrimental to the fatigue behaviour of the structure of which they form a part". In a recent investigation (Hamel, 1987) with 2 mm thick non-load carrying joints, it was shown that these joints were significantly reducing the fatigue limit. It was also concluded that these results might be an indication that the effect of non-load carrying attachments is increasing with the thickness.

EXPERIMENTAL PROCEDURES

The fatigue samples for the present investigation were fabricated from 301L/MT stainless steel. The mechanical properties and chemical composition of 301L/MT as provided by the alloy manufacturers are given in Table 1. The nugget diameters are given in Table 2. In this table like in the following of the text, the thickness is designated by two numbers where the first is the thickness of the short sheet and the second the base sheet. The width of the samples was 50 mm and the free length (between the grips) 195 mm.

Additional fatigue tests were also done with holed samples of the same material as a comparison. These samples had a 6.35 mm drilled hole in 3 mm thick, 47 mm wide plates. Most of the fatigue tests were performed on a 100 kN axial servo-hydraulic machine in load control mode. A 250 kN machine was used for some of the tests at higher loads. Many of the tests were performed under computer control. This eliminated the need of readjusting the control signal during the test providing a more constant load amplitude. All the tests were done at a load ratio R of 0. The tests are run until complete separation of the samples.

For every test, an extensometer was attached across the nugget on the stressed sheet side. The signal of the extensometer is used to monitor the compliance change during the test and to provide an early indication of crack initiation and to indicate when the crack reaches the opposite face of the sheet. The procedure is explained in details in a previous reference (Hamel, 1987) and is based on the presence of a minimum in the compliance vs number of cycles curve.

RESULTS AND DISCUSSION

The results of the fatigue tests are plotted on Fig. 2 to 5 for the non-load carrying joints and on Fig. 6 for the holed samples. On these curves the stress range (equal to S_{max} for R = 0) is plotted versus the number of cycles for initiation and fracture. It should be noticed that the criteria used to determine the "initiation" does not correspond to the nucleation of a small microscopic crack but to the moment when the thumbnail crack almost reaches the outside face of the sheet.

ISFM-R

TABLE 1 Mechanical Properties and Chemical Composition of 301L MT

a) Mechanical Properties

t (mm)	σ_y (.2% offset) (MPa)	σ_{UTS} (MPa)	elongation %
2	573	920	32.8
4.5	613	951	31.8

b) Chemical Composition

C	Si	Mn	P	S	Ni	Cr
.030	1.00	2.00	.040	.020	6.5-8.8	17.0-18.0

TABLE 2 Spot Diameters

t_s/t_b (mm)	spot diameters (mm)	
	required	measured
1.5/1.5	5.6	7.6
3/3	7.6	9.5
4.5/4.5	8.9	9.2
1.5/4.5	5.6	7.0

<u>1.5/1.5 mm Samples</u>

On Fig. 2, the endurance limit is not defined accurately and is indicated by a hatched area. Filled symbols at low stress correspond to samples from a different batch than those represented by open symbols. It was noticed that the indentation of the nugget of these samples had a different shape due to a different electrode (hemispherical rather than truncated cone). The overall thickness at the center of the spot was slightly larger (2.92 ±.01 vs 2.85 ±.04 mm) in spite of a deeper indentation (.108 ±.02 vs .064 ±.008 mm). The initial normalized compliance of these samples was also slightly smaller than for the other specimens (.975 ±.03 vs 1.02 ±.01 mm). This observation agree with previous data (Hamel, 1987) where higher compliance was correlated with higher endurance and where higher compliance was attributed to nuggets with large porosities, of smaller diameter or deeper indentation. It is then clear that the fact these specimens were from a different batch affected the data. On the other hand it is not completely clear which factor in the nugget geometry influenced the most the results. The additional tests with the new batch of samples were performed to confirm the determination of the fatigue limit based on a single specimen since it was known from other tests that a false fatigue limit might be caused by a specimen slightly overloaded during testing. Obviously, the new batch of samples didn't improve this aspect of the data.

One sample (2×10^6 cycles, 273 MPa on Fig. 2) exhibited a markedly higher endurance than the other samples. Post failure examination indicated that cutting fluid had penetrated between the two sheets before testing. The fatigue endurance improvement caused by oil has already been noticed by Overbeeke (1974).

Two samples (indicated by triangles) tested at high stress failed outside the gauge length due to cracks initiated on the edge of the sample. Large plastic deformation around the weld nugget of these samples reduce the effective stress intensity factor in this area and can promote fatigue initiation at the edge of the samples. These edges are sheared which leaves a rough finish and strain hardened material. In fact, many small cracks were observed along the edges. A third sample also failed outside the nugget area at a stress of about 220 MPa. This is much more amazing and no explanation can be found for it. However, post failure fractographic examination revealed fatigue crack initiation at the expected location on the nugget periphery.

3/3 and 4.5/4.5 mm Samples

On Fig. 3 and 4, the endurance limit is much better defined than for Fig. 2. All the failures occured in the nugget area and no cracks in the machined (milled) edges were observed. The initiation for the 3 mm samples at higher stress is not clearly defined because there was no minimum on the compliance vs number of cycles curve.

1.5/4.5 mm Samples

On Fig. 5, the endurance limit is indicated by a hatched area as in Fig. 2. The mean compliance of the unfailed samples within the limits of the hatched area is 1.04 ±.04 and 1.08 ±.05 for the failed samples. These data agree with previous discussion about the effect of compliance but the statistical significance is more questionable in this latter case.

Influence of thickness

On Fig. 7, the curves of Fig. 2 to 5 are superimposed along with other data taken from the literature. The three curves labeled "PLAIN" are for smooth specimens of 301L MT at R = -1 as provided by three manufacturers of the alloy. The curves labeled 2/2 from another reference (Hamel, 1987) correspond to non-load carrying spot welded joints of 2 mm thick SAE 950A HSLA steels. The curve labeled 1/1 from Defourny (1981, 1985) are for 1 mm thick non-load carrying samples of E380 HSLA steel tested at a load ratio R = .11.

Although there is a large scatter band for the endurance limit of the 1.5/1.5 and 1.5/4.5, the figure shows that the endurance limit and the high-cycle endurance decreases with the thickness. The 3/3 curve is within the scatter band of the 1.5/1.5 curve but its fatigue endurance was clearly inferior in the high-cycle region (10^5-10^6 cycles). The 4.5/4.5 specimens clearly had the lowest endurance. In every case, the fatigue limit is lower than the plain specimens[1]. This is the most important result of this investigation. The

---

[1] Some might argue that the plain specimens were tested at R = -1 rather than 0. This effect will be discussed later.

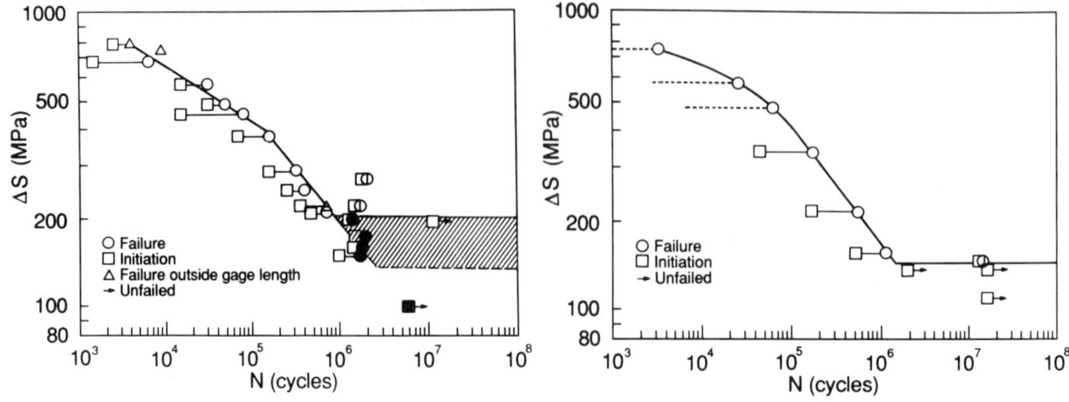

Fig. 2   SN curve for 1.5/1.5 specimens of 301L MT. Closed symbols designate samples from a different batch.

Fig. 3   SN curve for 3/3 specimens of 301L MT.

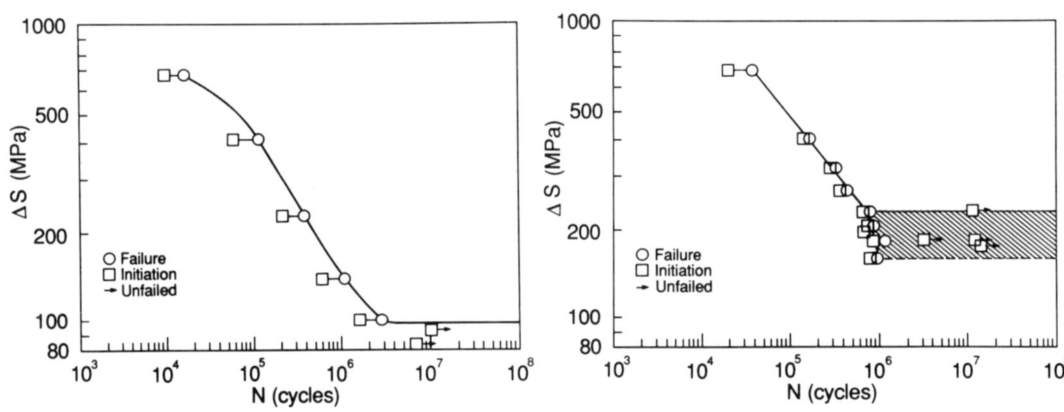

Fig. 4   SN curve for 4.5/4.5 specimens of 301L MT.

Fig. 5   SN curve for 1.5/4.5 specimens of 301L MT.

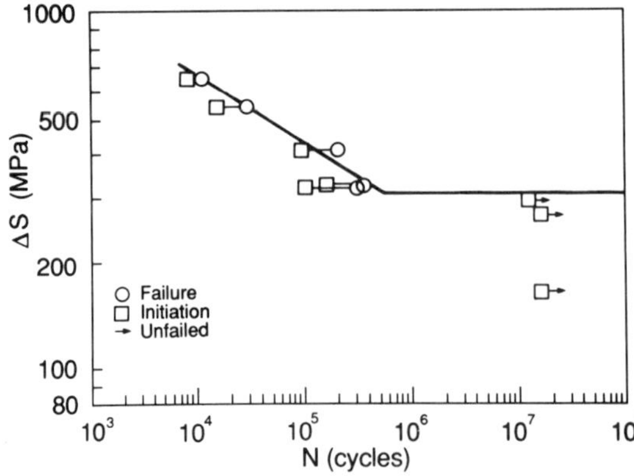

Fig. 6   SN curve for 301L MT samples with central drilled hole

presence of a non-load carrying attachement significantly reduces the fatigue endurance and the effect increases with the thickness of the attachement. In the case of 4.5/4.5 specimens, the fatigue limit is as low as about 10% of the static strength of 951 MPa.

The effect of thickness on the fatigue endurance of non-load carrying spot welded joints agrees with a more general observation (Gurney, 1979) that the fatigue endurance decreases with thickness in welded structures. This can be interpreted with the fracture mechanics approach. For a given stress, a crack in a piece of the same proportions but of larger dimensions will have a higher stress intensity factor. The geometry of the two sheets and the nugget of a spot welded joint form a crack-like defect. For a given geometry and stress, the stress intensity factor increases with the thickness.

Different thickness. The behaviour of the 1.5/4.5 specimens is very similar to the 1.5/1.5 and different from the 4.5/4.5. This clearly indicates that the stress (not the load) in the base plate and the thickness of the attachment are the controlling factors. The thickness of the base plate has a negligible influence. However, the endurance limit scatter band is slightly higher for the 1.5/4.5 specimens than for the 1.5/1.5 one. This might be attributed to a small influence of the relative thickness of the attachment and the base

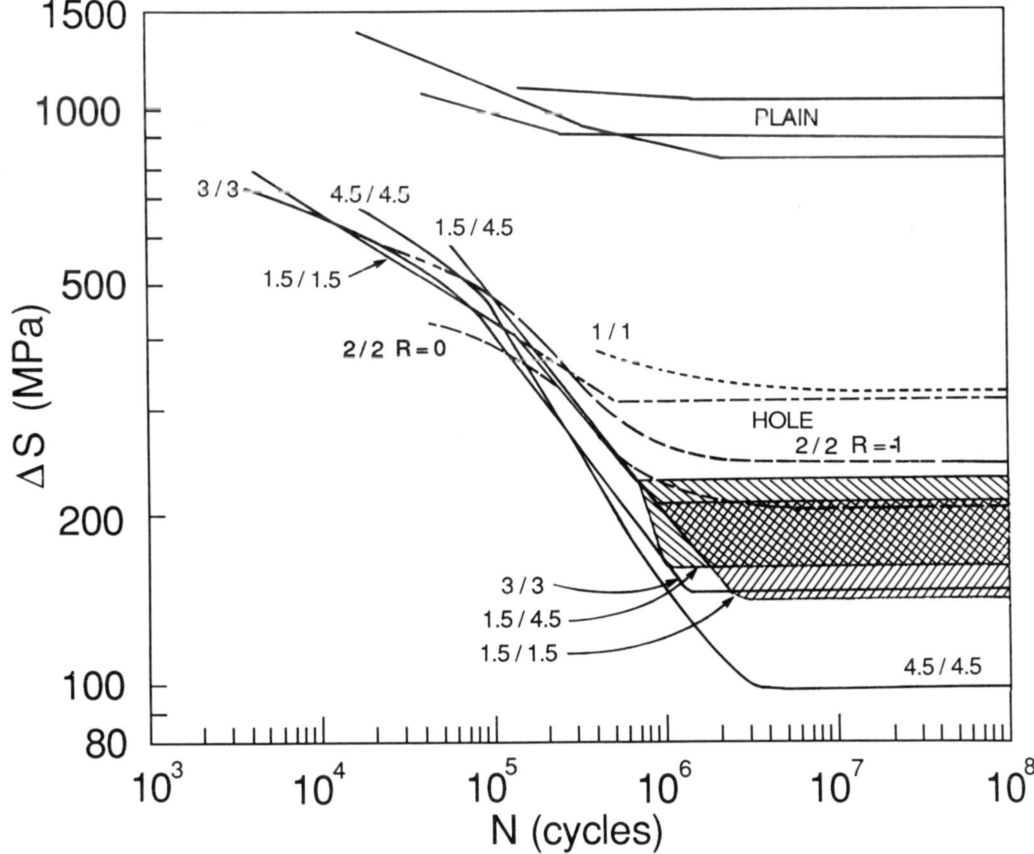

Fig. 7 Summary of all ΔS vs N curves for 301L MT (plain, holed and spot welded) and other data for non-load carrying spot welded joints.

plate. It could also be attributed to the fact that the measured spot diameter of the 1.5/4.5 specimens (7.0 mm) was slightly smaller than for the 1.5/1.5 specimens (7.6 mm). When comparing Fig. 2 and 5, it appears that the propagation phase is longer for the 1.5/1.5 specimens than for the 1.5/4.5 specimens. It has to be remembered that the method used to determine the "initiation" includes the phase to propagate the thumbnail crack to the opposite face of the sheet. That phase is of course longer for thicker plate. However, at that moment, the crack is larger for the thicker plate, has a higher stress intensity factor and will propagate faster to the final failure.

Spot diameter. In this study, the spot diameter was varied with the thickness to be the same as in actual structures based on the ·actually accepted standards. No test was done to investigate separately the effect of the plate thickness and of the nugget diameter. It is believed that the endurance depends on the joint stiffness which is a combined function on these two factors (thickness and nugget diameter).

Influence of Materials

On Fig. 7, curves from the literature (Defourny, 1981, 1985) for other non-load carrying joints from different material (HSLA steel) are also indicated. The endurance limit of the 1/1 specimens is the highest of all the data. For the 2/2, R = 0 curve, it is within the scatter band of the 1.5/1.5 and 1.5/4.5 results. These data agree with our previous discussion about the thickness effect and suggest the alloy has a less important effect than the thickness on the endurance limit. The tests of Sperle (1984) on 1.2 mm thick sheets showed a negligeable influence on the non-load carrying attachments. This agrees with the conclusion that the influence decreases with the thinner sheets. The reason why the influence was smaller for the 1.2 thick sheets of Sperle (1984) than for the 1 mm thick sheets of Defourny (1981, 1985) is not known. Some secondary factors not considered in this study (like the alloy or the specimen geometry) might be the explanation.

Influence of Load Ratio, R

The two dashed curves for the 2/2 samples indicate that the influence of R on the endurance limit is relatively small. This can be attributed to the fact that most of the life is spent in the short crack growth ·regime where the influence of R is small (Verreman,1986). For the same reason, it seems justified to compare the endurance limit of plain specimens at R = -1 and welded joints at R = 0 in terms of stress range (ΔS) rather than maximum stress even if a small influence of R is known to exist.

The situation is different at high stress where all the curves tend to their respective static strength. In this low-cycle regime, it is better to plot the curves in terms of S_{max} as shown on Fig. 8. Of course, the curves at R = 0, occupy the same positions as on Fig. 7 and only the curves for R ≠ 0 are moved. All the curves tend to their respective static strength values of 920 MPa for 301L, 480 MPa for the 2/2 HSLA specimens and 500 MPa for the 1/1 HSLA specimens.

All the curves for 301L MT all well grouped for less than 10^5 cycles. The reason why the curves for the 4.5 mm thick sheets are higher is not known exactly, and could simply be attributed to data scatter. As shown in Table 1, the yield and tensile strenghts are slightly higher than for the thinner sheets but the small difference is not believed to be large enough to explain the difference in fatigue life.

It is also significant that the order of the curves in the high-cycle regime is not changed. The conclusions stated earlier about the effect of non-load carrying spot welded joints are not affected and cannot be attributed to the fact that the data available for plain specimens were at R = -1. Moreover, as mentioned earlier, the fact that the two curves for the 2/2 samples are much closer on the ΔS than on the S_{max} (Fig. 8) curve justifies the comparison on the ΔS curve in high-cycle fatigue. Fatigue data were not obtained at $N < 10^4$ cycles for the 2/2 specimens but it is known that these curves will tend to the static strength of 480 MPa (960 for R = -1 on ΔS scale) at N = 1. Consequently the two curves will move off on the ΔS curve and will come closer on the S_{max} curve.

It can be observed that the curve for the samples with a hole, crosses the curves for non-load carrying spot welded samples at intermediate lives (10^4-10^5 cycles). This behaviour can be qualitatively explained by the model developed by Dowling (1979) and Socie (1984). The samples with a hole behave like a blunt notch. For such a notch, crack growth consumes the majority of the fatigue life at intermediate stress and crack nucleation dominates the endurance limit behaviour. On the other hand, the spot welded samples behave like sharply notched members. In this case short crack propagation consumes the majority of the fatigue life at all stress levels. The endurance limit is controlled by the behaviour of short non-propagating cracks.

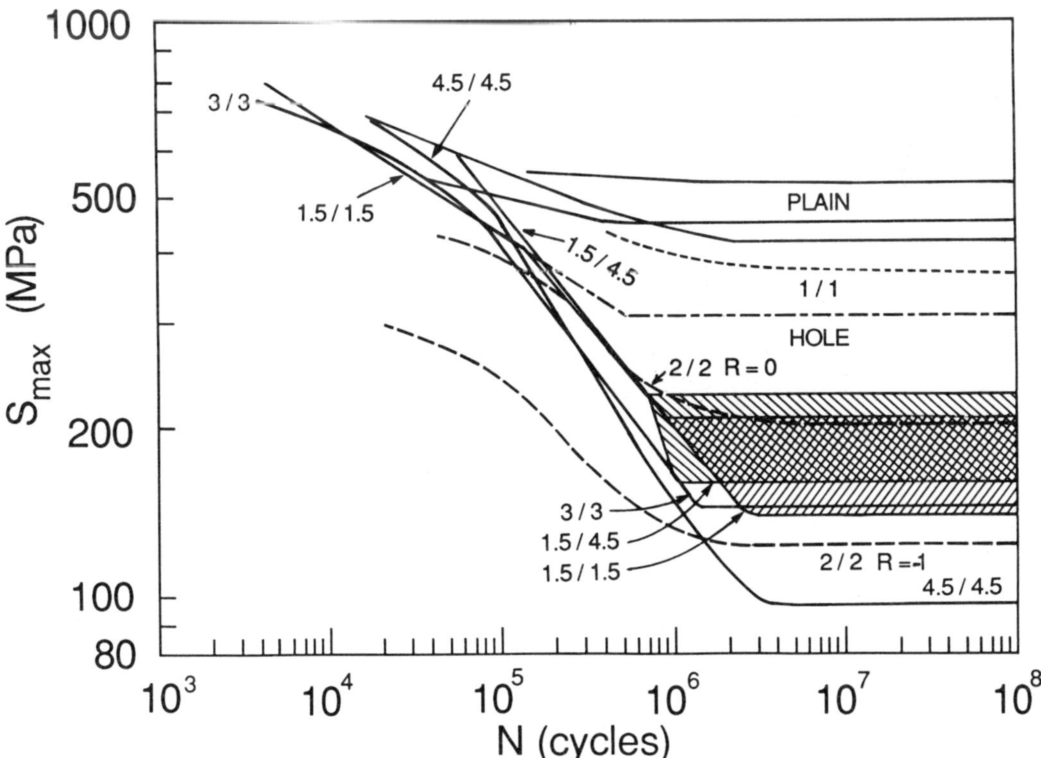

Fig. 8 Summary of all curves of Fig. 7 plotted on a S_{max} vs N graph

CONCLUSION

The following conclusions can be stated about the behaviour of non-load carrying spot welded joints.

1) The presence of a non-load carrying spot welded joint significantly reduces the fatigue limit of the base sheet to which it is attached

2) This life reducing effect significantly increases with the thickness of the attachment

3) The controlling factors are the stress in the base sheet and the rigidity of the attachment (determined by thickness of the attachment, size and quality of the spot). The thickness of the base sheet seems to have a negligible effect

4) The steel composition and mechanical properties seems to have a negligible effect at low stress

5) The fatigue endurance reducing effect of non-load carrying spot welded joints is eliminated at high stresses. The mechanical properties of the steel are important at high stresses

6) The fatigue endurance is better described by the ΔS parameter at low stress and by the S_{max} parameter at high stress.

7) The fatigue endurance is improved by the presence of oil between the sheets

From a practical point of view, the first two conclusions are obviously the most significant. Non-load or low-load carrying spot welded joints should be carefully considered in fatigue design especially for thick sheets. Not only the load carried by a spot but also the stress existing in the sheets should be considered. Additional research is however needed to quantify the possible synergetic effect of the stress level of the base sheet and the load transmitted by the spot.

ACKNOWLEDGEMENT

The authors would like to thank R. Lavallée who carried on the tests. We also thank Bombardier Inc. who manufactured the spot welded samples. Helpful discussions with J. Janssen of Bombardier Inc. were also greatly appreciated.

REFERENCES

Cooper, J.F. and Smith, R.A. (1985). The measurement of fatigue cracks at spot welds, Int. J. Fatigue, 7, 3, 137-140.
Cooper, J.F. and Smith, R.A. (1886). Fatigue crack propagation at spot welds. Metal Construction, June 1986, 383R-386R.
Davidson, J.A. (1983a). A review of the fatigue properties of spot-welded sheet steels. 1983 SAE Int. Congress and Exposition, Detroit, Mich., Technical Paper 830033.
Davidson, J.A. and Imhof, J.E., Jr. (1983b). A fracture-mechanics and system-stiffness approach to fatigue performance of spot-welded sheet steels, SAE Int. Congress and Expositions, Detroit, Mich., Technical Paper 830034.

Davidson, J.A. (1983c). Design related methodology to determine the fatigue life and related failure mode of spot welded sheet steels. 1983 Int. Conf. on Technology and Applications of HSLA Steels, Philadelphia, Penn., ASM Paper 8306-022.

Davidson, J.A. and Imhof, J.E., Jr. (1984). The effect of tensile strength on the fatigue life of spot-welded sheet steels. 1984 SAE Int. Congress and Exposition, Detroit, Mich., Technical Paper 840110.

Defourny, J., d'Haeyer, R. and Bragard, A. (1981). Experience of CRM in the resistance spot welding of high strength steel sheet. Proceedings of Welding of High Strength Steel Symposium, Coventry.

Defourny, J. and Bragard, A. (1985). Welding in the World, 23, 5/6, 100-122.

Dowling, N.E. (1979). Fracture Mechanics, ASTM STP 677, Smith, C.W., ed. ASTM, 247-273.

Gurney, T.R. (1979). Fatigue of Welded Structures, 2nd ed., Cambridge University Press, 66 and 123.

Hamel, F.G. and Masounave, J. (1987). The fatigue behaviour of non-load carrying spot welded joints. Submitted to Int. J. Fatigue.

Iwasaki, T. Tanaka J., Kabasawa, M. and Nagae, M. (1983). Welding Research Abroad, XXIX, 3, 38-47.

Orts, D.H. (1981) Fatigue strength of spot welded joints in HSLA steel", 1981 SAE Int. Congress and Exposition, Detroit, Mich., Technical Paper 810355.

Overbeeke, J.L. and Draisma, J. (1974). Metal Construction, 6, 213-219.

Pollard, B. (1974). Welding Research Supplement, 53, 8, 343s-350s.

Pook, L.P. (1975). Int. J. Fract., 11, 173-176.

Rivett, R.M. (1983). Assessment of resistance spot welds in low carbon and high strength steel sheet, 1983 SAE Int. Congress and Exposition, Detroit, Mich., Technical Paper 830126.

Socie, D.F., Dowling, N.E. and Kurath, P. (1984). Fracture Mechanics: Fifteenth Symposium, ASTM STP 833, Sanford, R.J. ed., ASTM, 284-299.

Sperle, J.O. (1984). Metal Construction, 16, 11, 678-679.

Verreman Y., Baïlon J.-P. and Masounave, J. (1986). Fatigue short crack propagation and plasticity induced crack closure at the toe of a fillet welded joint", The Behaviour of Short Fatigue Cracks, EGF Pub. 1, Miller, K.J. and De LosRios, E.R., Mechanical Engineering Publications, London, 387-404.

Wang, P.C., Corten, H.T. and Lawrence, F.V. (1985). A fatigue life prediction method for tensile-shear spot welds. 1985 SAE Int. Congress and Exposition, Detroit, Mich., Technical Paper 850370.

EXPERIMENTAL STUDIES ON FATIGUE CRACK PROPAGATION
IN TUBULAR T-JOINTS

K.Munaswamy[*],P.G.Williams[*],A.S.J.Swamidas[*],M.Arockiasamy[**] and O. Vosikovsky[***]
* Faculty of Engineering and Applied Science
Memorial University of Newfoundland, St. John's, Canada
** Dept. of Ocean Engg., Florida Atlantic Univ., Boca Raton, Florida,U.S.A.
*** Physical Metallurgy Research Laboratories, CANMET, Ottawa, Canada

ABSTRACT

The paper presents experimental results of the fatigue tests on unstiffened and stiffened tubular joints. The experimental study gives results of static and fatigue tests on one unstiffened and two stiffened tubular T-joints under axial loading mode. Alternating current field measurement (ACFM) technique was employed in the monitoring of crack growth. Stress distributions and crack profiles around the brace chord intersection are presented. Also crack growth curves for the hot spot locations (i.e., saddle point) of all the three joints are given. In addition the test facility, instrumentation and test procedures are described.

KEY WORDS

Fatigue; Tubular T-Joint; Stiffened and Unstiffened; Axial Loading; ACPD; Crack Growth

INTRODUCTION

In offshore structures, which are generally of tubular construction, the intersection of the chord and the brace induces high local stresses adjacent to the weld; consequently fatigue damage will generally occur in this region. Therefore the study of fatigue crack growth around the weld toe region of the tubular joint is extremely important. Since experimental and analytical investigation of this problem using large prototype specimens are both costly and time consuming, much of the reported work has been carried out with idealized welded plate specimens and small scale tubular joints. Most of the earlier research programs were concerned with the determination of the influence of weld profile, plate thickness, corrosive environment and cathodic protection on the fatigue lives of welded T-plates, cruciform and small/medium scale tubular joints. These investigations were mostly sponsored and supported by ECSC and UKOSRP. Details of these research programs can be seen in the conference proceedings edited by Noordhoek and de Back (1987).

Canadian offshore steel research is of fairly recent origin. Fatigue studies on offshore steels in Canada was initiated and sponsored by the Department of Energy Mines and Resources, Canada (Thomson and Tyson, 1987). Under this program fatigue strength of T-plate and pipe-plate specimens have been determined considering the effect of corrosion and plate thickness. Also fracture mechanics approach is being explored in determining the fatigue lives of these joints (Vosikovsky and others, 1984,1985,1987; Lambert and others, 1987; Bell and others, 1987; Burns and others, 1987). Finite element technique has been used to determine stress concentration factors and through-thickness stress distribution (Bhuyan and others, 1985, 1986; Munaswamy and others, 1986). As part of the overall Canadian program the investigation of fatigue behaviour of stiffened and unstiffened tubular T-joints under different environmental conditions and loading modes was undertaken by the Physical Metallurgy Research Laboratories, CANMET, Dept. of Energy Mines and Resources, through a contract to AMCA International. Under this program twelve tubular T-joints are to be tested under constant amplitude loading. Six test specimens are to be tested at the University of Waterloo under in-plane bending (IPB) and out-of-plane bending (OPB) loading modes. The other six specimens are to be tested at the Memorial University of Newfoundland under axial loading mode, both in air and in seawater with cathodic protection. The first specimen in each of the series is an unstiffened one. The specimens consist of a 457 mm diameter brace welded to a 914 mm dia chord; both have the same wall thickness of 19 mm. The two ring stiffeners are placed symmetrically on either side of the axis of the brace, 250 mm apart. The thickness and the depth of stiffener are 19 mm and 100 mm, respectively.

This paper presents the test facility and the corresponding instrumentation utilized at Memorial University and some of the salient results of unstiffened joint TA1 (tested at a maximum hot spot stress range of 250 MPa) and the two subsequent stiffened joints TA2 and TA3 (tested at maximum hot spot stress ranges of 250 MPa and 160 MPa, respectively) under constant amplitude axial loading.

TEST FACILITY

The self-straining test rig was designed by Prof. U.H.Mohaupt of the University of Waterloo. The rig was designed to perform axial load tests, with the axis of both the chord and brace lying horizontally. To avoid some of the alignment and restraint problems, special split end rings were designed to hold the flanged ends of the chord of tubular specimen. In order to prevent the development of fixity at the loaded end, load was applied through a double-swivel joint; and to allow easy rotation and thus prevent the development of end restraints, both ends of the chord were provided with cam followers. The double-swivel and the cam followers were also fabricated by the University of Waterloo. The test frame and the end plates were fabricated by EASTEEL Industries Ltd., St.John's. The remaining parts were made and assembled by the Technical services, Memorial University of Newfoundland. The assembled test rig with the specimen in place is shown in Fig. 1.

INSTRUMENTATION FOR TESTING AND DATA ACQUISITION

The fatigue loading system, manufactured by MTS, was used to apply the axial load on the brace. The strain gauge and ACFM (alternating current field measurement) data were acquired using the MINC-11 PDP system. Two Keithley 100 channel low voltage scanner units were used to switch the data lines from strain gauges and ACPD probes to the computer.

STRESS ANALYSIS OF STATIC TESTS

In order to study the stress distribution at critical regions of the joint during both static and fatigue tests, the specimens were extensively instrumented with strain gauges, both on the inside and outside surfaces of the chord and on the outside surface of the brace. The schematic diagram of the test specimen showing the nomenclature for locating the strain gauges is shown in Fig. 2. One quarter ($0°$ to $90°$) of the chord surface was mounted with three rows of triangular rosette strain gauges. The remaining three quarters of the chord surface was also sparsely instrumented with strain gauges to check the symmetry of loading on the test specimen. Strain gauges were also mounted on the inside surface of the chord, exactly below those on the outside surface. The first two rows of gauges along the weld toe were positioned according to the guidelines given by Pozzolini (1981), so that on linear extrapolation the nominal hot spot stresses at the weld toe could be obtained. The schematic arrangement of strain gauges on the unstiffened joint is shown in Fig. 3(a).

The specimen was loaded incrementally up to 80% of the desired maximum load, taking strain gauge readings to observe the linearity and then decreased to zero load to check and correct for any residual strains. This procedure was repeated 3 to 4 times to shakedown the plastic strains caused by the superposition of the applied strains and the initial residual strains. Using the last of the preloading cycles, the maximum load to obtain the specified hot spot stress range was decided. The initial radial stress variation along the radial line at saddle point are shown for the three specimens TA1 to TA3 in Figs. 4, 5 and 6. The variation of the radial membrane and bending stresses are also shown in these figures. It is observed that the radial bending stresses become negative almost at the same point for both the stiffened joints; in addition the tensile membrane stresses increase in a linear manner according to the applied stress range at the weld toe.

The radial stress distribution along the weld toe obtained by extrapolation and the stresses at the various gauge locations, normalized with respect to the maximum hot spot stress are shown in Figs. 7, 8 and 9. It is seen that the normalized stress distribution is almost the same for the two stiffened joints . During fatigue tests strain gauge readings were monitored for some of the critical locations. The radial strain variation in the unstiffened joint for various increasing number of load cycles is given in Fig. 10. As the test progresses, the radial strain decreases and even becomes compressive when the crack extends through the thickness.

Earlier initiation of crack, on the side where strain gauge 31 was located was indicated by the decreased strain in that gauge; in addition the crack initiation was also detected through the crack microgauge. Detailed analysis of the test results are given in the progress report (Munaswamy and others, 1987).

CRACK SHAPE MONITORING DURING FATIGUE TESTS

The initiation and subsequent growth of fatigue cracks from the toe of the weld through the thickness of chord material was carefully monitored using alternating current field measurement (ACFM) technique (otherwise known as alternating current potential difference - ACPD method) and the crack microgauge. To obtain the crack depth measurements active and reference probe pairs are required at each location. The active pair spans the weld toe and the reference pair is located on the same radial line perpendicular to the weld toe. The radial distances between the active pair and reference pair were kept at 10 mm. Such probe pairs were distributed along the weld toe in the transverse direction at approximately 6 mm increments near the hot spot location. Altogether 60 such active and reference probe pairs were provided on either side of the brace for unstiffened joint and 50 for stiffened joint. The schematic diagram for the positioning of these probes are given in Fig. 3(b).

For determining the crack depth from the monitored crack microgauge readings, calibration tests were conducted on six T-plate specimens. The measured crack depths (from beach marks on T-plate) and crack depths obtained using ACFM technique were curve fitted using a second order polynomial regression analysis. These regression coefficients were used in the subsequent analyses. A more detailed discussion of this ACFM technique and the calibration procedures used in the study are given by Munaswamy and others (1987).

DISCUSSION

During the fatigue tests, ink stain and beach marks were introduced at three to five crack depths. Figures 11, 12 and 13 show the crack profiles indicated by beachmarks and ink stains and the predicted crack profiles by ACPD method. As can be seen, the ACPT technique accurately predicts the crack depth in the central parts, near the maximum depth. Near the crack-surface intersection the ACPD overpredicts the depth.

Analysis of test results shows that variation of radial stresses along the radial line at the hot spot location for all the three joints is similar, but the stress distribution around the weld toe between stiffened and unstiffened joints are distinctly different, viz., a more flattened distribution at hot spot location in the case of stiffened joints. Figures 14 to 16 give a detailed plot of the stresses and crack development for the three specimens. Five to seven major cracks initiated at the weld toe over a length exceeding the width of the stress peak. The cracks coalesced into a single crack when the maximum depth exceeded ~2 mm. In stiffened joints (Fig. 15,16) the surface length of the single crack, before it penetrated the wall thickness was limited by the spacing of the stiffeners.

The crack growth through the wall thickness at the hot spot location as a function of number of cycles, is shown in Fig. 17, the variation in fatigue crack growth rates is presented in Fig. 18. As can be seen, the stiffeners have little effect on fatigue crack development. The fatigue crack growth and crack growth rate curves for TA1 and TA2 joints tested at the same stress range are close to each other. The stiffened joint TA3, tested at lower stress range exhibited longer life and lower crack growth rates.

The fatigue crack growth rate curves in Fig. 18 show a distinct pattern. The first phase associated with the coalescence of initial single cracks (up to ~2 mm depth) is characterized by a sharp increase in growth rate. Thus initiation and coalescence of single cracks accounts for 20 to 40% of fatigue life for crack breakthrough (wall penetration). Phase II, characterized by an almost constant growth rate, extends for up to ~70% of the wall thickness, and consumes about 30% of the life. During the remainder of the life, before wall penetration, the growth rate in most cases gradually decreases.

The final results from stress distribution and fatigue life measurements are summarized in Table 1. Comparison of stress concentration factors (SCF = hot spot stress/nominal stress in the brace) shows that the main beneficial effect of stiffeners is reflected in reduction of SCF by a factor of ~3.5. Thus the load-bearing capacity of the joint at the same hot spot stress is increased by the same factor (compare loads for TA1 and TA2 joints).

Fatigue lives for crack initiation (defined by crack depth equal to 0.5 mm) crack breakthrough (wall penetration) and the end of the experiment (defined by a loss of load-bearing capacity of the joint) are significantly shorter for the stiffened joint TA2 than the unstiffened TA1. However, the differences between the lives are within the limits of expected scatter from tests of large welded joints. Slightly higher hot spot stress range of TA2 joint also obviously contributed to life reduction. Thus the results indicate that the hot spot stress range, successfully used as stress parameter in fatigue life prediction of unstiffened joints, will be equally applicable for stiffened joints, and standard S-N curves can be used for life calculations.

CONCLUSIONS

The following tentative conclusions on effects of internal stiffeners on fatigue behaviour of axially loaded tubular joints can be drawn from the partially completed test program:

The stiffeners reduce the stress concentration factor by a factor of ~3.5 and thus proportionally increase the fatigue-load-bearing capacity of the joint.

The hot spot stress range appears to be a valid stress parameter for stiffened joint fatigue life prediction using standard S-N curves.

The ACPD technique combined with beachmarking can accurately monitor fatigue crack development.

Initiation and early growth of fatigue cracks, almost up to wall penetration, are little effected by the stiffeners.

ACKNOWLEDGEMENTS

The authors would like to thank Dean G.R. Peters and Associate Dean Dr. T.R. Chari, of the Faculty of Engineering and Applied Science and Dr. I. Rusted, Vice President, for their continued interest and active support. The fourth author would also like to acknowledge the encouragements of Prof. S. Dunn, Chairman, Ocean Engineering Dept. and Dr. T. Shoup, Dean of Engineering, Florida Atlantic University, Boca Raton. The support of this investigation by the AMCA-DSS Grant Nos. 235Q.23440-3-9276 and BN 35 is gratefully acknowledged. Special thanks are due to Mr. M.J. Pates of Materials Technology Centre, a division of Arctic Canada Limited for his constant encouragement and careful project management.

REFERENCES

Bell, R., Vosikovsky, O., Burns, D.J. and Mohaupt, U.H. (1987). A fracture mechanics model for life prediction of welded plate joints. Proceedings of the 3rd International ECSC Offshore Conference on Steel in Marine Structures, Delft, The Netherlands, 901-910.

Bhuyan, G.S., Arockiasamy, M. and Munaswamy, K. (1985). Analysis of tubular joint with weld toe crack by finite element methods. J. Comm. in Applied Numerical Methods, 1, 325-331.

Bhuyan, G.S., Arockiasamy, M., Munaswamy, K. and Vosikovsky, O. (1986). Finite element analysis of cracked and uncracked tubular T-joint. Canadian J. of Civil Engineering, 13, No3,261-269.

Burns, D.J., Lambert, S.B. and Mohaupt, U.H. (1987). Crack growth behaviour and fracture mechanics approach. Proceedings of the 3rd International ECSC Offshore Conference on Steel in Marine Structures, Delft, The Netherlands, 137-160.

Lambert, S.B, Mohaupt, U.H., Burns, D.J. and Vosikovsky, O. (1987). Simulation of fatigue behaviour of tubular joints using a pipe-to-plate specimen. Proceedings of the 3rd International ECSC Offshore Conference on Steel in Marine Structures, Delft, The Netherlands, 489-500.

Munaswamy, K., Williams, P.G. and Swamidas, A.S.J. (1987). Progress report on fatigue test of tubular joint 1. Submitted to CANMET.

Munaswamy, K.,Bhuyan, G.S., Swamidas, A.S.J. and Arockiasamy. (1986). Experimental and analytical studies on fatigue of stiffened and unstiffened tubular T-joints., Proceedings of Offshore Technology Conference, OTC 5308, 153-161.

Noordhoek, C. and de Back, J. (1987). Ed. Proceedings of the 3rd International ECSC Offshore Conference on Steel in Marine Structures (SIMS '87), Delft, The Netherlands, Elsevier Science Publishers.

Pozzolini, P.F. (1981). Tests on tubular joints. Plenary sessions, Steel in Marine Structures, Paris, France.

Thomson, R. and Tyson, W. (1987). An overview of the national Canadian offshore steels research programme. Proceedings of the 3rd International ECSC Offshore Conference on Steel in Marine Structures, Delft, The Netherlands, 29-38.

Vosikovsky, O. and Rivard, A. (1984). Fatigue crack propagation in welded plate T-joints. Interim Report, ERP/PMRL, CANMET, Ottawa.

Vosikovsky, O., Bell, R., Burns, D.J. and Mohaupt, U.H. (1985). Fracture mechanics assessment of fatigue life of welded plate T-joints, including thickness effect. Proceedings of 4th international Conference on Behaviour of Offshore Structures, Delft, The Netherlands, 453-464.

Vosikovsky, O., Bell, R., Burns, D.J. and Mohaupt, U.H. (1987). Effects of cathodic protection and thickness on corrosion fatigue life of welded plate T-joints. Proceedings of 3rd International ECSC Offshore Conference on Steel in Marine Structures, Delft, The Netherlands, 787-798.

TABLE 1

a Parameter	Unstiffened Joint TA1	Stiffened Joint TA2	Stiffened Joint TA3
Max. Hot spot stress range	248 MPa	251.725 MPa	161.93 MPa
SCF	24.15	6.3847	7.0735
Load Ratio	0.16	0.05	0.05
Max. Load	315 kN	1085 kN	630 kN
Crack initiation (10.5 mm)	90-100 k cycles	100-120 k cycles	360-380 k cycles
Crack Breakthrough	665 k cycles	411.65 k cycles	1348.87 k cycles
End of Experiment	890 k cycles	520.465 k cycles	1766.87 k cycles

FIG 1. TEST FRAME

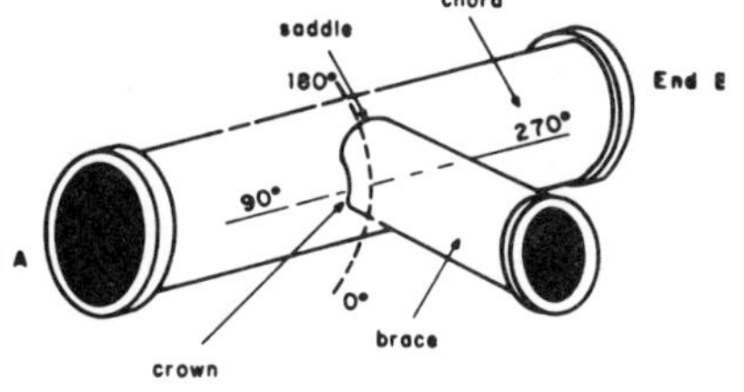

Fig. 2 Dimensions of tubular T-joints used in the current research program

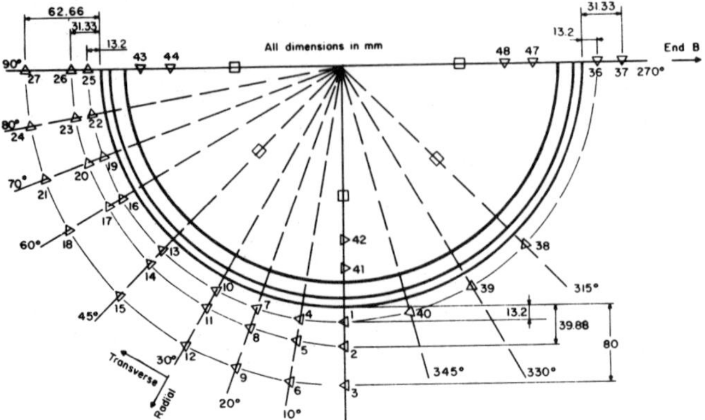

Fig. 3(a) Plan view of chord surface, strain gauge locations and orientations

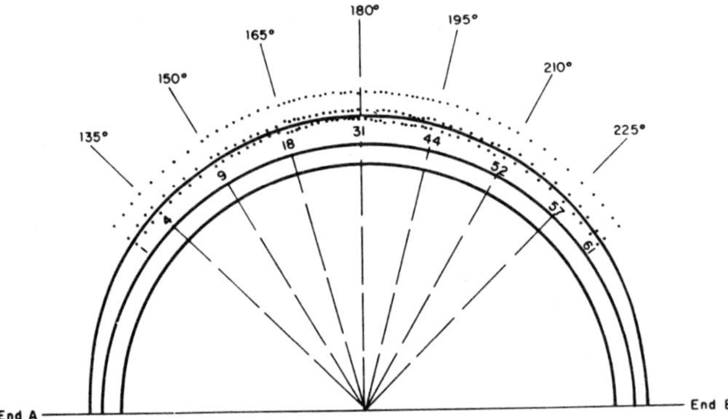

Fig. 3 (b) Plan view of chord surface – ACPD Probe locations

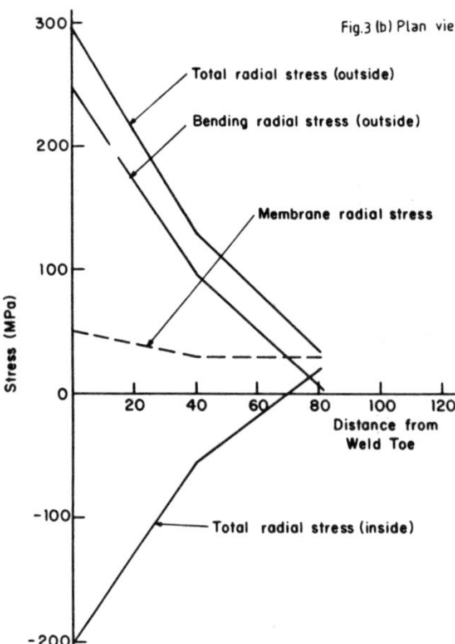

Fig. 4   Radial surface stress variation along the outside and inside radial line at saddle point – maximum load of 315.0 kN

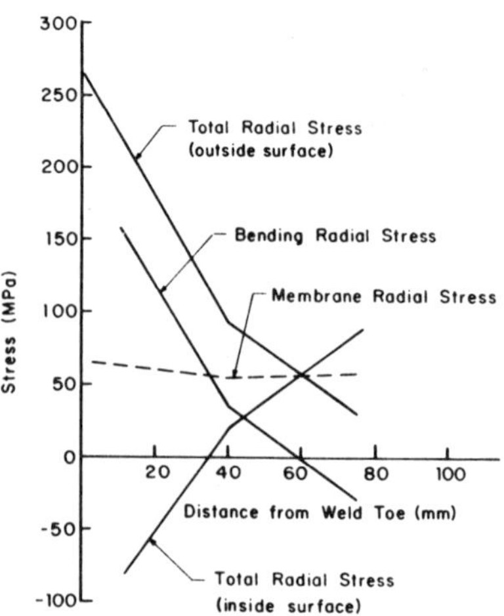

Fig. 5   Radial surface stress variation along the outside and inside radial line at saddle point – max. load 1085 kN (stiffened joint – TA2)

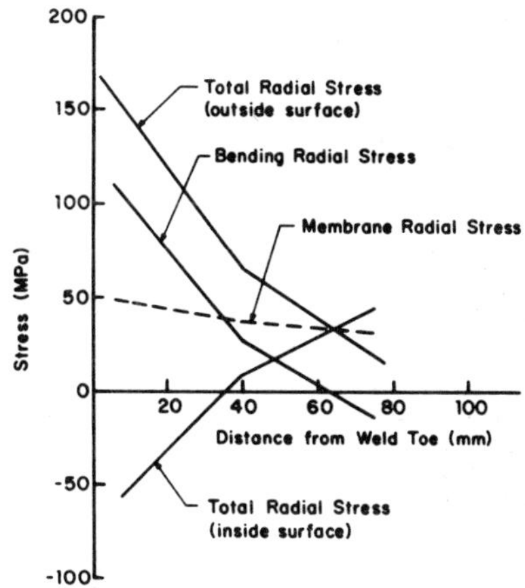

Fig. 6 Radial surface stress variation along the outside and inside radial line at saddle point – max. load 630 kN (stiffened joint – TA3)

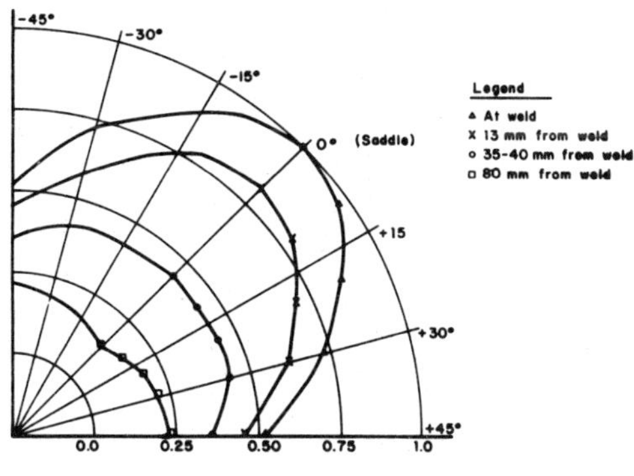

Fig. 7 Polar plot of radial stresses along a quadrant of ring on the outside of the chord – specimen TA1 subjected to axial tensile load.

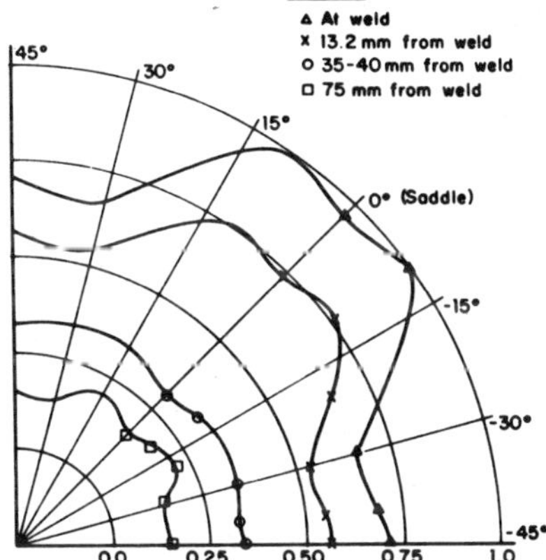

Fig. 8 Polar plot of radial stresses along a quadrant of ring on the outside of the chord – specimen TA2 subjected to axial tensile load

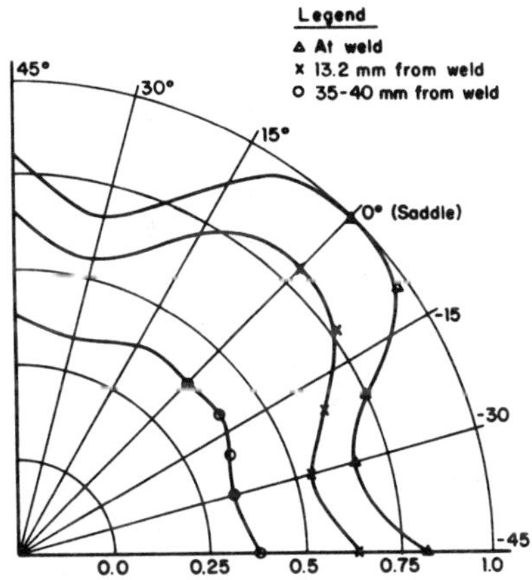

Fig. 9 Polar plot of radial stress along a quadrant of ring on the outside of the chord – specimen TA3 subjected to axial tensile load

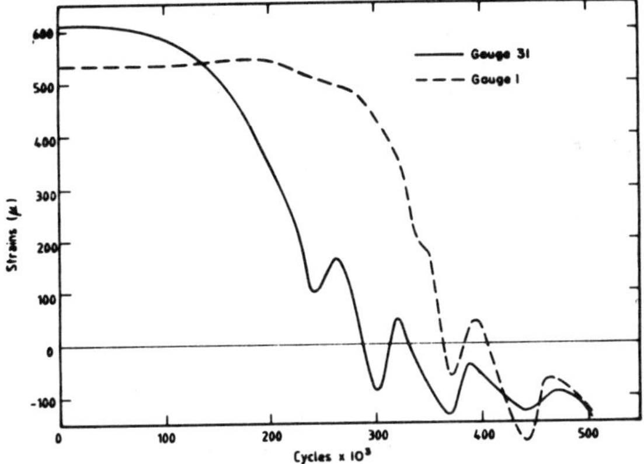

Fig. 10 Radial strain on the chord outside surface at mean load of 162.0 kN vs. no. of cycles

FRACTURE MECHANICS

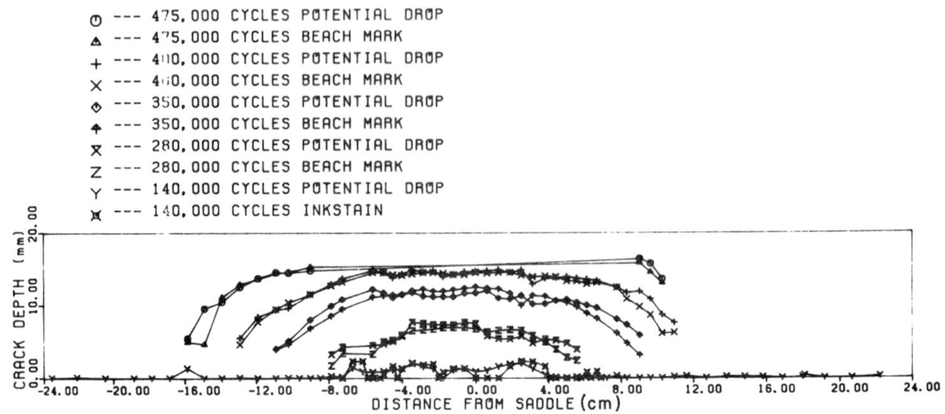

FIG 11. CRACK DEPTH Vs. POSITION – POTENTIAL DROP AND BEACH MARK COMPARISION

(UNSTIFFENED JOINT – TA1)

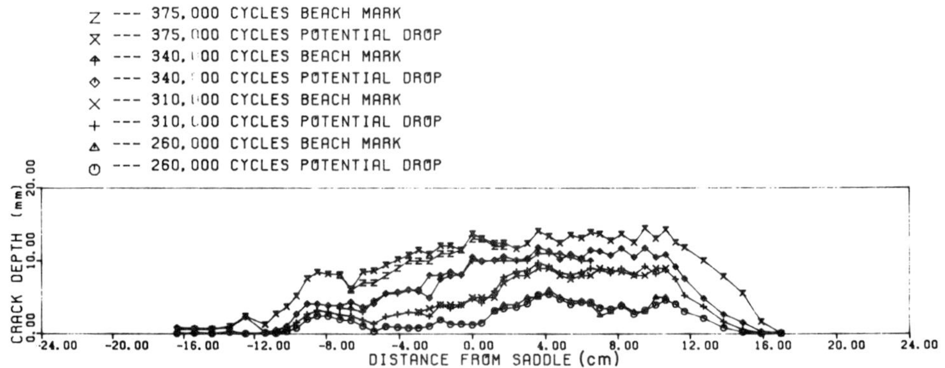

FIG 12. CRACK DEPTH Vs.POSITION – POTENTIAL DROP AND BEACH MARK COMPARISION

(STIFFENED JOINT – TA2 side 2)

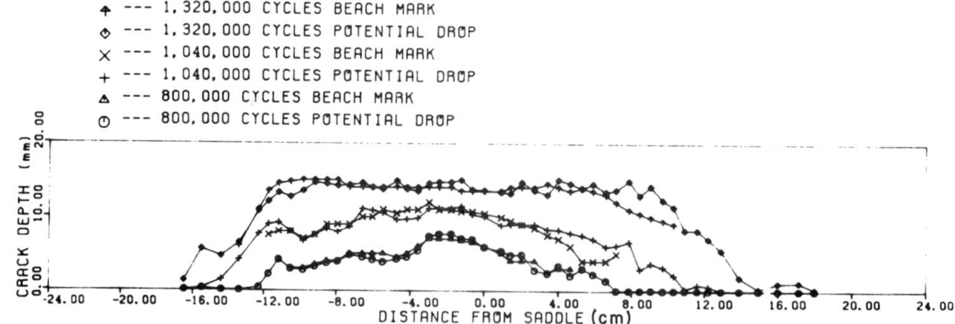

FIG 13. CRACK DEPTH VS. POSITION – POTENTIAL DROP AND BEACH MARK COMPARISION

(STIFFENED JOINT – TA3)

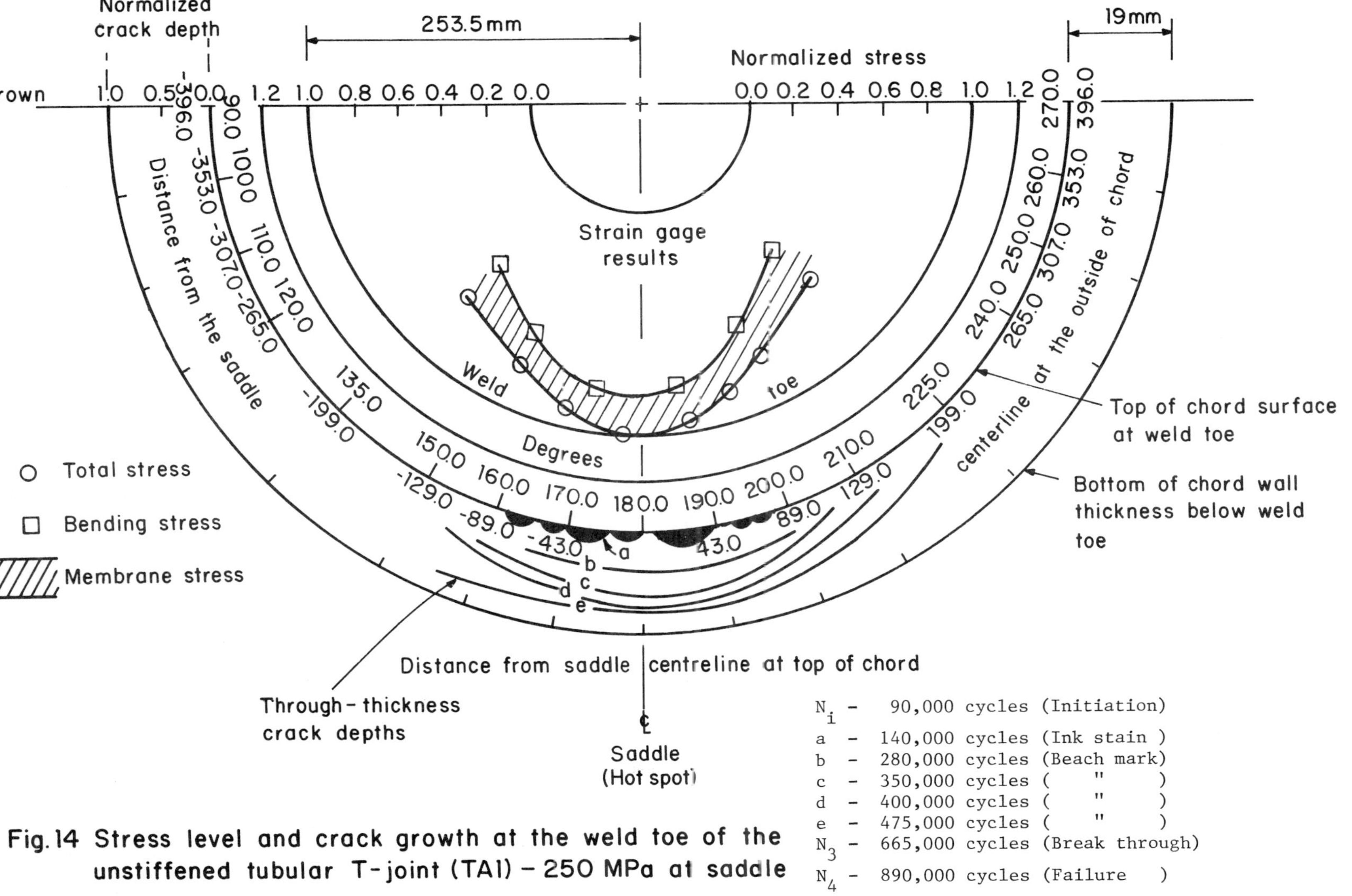

Fig.14 Stress level and crack growth at the weld toe of the unstiffened tubular T-joint (TA1) – 250 MPa at saddle

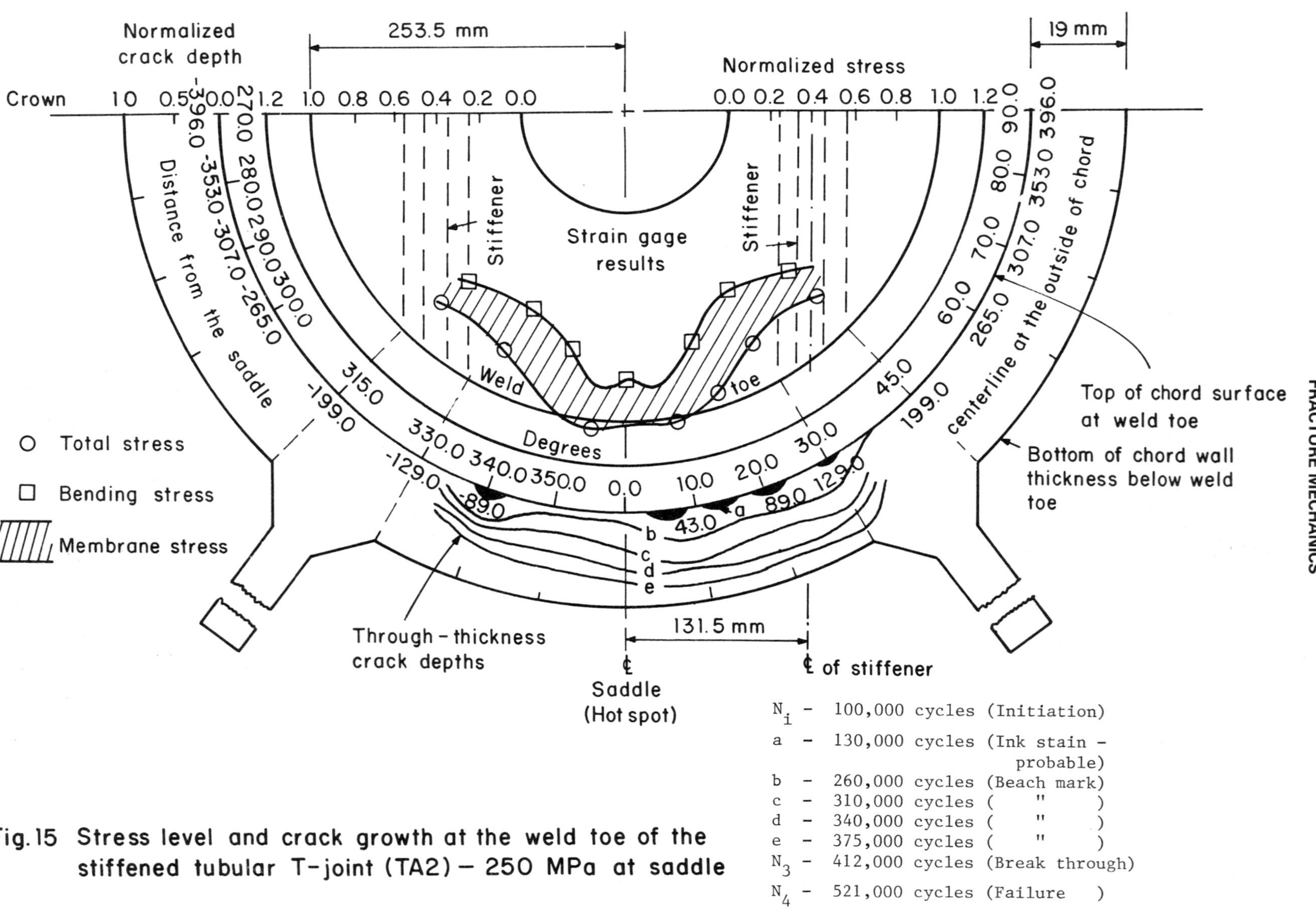

Fig.15 Stress level and crack growth at the weld toe of the stiffened tubular T-joint (TA2) – 250 MPa at saddle

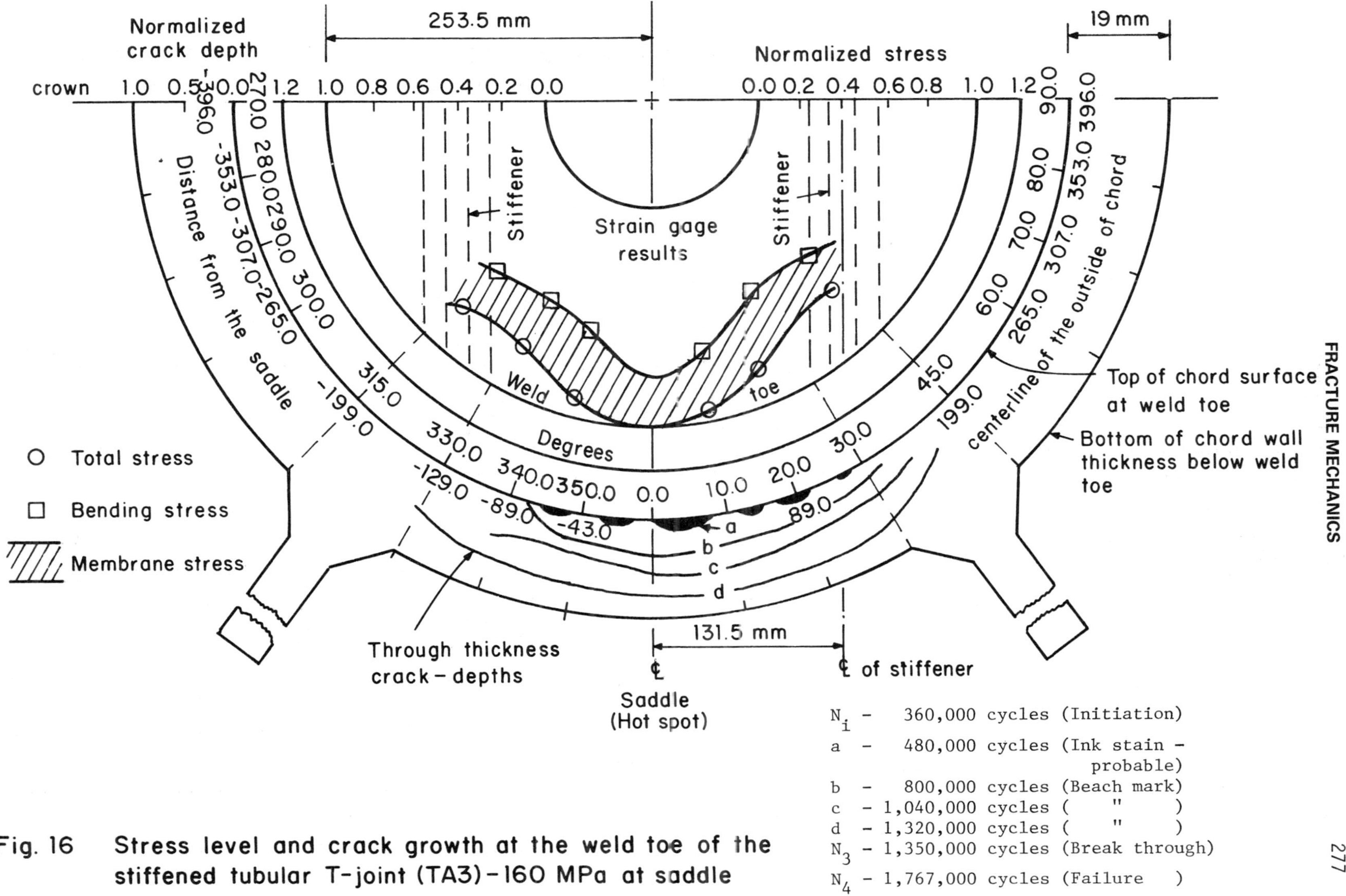

Fig. 16 Stress level and crack growth at the weld toe of the stiffened tubular T-joint (TA3)–160 MPa at saddle

FRACTURE MECHANICS

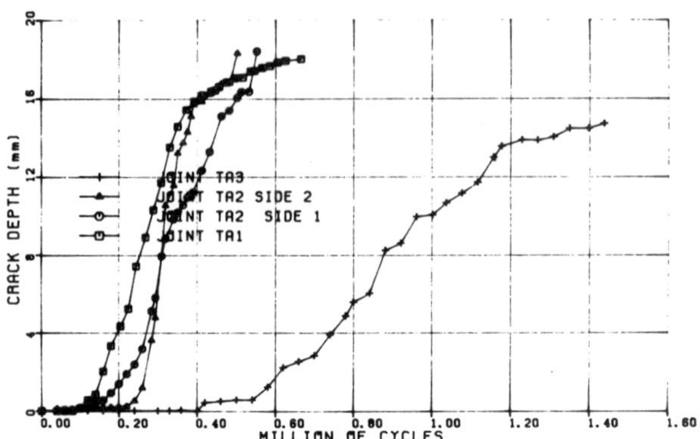

FIG 17. CRACK GROWTH CURVES FOR THREE JOINTS
AT SADDLE POINTS

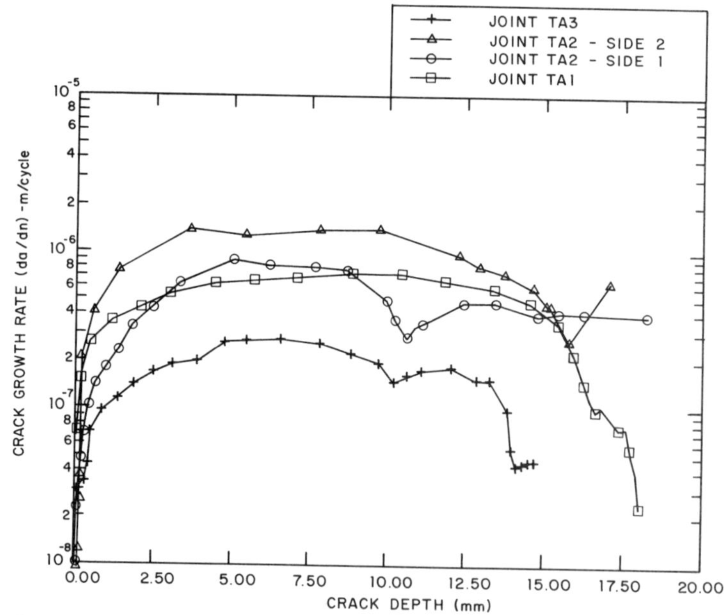

Fig. 18 CRACK DEPTH VS. CRACK GROWTH RATE FOR JOINTS TA1, TA2 & TA3

SESSION 4:   MATERIAL DEVELOPMENT AND MICROSTRUCTURE/
PROPERTY RELATIONSHIP

CHAIRMAN: G. BELLAMY
Ontario Hydro
Toronto

THE INFLUENCE OF MICROSTRUCTURE ON FRACTURE TOUGHNESS LEVELS IN
A357 ALUMINUM ALLOY

Carolyn W. Meyers
Assistant Professor
George W. Woodruff School of Mechanical Engineering
Georgia Institute of Technology
Atlanta, Georgia 30332

ABSTRACT

Cast aluminum alloy A357 is used in structural applications when treated to the T6
temper. The first part of the protocol, the solution heat treatment, produces
measurable changes in the silicon-rich eutectic microconstituent. This is a study
of how these changes affect fracture toughness levels. The fracture toughness
levels, determined using the chevron-notched short bar specimen, for a range of
solution heat treatments reflect the morphological changes in the silicon-rich
microconstituent. Through SEM analysis, mechanistic explanations are proposed.

KEYWORDS

Micromechanisms, fracture toughness, fibrous fracture, void coalescence,
progressive fracture

INTRODUCTION

Aluminum alloy A357 possesses excellent castability. When treated to the T6
temper, this cast alloy is noted for its unusual combination of moderate strength
and good ductility. Yet the use of alloy A357 in fracture critical applications
in the airframe industry is not as widespread as might be expected. According to
Oswalt and Lii (1984), design property data are insufficient and unreliable for
the usage of castings in structural applications. Because the thermal processing
parameters of the T6 temper (temperature and especially the duration of the heat
treatments) cover a range of values, much scatter is observed in reported tensile
and fracture data on alloy A357.

Previous research (Meyers, 1984, 85, 86) on aluminum alloy A357 documented the
effects of the first stage of the T6 temper treatment, the solution heat
treatment. The duration of this treatment changes the characteristics of the
major microconstituent - the silicon-rich structures in the eutectic. These

microconstituent changes in turn strongly influence the subsequent tensile property levels attainable in A357 alloy. Such characteristics of the silicon-rich structures as their average size, numerical density, spacing and morphology vary with the duration of the solution heat treatment. The silicon-rich structures undergo a coarsening process as the time of the solution heat treatment increases. The magnitudes of the changes in the characteristics of the silicon-rich structures have been found to be measurable and predictable. More importantly, the effects of changes in these characteristics on tensile properties can likewise be significant and predictable. Since tensile and fracture behavior are both strongly microstructure dependent, it is reasonable to expect fracture toughness levels to be influenced by the duration of the solution heat treatment and the ensuing changes in the silicon-rich structures. While fracture mechanics provides a quantitative tool for predicting fracture behavior, as Wells (1985) suggests, a truly integrated approach to understanding the fracture process is needed. This approach must include the examination of the macro and microstructural features of the alloy. This is the object of this paper: to study the extent of the influence of a dominant microconstituent, the silicon-rich structures in the eutectic of A357 alloy, on subsequent fracture toughness levels.

BACKGROUND

The progressive fracturing process, simple and macroscopic in approach, dominates current fracture mechanics concepts. Although aluminum alloys such as A357 often exhibit catastrophic brittle failure, the fracture surfaces are extensively covered by voids. These voids, indicative of fibrous fracture, necessitate the modification of conventional fracture mechanics concepts to include a basic understanding of the micromechanics of the fracture process and specifically, of the fibrous fracture mode. But the mechanical details of progressive fracture including void growth, cleavage, and grain boundary separation are not as fully understood as the strict mechanics of the macroscopic fracture process.

The complex microstructures associated with high strength aluminum alloys like alloy A357 commonly include several microconstituents. According to Hahn and Rosenfield (1975) among these are large size inclusions, intermediate size dispersoids and fine scale aging precipitates. The respective amounts and morphologies of these microconstituents appear to influence fracture toughness levels in high strength wrought aluminum alloys (Chen and Knott: 1981). The presence of these microconstituents and their respective characteristics account for the two observed extremes in fracture behavior. Microvoid coalescence occurs in the soft matrix around non-metallic inclusions and fast shear in the regions of the shear strain localization.

Aluminum alloy A357, a ternary alloy of aluminum, silicon, and magnesium, possesses a microstructure consistent with both of these extremes in fracture behavior. As shown in Fig. 1, the as-cast microstructure consists of primary aluminum-rich solid solution dendrites outlined by a divorced eutectic microconstituent. Embedded in the latter are silicon-rich platelets. When treated to the T6 temper, and chemically modified these silicon-rich platelets become more discrete and appear globular in morphology. Not apparent in Fig. 1 is the presence of Mg_2Si within the dendrites formed as a result of the tempering treatment. The silicon-rich structures after tempering are sources for microvoid coalescence as they decohere and as some of these structures crack; they also provide sources for strain localization along with the Mg_2Si precipitates. An example of the resulting fibrous fracture associated with this alloy is shown in Fig. 2.

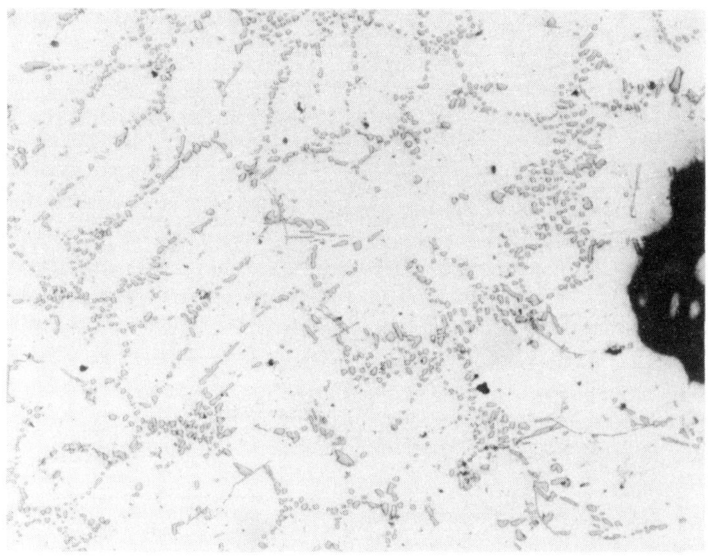

Fig. 1. As Cast Microstructure of A357 Alloy 135X.

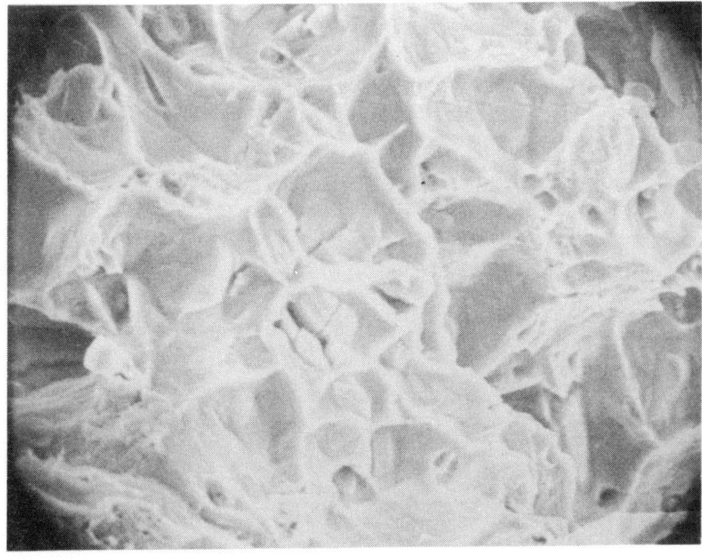

Fig. 2. Fibrous Fracture of A357 Alloy.

The effect of these silicon-rich structures on the progressive fracture process is are debatable. Several researchers (Garrett and Knott, 1978; Chen and Knott, 1981, Van Stone and Psioda 1975, Sailors, 1976) propose that larger particles nucleate voids preferentially. Therefore, microconstituent size should control the progressive fracture process. For continued plastic deformation, void growth occurs by dislocation slip in the ligaments between the voids. These ligaments which are proportional to the void spacing control void growth. Both of these events - void nucleation and void growth require large strains and depend on the interactions of the constituents of the microstructure.

Traditionally, predictions of fracture toughness have been based on two approaches: continuum mechanics and dominant microstructural mechanism. For the former approach fracture toughness depends on elastic modulus, yield strength, strain hardening exponent and effective fracture ductility (Weiss, Kasai and Sieradzki, 1976). The mechanics involved is well developed and will not be discussed here as this is not the focus of this paper. Most of these approaches include a basic length parameter representing the dimensions of the spacings of inclusions to account for local strain at the crack tip.

An alternative approach is that resistance to fracture is controlled by void growth and coalescence as proposed by Hahn and Rosenfield (1975). Their basic length parameter is the spacing between intermetallic inclusions.

Chen and Knott (1981) working with the 7XXX series of aluminum alloys sought to combine the void initiation and void growth stages in their expression for K_{IC}:

$$K_{IC} = \left[\frac{mb}{10A} E\, \sigma_c\, \sigma_y\, n^2\, \frac{\lambda}{d} \right]^{1/2} \tag{1}$$

is which
- m = constant of the order of 1.0
- b = Burgers vector
- A = proportionality constant
- σ_c = cohesive strength
- λ = particle spacing
- d = particle diameter

As is evident, the fracture toughness depends on the ratio of inclusion (or particle) spacing to its size.

The role of the silicon-rich structures in influencing fracture toughness levels in aluminum alloy A357 was examined by Meyers and Lyons (1987). Their results indicate that fracture toughness in A357 alloys is more strongly affected by the number $(N_\ell)_{ID}$ and the spacing (MFP) of the silicon-rich structures than by the dendrite arm spacing, DAS. The specific relationships are:

$$K_{ICSR} = 47.6 - 0.41\ \text{DAS} - 116.5\ (N_\ell)_{ID} \tag{2}$$

$$\text{MPa}\ \sqrt{m} \qquad\qquad (\mu m) \qquad\qquad (\mu m)$$

and

$$K_{ICSR} = 42.8 - 0.40\ (\text{DAS}) + 1.62\ \text{MFP} \tag{3}$$

$$\text{MPa}\ \sqrt{m}$$

in which K_{ICSR} = plane strain fracture toughness using the chevron notch short rod specimen

Since the fracture surface in A357 alloys appears macroscopically brittle but is dimpled microscopically, progressive fracture and fast shear are evidenced. The fracture path is primarily through the interdendritic material, the eutectic constituent. It then follows that the controlling microstructural feature in A357 alloys may well be the dendrite in that its size limits the eutectic constituent. Vorren, Evenson and Pederson (1980) working with Aℓ-7Si-Mg alloys report a decrease in fracture toughness with increasing dendrite arm spacing (DAS). Their results are also consistent with those reported by Oswalt and Misra (1982) as well as Meyers, Saigal and Berry (1983). The larger DAS really implies larger more widely spaced silicon-rich structures for a given volume fraction. This paper extends the work of Meyers and Lyons (1987) to examine progressively larger silicon-rich structures produced by extended solution heat treatment and therefore, coarsening.

EXPERIMENTAL PROCEDURE

Cylindrical specimens were cast from A357 bar stock in the research foundry in the George W. Woodruff School of Mechanical Engineering. Room temperature steel permanent molds preheated to 100°C were used. To vary the average sizes of the silicon-rich structures, the chevron notch shortrod (CNSR) test blanks were solutionized at 540°C for up to 100 hours. At ten hour intervals selected samples were removed from the heat treatment furnace, water quenched and artifically aged. The aging treatment was constant for all samples: 155°C for four hours. CNSR test specimens, shown in Fig. 3 were then extracted from the blanks.

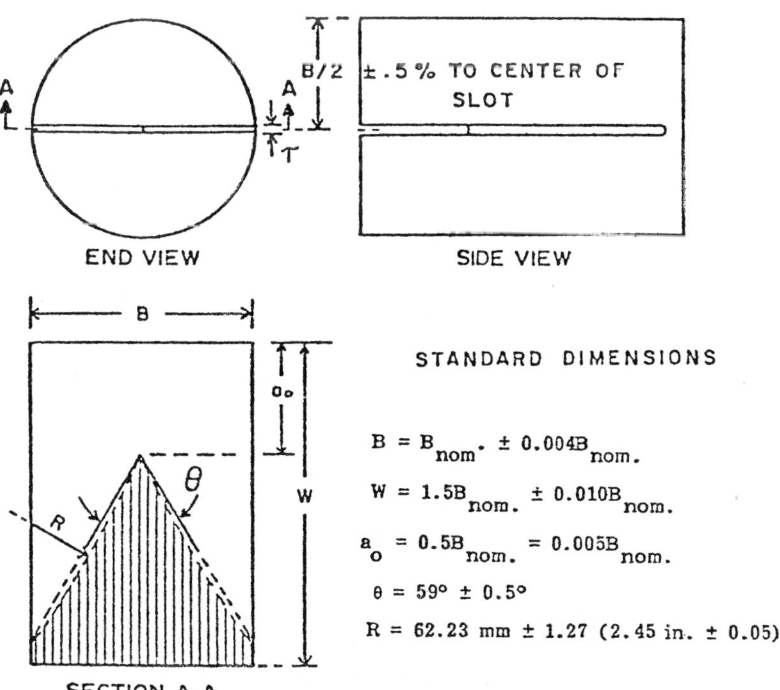

Fig. 3. Short-Rod Specimen

After testing on the Fracjack II, the specimens were broken completely to allow examination of the fracture surfaces using scanning electron microscopy. Additionally, the samples were sectioned perpendicular to the crack plane to allow the observation of the path of the advancing crack.

Metallographic specimens were extracted from each short rod blank. Using point counting techniques on a light optical microscope, the following average characteristics of the silicon-rich structures were determined for each solution heat treatment time:

N_ℓ: bulk lineal density

$(N_\ell)_{ID}$:   interdendritic lineal density

$\dot{A}$:   average area

$\dot{d}$:   mean tangent diameter

$\dot{L}_2$:   mean intercept length

MFP: mean free path

DAS: secondary dendrite arm spacing

AR: derived aspect ratio.

Computer assisted linear regression analyses were used to gauge the most accurate predictor of fracture toughness levels. The aforementioned average silicon rich structure characteristics were the independent variables and the short rod fracture toughness values were the dependent variables.

RESULTS AND DISCUSSION

To study the influences of microstructure on fracture toughness in aluminum alloy A357, one microstructural feature - the silicon-rich eutectic microconstituent - was isolated. Previous research (Meyers, 1985) indicated the characteristics of the silicon-rich structures varied with the duration of the solution heat treatment stage of the T6 temper treatment. The solution heat treatment time dependencies of the characteristics of the silicon-rich structures such as their numerical densities and average sizes further indicated a coarsening like process described by

$$R = B_1 + B_2 \ (V_V)t^{1/3} \tag{4}$$

where R is the average equivalent radius of the eutectic silicon-rich structures,
     B_1 and B_2 are constants,
     (V_V) is the volume fraction silicon-rich structures, and
     t is the solution heat treatment time.

Accompanying this process is a concurrent decrease in the numerical densities of the silicon rich structures. The average spacings or mean free paths between these structures increased, for a constant volume fraction.

The results of identically treated miniature tensile tests show that ultimate strength and percent uniform elongation exhibit the same solution heat treatment time dependency as expressed by the size parameter in Equation 4. As expected when microstructural features were analyzed with these two tensile properties, the size parameters (descriptive of the silicon-rich structures) were the most accurate microstructural predictors of ultimate strength and percent uniform elongation for unmodified A357 alloy.

These same descriptive parameters of the silicon-rich eutectic structures for each solution heat treatment were used as independent variables in regression analysis with the corresponding short rod fracture toughnesses for each test condition. These results appear in Table 1.

Initially, the variation in fracture toughness with dendrite arm spacing predicted by previous researchers (Vorren,, Evenson and Pedersen, (1980) and others) was checked using the short rod test.

These results are depicted in Fig. 4. The linearized relationship is

$$K_{ICSR} = -0.18(DAS) + 47.6 \qquad (5)$$
$$(mPa \sqrt{m}) \qquad (\mu m)$$

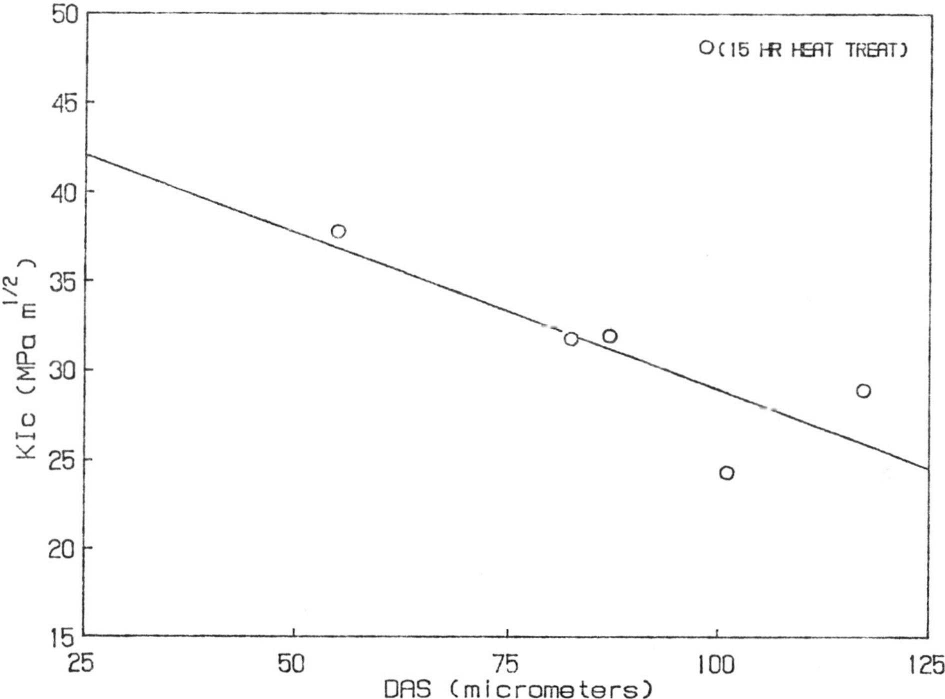

Fig. 4. Aluminum Secondary Dendrite Arm Spacing vs. Fracture Toughness.

Since cracks in aluminum alloy A357 appear to propagate mainly through the eutectic microconstituent, the finer distribution of this microconstituent causes higher fracture toughness levels.

The deleterious effects of the larger dendrite arm spacing due to slower solidification rates on fracture toughness levels can be offset by longer solution heat treatment times. Local yielding in ductile materials relaxes the peak stress at a crack tip to a value approaching the material's yield strength, σ_y. This creates a zone of plastic deformation which for plane strain conditions is given to a first approximation by

$$r_p \simeq \frac{1}{6\pi} \left(\frac{K_I}{\sigma_y} \right)^2 \tag{6}$$

in which r_p is the plastic zone radius and
       K_I is the opening mode stress intensity factor.

Using typical values of K_I = 25 mPa √m and σ_y = 290 mPa for A357–T6, the radius of the plastic zone was calculated by Lyons (1987) to be 0.39 mm.

The deformation in this plastic zone generates a region of extensively deformed material near the crack surface. Because the material in this region was highly strained, it did not polish as well as material far from the crack. The size of this "process region" was measured with a calibrated test line to be approximately

TABLE 1.   Linear Regression Results of Size, Shape and Spacing
            Parameters with Toughness. $K_{IC})_{SR}$ = mx + b.

Parameter x	Slope m	Intercept b	Correlation Coefficient	Comments
A	3.42	22.59	0.94	Casting 1
A	2.15	13.86	0.79	Casting 2
$(\bar{L_2})_{ID}$	23.23	13.19	0.88	Casting 1
$(L_2)_{ID}$	27.28	-3.69	0.64	Casting 2
AR	0.90	31.23	0.20	Casting 1
AR	5.34	2.41	0.73	Casting 2
1/AR	-25.91	41.61	0.26	Casting 1
1/AR	-129.00	55.20	0.74	Casting 2
d	3.90	20.35	0.76	Casting 1
d	3.49	8.46	0.76	Casting 2
$d^{1/2}$	15.11	5.98	0.78	Casting 1
$d^{1/2}$	16.16	-10.14	0.79	Casting 2
MFP	1.75	0.21	0.94	Casting 1
MFP	1.43	13.43	0.89	Casting 2
MFP/d	11.40	10.50	0.75	Casting 1
MFP/d	16.66	-1.67	0.82	Casting 2
$(N_\ell)_{ID}$	-112.39	48.75	0.93	Casting 1
$(N_\ell)_{ID}$	-345.31	33.61	0.70	Casting 2
$Time^{1/2}$	3.31	24.60	0.96	Casting 1
$Time^{1/2}$	2.13	21.60	0.91	Casting 2

0.1 mm. This experimentally determined plastic zone size is of the same order of
magnitude as the theoretical prediction.   Because it was difficult to accurately
determine the exact boundary of the process region with the microscope, the
experimental value is less than that predicted by Equation 6.

The critical crack length was approximated from short rod samples whose tests were
interrupted when the applied load reached its critical value. After sectioning
perpendicular to the crack plane, the depth of the crack was measured and related
to the actual crack length by the geometry of the chevron notch. The average
critical crack length was 12 mm.   One criterion for the applicability of LEFM is
that the crack length be at least 25 times the radius of the plastic zone.
Comparing the experimental results we see:

$$a_c = 12 \text{ mm} > 25 \ r_p = 9.8 \text{ mm} \tag{7}$$

Thus, this criterion is satisfied.

The plane strain fracture toughness values   obtained in this research ranged from
22.4 to 38.0 MPa$\sqrt{\text{m}}$. Published values for A357-T6, as determined by the Compact
Tension fracture toughness test, range   from 27.9 to 30.1 MPa$\sqrt{\text{m}}$. Valid fracture
toughness values which exceeded this range are attributed to the differences in
the solution heat treatment times and temperatures. Other sources of discrepancy
include chemical modification of the commercial castings, which was not done here.

In order for short bar toughness values to be considered valid, the crack must be
maintained in the chevron slot plane. Examinations of the fracture surface and of
interrupted tests indicated that this condition was met up to and beyond the point
where the growing crack achieved its critical crack length during each test.
Significant deviation from the intended plane occurred in a few samples as the
crack size increased beyond the critical size.

Since cracks appear to propagate primarily through the eutectic, it is concluded
that this microconstituent has a lower fracture toughness than the aluminum
matrix. This decreased fracture toughness is due to the presence of the hard,
brittle silicon-rich structures in the eutectic, which nucleate voids as they are
enveloped by the plastic strain region at the tip of an advancing crack. Fracture
toughness is therefore improved by decreasing the density and increasing the
spacing between these structures with extended solution heat treatments.

The plastic zone radius, as calculated by Lyons (1987) was 490 μm. Using a
typical as-cast bulk lineal density of 0.04 particles per μm, roughly 20 silicon-
rich structures were enveloped by the plastic zone along a line extended from the
advancing crack. After solution heat treatment for 30 hours, the bulk lineal
density decreased to 0.02 particles per μm, so that roughly 10 silicon-rich
structures were enveloped. Thus, fewer particles within the plastic zone which
could potentially nucleate voids corresponds with higher fracture toughness
levels.

This is verified by the stereological data extracted from samples subjected to
extended solution heat treatments.   Of the silicon-rich structure parameters
monitored, the lineal density of the silicon-rich structures $(N\ell)_{\text{ID}}$ and the mean
free path (MFP) between them were the most accurate predictors of fracture
toughness levels. Based on lineal density, fracture toughness levels can be
approximated by

$$K_{ICSR} = -112.9 \; (N_\ell)_{ID} + 48.75 \tag{8}$$

(MPa√m) (μm^{-1})

These results are shown in Fig. 5 for fine (casting 1) and coarse (casting 2) dendrite arm spacings. It appears that fracture toughness levels are more strongly affected by the interdendritic microconstituent than by the dendrite size itself. The effect of the finer dendrite size is to raise the base level of fracture toughness in general but not to influence the response to extended solution heat treatment.

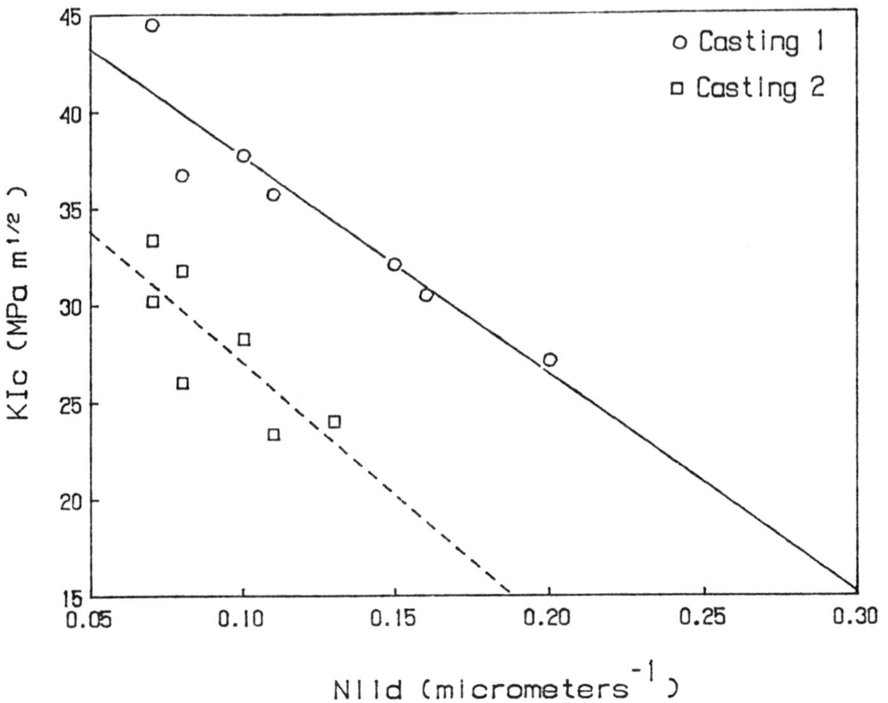

Fig. 5. Fracture Toughness vs. Interdendritic Lineal Density.

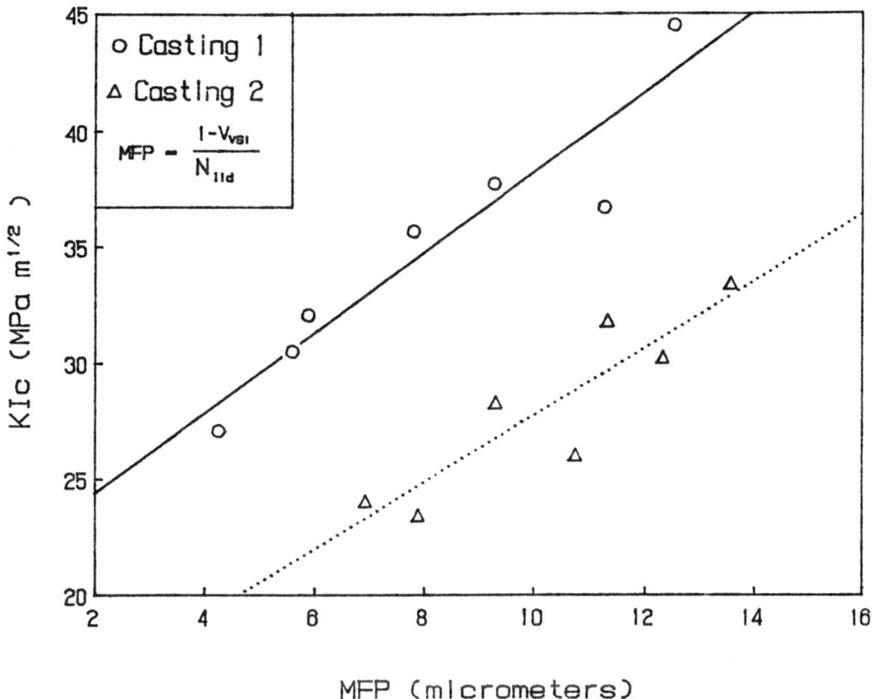

Fig. 6. Mean Silicon-Free Path vs. Fracture Toughness.

The extended solution heat treatments resulted in larger spacings and therefore larger ligaments of the higher toughness aluminum-rich solid-solution material between voids.   The best empirical relationship obtained between fracture toughness and the mean free path was

$$K_{ICSR} = 1.75 \ (MFP) + 21.0 \tag{9}$$
$$(MPa\sqrt{m}) \qquad (\mu m)$$

This influence of particle spacing on fracture toughness levels is shown in Figure 6.

These experimental results support the predictive equations presented by Sailors for steels (1976) and by Chen and Knott (1981) for wrought aluminum alloys. Both of those theoretical relationships predict that fracture toughness increases with the square root of the particle spacing.   However, taking the square root of the experimentally determined spacings failed to improve the correlation with fracture toughness.   This discrepancy resulted in part from the microstructural differences between cast A357 and steel or wrought aluminum alloys.

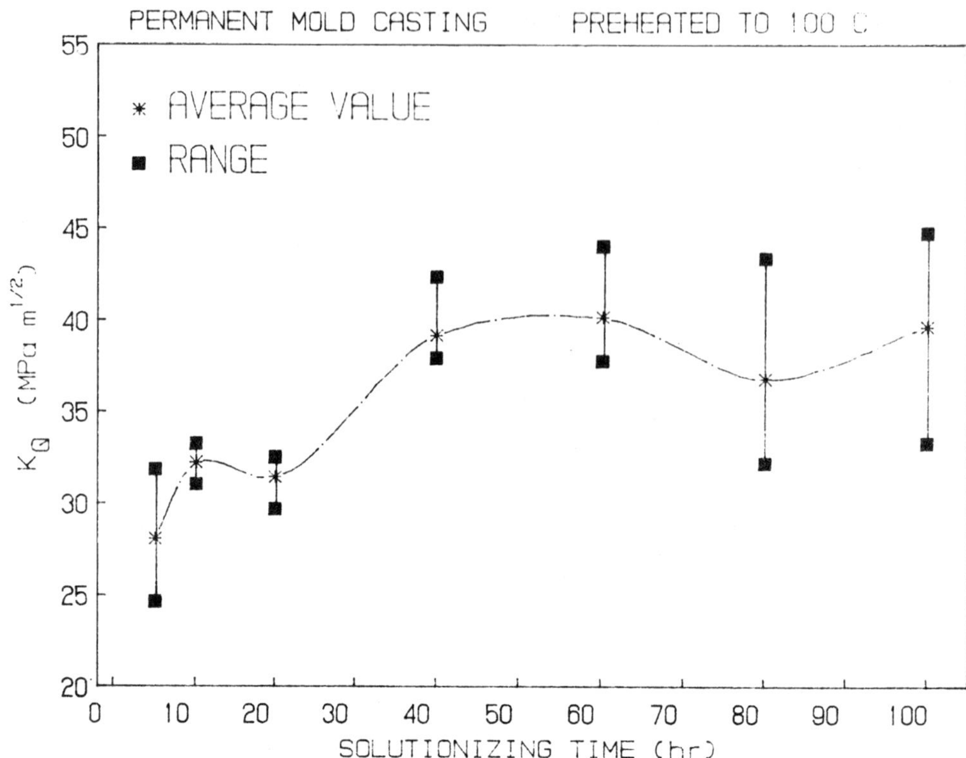

Fig. 7. Fracture Toughness vs. Solutionizing Time.

Extended heat treatments for up to 100 hours (Fig. 7) produce a continuation of the trends already noted. That is, as the silicon-rich structures increase in spacing and decrease in numerical density, fracture toughness levels continue to rise. This provides further indication that fracture toughness levels in A357 alloys strongly depends on the aluminum-rich matrix ligament size between silicon-rich eutectic structures.

As stated in the background, previous research by Meyers (1986) indicated that silicon-rich structure size (and therefore spacing) and density significantly affect ultimate strength and percent uniform elongation levels in A357 alloys. The reported influences of these same silicon-rich structure parameters on the fracture toughness levels would suggest a relationship between fracture toughness and these two mechanical properties. Further research is focused on exploring this connection.

CONCLUSIONS

On a broad basis dendrite arm spacing is critical in determining fracture toughness levels in A357 aluminum alloy. Dendrite arm spacing controls the scale of the eutectic microconstituent. This, in turn, influences the sizes and spacings of the silicon-rich eutectic structures. These two parameters change markedly with the duration of the solution heat treatment. The relatively steady increase in the fracture toughness levels of aluminum alloy A357 as the solution heat treatment time increases can be directly attributed to the increase size of the ligaments between the silicon-rich structures. This implies that the extent of void coalescence by internal necking increases with extended solution heat treatment. Ongoing research involves the measurement of void and ligament sizes for the different heat treatments and the determination of the existence of a critical limit on the silicon-rich structure parameters above which fracture toughness levels would be unaffected.

ACKNOWLEDGEMENTS

The author expresses her gratitude to the National Science Foundation which supported this research, to Mr. Jed Lyons for his experimental work, to the George W. Woodruff School of Mechanical Engineering for the use of facilities and equipment and to Miss Melinda A. Wilson for the preparation of this manuscript.

REFERENCES

Chen, C. Q. and Knott, J. F., "Effects of Dispersoid Particles on Toughness of High-Strength Aluminum Alloys", _Metal Science,_ Vol. 15, Aug (1981), pp. 357-364.

Firrao, D. and R. Roberti, "Interrelation Among Microstructure, Crack-Tip Blunting, and Ductile Fracture Toughness in Mild Steels", AFR Vol. 2-R, pp. 1311-1319.

Gangulee, A., and Gurland, J., "On the Fracture of Silicon Particles in Aluminum-Silicon Alloys", _Trans. AIME_ 239, (1959), 269-272.

Garrett, G. G., and Knott, J. F., "The Influence of Compositional and Microstructural Variations on the Mechanism of Static Fracture in Aluminum Alloys", _MET Trans._, 9A, Sep. (1978), pp. 1187-1200.

Knott, J. F., "Microstructure and Crack-Tip Fracture Processes in Metal", Paper 9, The Metals Society Conf. on "The Mechanics and Physics of Fracture", Churchill College, Cambridge, January 6-8, 1975, pp. 86-94.

Knott, J. F,., "Micromechanisms of Fibrous Crack Extension in Engineering Alloys", Metal Science, Aug. (1980), pp. 327-336.

Hahn, G. T., and Rosenfield, A. R., "Metallurgical Factors Affecting Fracture Toughness of Aluminum-Alloys", _Metallurgical Transactions_, Vol. 6A, April (1975), pp. 653-667.

Meyers, C. W., Saigal, A., Berry, J. T., "Fracture Related Properties of Aluminum A357-T6 Cast Alloy and Their Interrelation with Microstructure", _AFS Transactions_, 83-35, pp. 281-288.

Meyers, C. W., "Solution Heat Treatment Effects in A357 Alloys", _AFS Transactions_, Vol. 93, (1985), pp. 741-750.

Meyers, C. W. "Solution Heat Treatment Effects on Ultimate Strength and Percent Elongation in A357 Aluminum Alloy", _AFS Transactions_, Vol. 94, (1986), pp. 511-517.

Meyers, C. W. and Lyons, J. S., "Fracture Toughness-Second Phase Particle Interactions in A357 Alloys", paper at TMS Annual Meeting, Denver, Colorado, (1987).

Meyers, C. W., "The Microstructure-Mechanical Properties Interrelationships in A357-T6 Cast Aluminum Alloys", Ph.D. Thesis, Georgia Institute of Technology, School of Chemical Engineering, (August 1984).

Misra, M. S., and Oswalt, K. J., "Aging Characteristics of Ti-Refined A356 and A357 Aluminum Castings", _AFS Transactions_, 90 (1982), pp. 1-19.

Oswalt, K. J., and Lii, Y., Northrup Corporation Report Number NOR 84-109, October 1984.

Sailors, R. H., "Relationship Between Tensile Properties and Microscopically Ductile Plane-Strain Fracture Toughness," _Properties Related to Fracture Toughness_, ASTM STP 605, American Society for Testing and Materials, 1976, pp. 34-61.

Van Stone, R. H., and Psioda, J. A. "Discussion of Metallurgical Factors Affecting Fracture Toughness of Aluminum Alloys", _Metallurgical Transactions A_, Vol. 6A, April (1975), pp. 668-670.

Vorren, O., Evensen, J. D., and Pedersen, T. B., "Microstructure and Mechanical Properties of AlSi (Mg) Casting Alloys", Metallurgical Research and Development Center Report, Norway.

Weiss, V., Kasai, Y., and Sieradzki, "Microstructural Aspects of Fracture", _Properties Related to Fracture Toughness,_ ASTM STP 605, (1976), pp. 16-33.

Wells, Joseph M., "Synergism of Microstructure, Mechanisms and Mechanics in Fracture," Journal of Metals, Vol. (1985) pp. 58-64.

CONTROL OF MICROSTRUCTURE TO INCREASE THE TOLERANCE
OF ZIRCONIUM ALLOYS TO HYDRIDE CRACKING.

C.E. Coleman, S. Sagat and K.F. Amouzouvi*

Chalk River Nuclear Laboratories, Metallurgical Engineering Branch,
Chalk River, Ontario KOJ 1JO, Canada.
*Whiteshell Nuclear Research Establishment, Materials Science Branch,
Pinawa, Manitoba ROE 1LO, Canada.

ABSTRACT

The microstructure of Zr-2.5 Nb has been altered in three ways in attempts to
increase the alloy's tolerance to delayed hydride cracking, namely:

- breaking up the β-phase reduces diffusivity of hydrogen and decreases crack
  velocity,

- a gettering element (yttrium) reduces susceptibility to cracking although the
  yttrium alloy has low toughness and poor corrosion resistance, and

- reducing the number of basal plane normals in the main stressing direction
  improves resistance to crack growth.

KEYWORDS

Zirconium   alloys;   delayed   hydride   cracking; β-phase;   yttrium getter;
crystallographic texture.

INTRODUCTION

Zirconium alloys are susceptible to hydrogen embrittlement when brittle hydride
precipitates are present. Two distinct phenomena are observed:

- a time dependent process called delayed hydride cracking (DHC), and
- a reduction in toughness.

This paper will be mostly concerned with DHC. In this form of cracking, hydrogen
preferentially dissolves from hydrides in the zirconium alloy matrix and migrates
up the stress gradient to the tip of a stress concentrator, such as a surface
defect, then reprecipitates. If the stress is large, at some critical condition
(perhaps hydride size) the first hydride will crack, the process is then repeated
and the crack grows in a series of steps. The life history of a crack is shown
schematically in Fig. 1 in which crack velocity, V, is plotted as a function of
stress intensity factor, K_I. Between the values of K_I, called K_{IH} and K_{IC},
cracks are stable and grow by DHC.   The DHC region is characterised by a

threshold value, K_{IH}, below which cracks do not grow and a crack velocity, V_c, which is nearly independent of K_I. Above K_{IC} cracks are unstable.

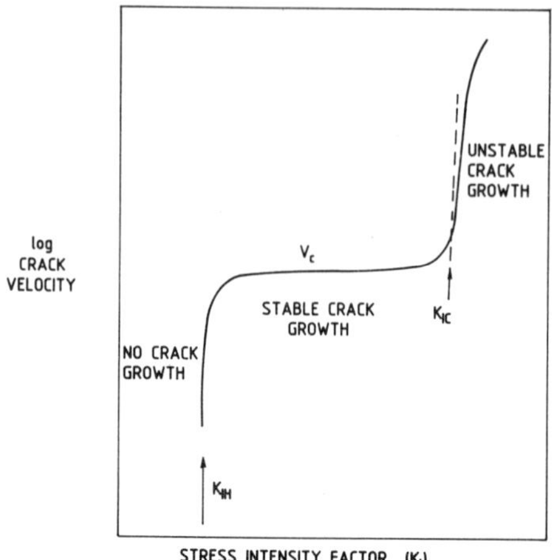

Fig. 1. Schematic diagram of the dependence of crack velocity on stress intensity factor.

When evaluating a component for its probability of failing by DHC we must consider several parameters:

- hydrogen concentration; hydrides are necessary to provide a source of hydrogen and the agent for embrittlement. Thus, to prevent DHC, the hydrogen concentration must not exceed the solubility limit,

- tensile stress; stress is necessary to fracture the hydrides and to induce the hydrides to precipitate during cooling with their plate normal parallel to the stress; such hydrides, called "reoriented hydrides", are very harmful because they provide an easy route for crack propagation and thus reduce both K_{IH} and K_{IC}. The larger the stress the higher the probability of reorienting and fracturing the hydrides.

- flaws; at moderate stresses flaws are needed to provide stress intensification, the critical size depending on stress and K_{IH},

- time; time is required for the hydrogen to diffuse to the flaw and for the hydride to grow. The rate of the attainment of the critical condition for cracking is a complicated function of temperature and loading history.

- microstructure; crystallographic texture, phase distribution and grain size and shape all affect cracking.

In previous papers we have specified methods for preventing cracking by controlling the hydrogen concentration, the tensile stresses and the size and number of flaws (Cheadle, Coleman and Ambler, 1987) and have recommended appropriate loading and temperature manoeuvres to minimise cracking (Coleman and co-workers, 1985). Thus these solutions involve the design, fabrication, construction and operation of components. In this paper we discuss possible methods of changing the microstructure of zirconium alloys to increase their tolerance to DHC. Desirable properties are high values of K_{IH} and low values of

V_c; in extruded and cold-worked Zr-2.5 Nb alloy and Zircaloy-2 (essentially a Zr-1.5 Sn alloy) the minimum value of K_{IH} is about 5 MPa√m (Cheadle, Coleman and Ambler, 1987) and reference values of V_c are shown in Fig. 2 as a function of temperature.

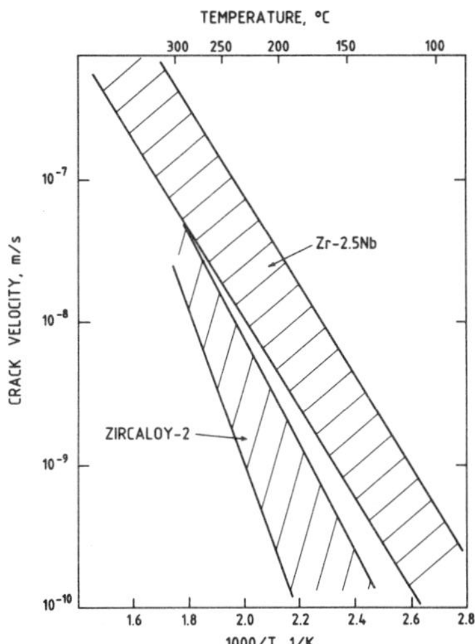

Fig. 2. Temperature dependence of crack velocity in cold-worked Zr-2.5 Nb and Zircaloy 2. Lines represent typical range of values.

THE ROLE OF MICROSTRUCTURE

The principles of the improvements in microstructure depend on reducing the mobility of the hydrogen and causing the hydride to precipitate in a less damaging form.

Mobility of Hydrogen

The flux of hydrogen to the crack tip is proportional to the diffusivity of hydrogen, D, the amount of hydrogen in solution, C, and the stress gradient, S, (Dutton and co-workers, 1977). Both D and C have temperature dependencies through their activation energies while S is controlled by the yield stress. In zirconium alloys, second phases may affect mobility. In Zr-2.5 Nb, depending on heat-treatment, a β-phase may be present. The diffusion coefficient and solubility of hydrogen in β-phase is greater than that in the α-phase (Sawatzky and co-workers, 1982) and if the β-phase is continuous it provides an easy pathway for hydrogen movement. When the alloy is in the extruded and cold-worked condition the β-phase is in the grain boundaries surrounding the α-grains. The presence of this grain boundary β-phase is partially responsible for greater crack velocities in Zr-2.5 Nb than in Zircaloy-2, which contains no β-phase, Fig. 2. In this paper we examine the effect on V_c of decomposition and redistribution of the β-phase.

A second way to affect mobility would be to add an alloying element that had a more stable hydride than zirconium and thus getter the hydrogen. If the alloying element had a low solubility in zirconium and did not form intermetallic compounds, it would be present in the form of discrete particles that would act as internal sinks but not otherwise interfere with the alloy. For nuclear applications the getter should also have a reasonably low capture cross-section for thermal neutrons. Yttrium appears to fulfill all three requirements:

- yttrium hydride has a free energy of formation of between -75 and -92 kJ/gm.atom of hydrogen whereas that of zirconium hydride is about -62 kJ/gm.atom of hydrogen (Mueller, Blackledge and Libowitz, 1968),

- the maximum solubility of yttrium in β-zirconium is 2 to 4 at%. The transformation from the β-phase to the α-phase is by a peritectoid reaction at about 0.6 at% yttrium at 1160 K. Below about 970 K most of the yttrium is out of solution and exists as particles of yttrium in a matrix of α-zirconium (Elliot, 1965),

- the macroscopic capture cross-section for thermal neutrons of yttrium is 3.8 m^2/m^3. Adding 1 at% yttrium would raise the cross-section of zirconium from 0.78 m^2/m^3 to 0.81 m^2/m^3.

Pieces of yttrium have been shown to be effective getters in nuclear fuel (Spalthoff and Wilhelm, 1969). Zirconium alloys with greater than 5 at% yttrium have been proposed (Patent Specification) as getters for residual gasses in hermetically sealed containers although the alloys also reacted strongly with oxygen. In this paper we report on the ability of small yttrium additions to increase the tolerance of Zr-2.5 Nb to DHC.

Less Harmful Hydrides

During DHC, hydrides at the crack tip are often reoriented. If reorientation is difficult then cracking is difficult. Hydrides tend to reorient easily if the grains are equiaxed and if their basal plane normal is parallel to a tensile stress (Cheadle, Coleman and Ipohorski, 1984). In cold-worked materials the grains are often elongated but the basal planes tend to be oriented in similar directions. If the crystallographic texture can be arranged so that the basal plane normal of each grain is nearly perpendicular to the direction of the major tensile stress then DHC will be reduced. For example, in extruded and cold-worked tubes of Zr-2.5 Nb the resolved fraction of basal plane normals in the circumferential direction, F_c, is usually high, about 0.6, Fig. 3a. Thus the basal plane normals tend to be parallel to the hoop stress, the highest stress in a pressurised closed end tube. Consequently, cracking is relatively easy on planes with normals parallel to the circumferential direction. Conversely, in the longitudinal direction , the resolved fraction of basal plane normals, F_L, is usually low, < 0.1, and cracking is very difficult (Coleman, 1982). In this paper we report progress on exploiting this effect of crystallographic texture.

EXPERIMENTAL METHODS

Materials

For the experiments to evaluate the importance of the β-phase, we used Zr-2.5 Nb alloy from CANDU pressure tubes. The initial microstructure, Fig. 4, consisted of α-phase plates surrounded by a thin layer of β-phase. To change the distribution of β-phase, the material was homogenized at temperatures up to 670 K for up to 10 days.

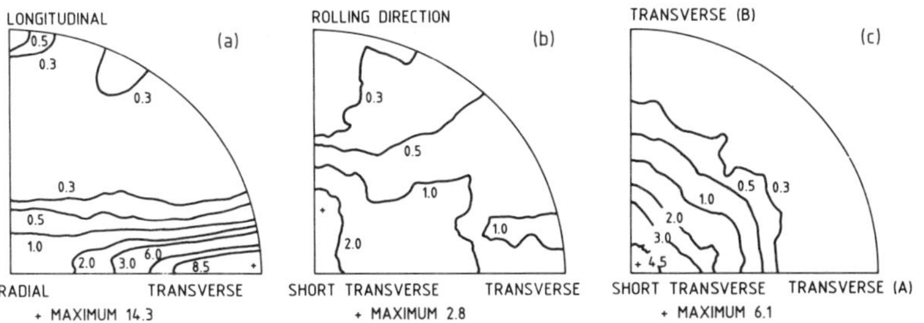

MATERIAL	RESOLVED FRACTION OF BASAL PLANE NORMALS		
	RADIAL/SHORT TRANSVERSE	TRANSVERSE (A)	LONGITUDINAL/ROLLING DIRECTION/TRANSVERSE (B)
PRESSURE TUBE	0.32	0.62	0.06
Zr-2.5Nb-1.4Y	0.47	0.32	0.21
CROSS-ROLLED PLATE	0.67	0.16	0.17

Fig. 3. Crystallographic texture of test materials (a value of 1.0 represents random orientation).

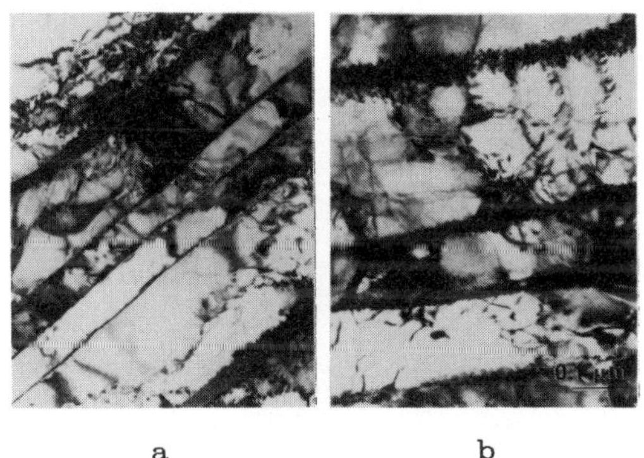

Fig. 4. Transmission electron micrographs of Zr-2.5 Nb pressure tube:
a - no heat-treatment
b - 72 h at 670 K.

A plate, 43 mm thick, 89 mm wide, and 600 mm long, of Zr-2.5 Nb containing 1.4 at% yttrium was used for the experiments to assess the efficacy of yttrium as a hydrogen getter. The material was vacuum arc melted and the resulting 100 mm square ingot was hot forged at 1270 K to 890 mm square. This was then hot rolled at 1100 K to a thickness of 56 mm and annealed at 970 K for 30 minutes. The final treatment was warm rolling. The microstructure was similar to that of pressure tubes except for longitudinal stringers of yttrium-rich particles 4 to 8 μm wide and elongated in the rolling direction up to 20 μm (Fig. 5). The basal plane normals were weakly concentrated in the thickness direction, Fig. 3b.

To investigate texture effects a Zr-2.5 Nb alloy ingot was forged at 1300 K to 240 mm diameter, reheated to 1300 K for 30 minutes and water quenched. A 25 mm slice was warm cross-rolled at 520 K to 18 mm, annealed at 1030 K for 30 minutes then warm cross-rolled at 520 K to 13 mm. The microstructure consisted of a uniform basketweave structure and when hydrides were present they tended to be parallel with the rolling plane (Fig. 6). The basal plane normals were highly concentrated in the thickness direction, Fig. 3c.

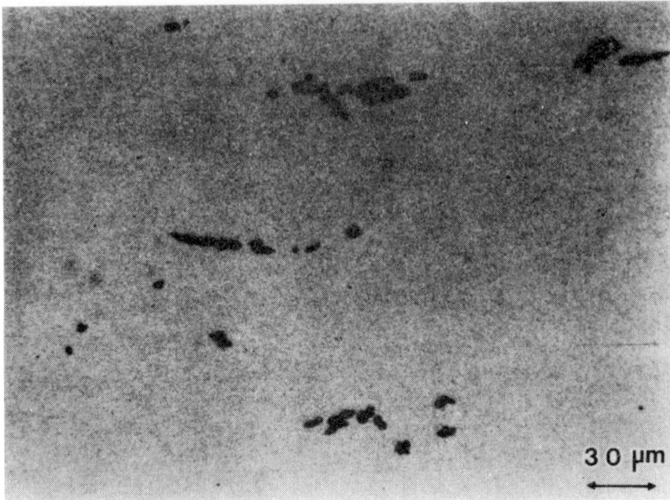

Fig. 5.   Yttrium rich particles in Zr-2.5 Nb with 1.4 at% yttrium.

Fig. 6.   Microstructure and hydride distribution in the cross-rolled Zr-2.5 Nb
plate on a plane with normal parallel to the rolling direction.

Hydrogen was added to some specimens by absorbing hydrogen gas for 4 h or less at
620 or 670 K, then the specimens were homogenized. Except for the studies of the
β-phase, specimens were homogenized at 670 K for 72 h.

The tensile properties at room temperature of the three test materials are
summarized in Table 1.

TABLE 1 Tensile Properties of Test Materials at Room Temperature

MATERIAL	ORIENTATION	HEAT-TREATMENT	0.2% YS (MPa)	UTS (MPa)	TOTAL ELONGATION %
Pressure Tube	Transverse	none	775	872	12 (a)
		24 h at 670 K	815	862	15
		72 h at 670 K	803	858	15
		96 h at 670 K	818	857	15
		168 h at 670 K	797	842	15
Zr-2.5 Nb-1.4 Y	Short Transv.	none	392	595	5 (a)
	Transverse	none	520	604	8
	Rolling Dir.	none	523	602	14
Rolled Plate	Short Transv.	none	795	871	24 (b)
	Transv. (A)	none	582	700	15 (c)
	Transv. (B)	none	573	695	14

	(a)	(b)	(c)
Specimen Gauge Length (mm):	16	5	19
Diameter (mm):	2.9	2.3	4.1

Specimens and Testing Techniques

Two types of specimens were used: cantilever beams and compact tension.

Cantilever beams were 38 mm long, 3 mm wide and 4 mm deep. A 0.5 mm notch was machined in the centre of the specimen and 0.1 mm deep side grooves were used to guide the crack. The specimens were loaded in pure bending and K_I was calculated from standard formulae (Brown, 1966). Cracking was detected using acoustic emission (Sagat, Ambler and Coleman, 1986). Before loading, the specimens were thermally cycled to 570 K and then cooled to the test temperature. The average crack velocity was obtained from the actual crack length and the time of stable growth. The threshold stress intensity factor, K_{IH}, was obtained by reducing load in small steps until cracking stopped for a minimum of 3 days. The corresponding K_I was assumed to be K_{IH}. Crack velocity was measured at temperatures ranging from 370 to 520 K and at a nominally constant stress intensity factor K_I of 17 MPa$\sqrt{m}$.

Compact tension specimens, 17 mm wide and 3.8 mm thick, were precracked by fatigue then loaded in tension at a K_I of about 6 MPa$\sqrt{m}$. The load on the specimen was increased by small increments (generally about 1 MPa$\sqrt{m}$), to estimate the V - K_I curve. Some specimens were tested isothermally at 370 and 470 K after cooling while others were cycled between 290 and 570 K with holding periods of 16 h at 370 K. Both acoustic emission and potential drop were used to monitor cracking.

Cantilever beam specimens were used to evaluate heat-treatment and texture effects while the compact tension specimens were also used for heat-treatment experiments as well as for the evaluation of yttrium as a getter. Some specimens were water quenched to refine the hydrides and enhance cracking. A valid fracture toughness, K_{IC}, of the Zr-2.5 Nb-1.4 Y alloy was determined using the ASTM E 399-81 standard test method.

RESULTS

The Effect of Heat-treatment on DHC

Crack velocities were obtained, using cantilever beam specimens, at various temperatures for Zr-2.5 Nb pressure tube material in the following conditions:

- as-received, 20% cold work,
- homogenized at 670 K for 24 h,
- homogenized at 670 K for 96 h with 0.5 to 1 at% hydrogen added.

Tensile properties after heat-treatment are included in Table 1 and the microstructures are shown in Fig. 4. The measured crack velocities are plotted versus temperature in Fig. 7. The lines represent linear regression of a standard Arrhenius plot.

In the second experiment the compact tension specimens containing 0.45 at% hydrogen were homogenised at 660, 620 and 580 K for 10 days then quenched into ice water to produce a distribution of small hydrides. Crack velocity at 470 K was shifted to lower values as the homogenization temperature was increased, but K_{IH} was unchanged, Fig. 8.

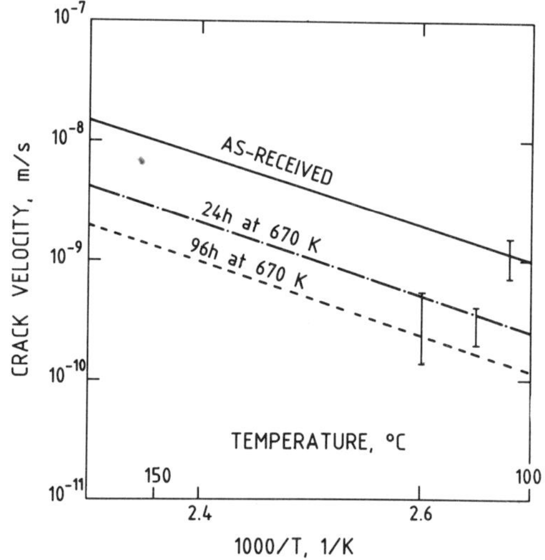

Fig. 7.   Effect of homogenization on crack velocity in Zr-2.5 Nb.   Bars represent range of data.

Yttrium Getter

Isothermal DHC tests were performed in air and in argon, at 370 and 470 K, on the as-received material with yttrium addition (containing 0.07 at% hydrogen) and on the material with 0.9 and 1.34 at% hydrogen added. The as-received yttrium alloy was also subjected to thermal cycling up to 570 K, at the rate of one cycle a day to further encourage DHC. During these thermal cycles the temperature was held at 370 K for 16 h.

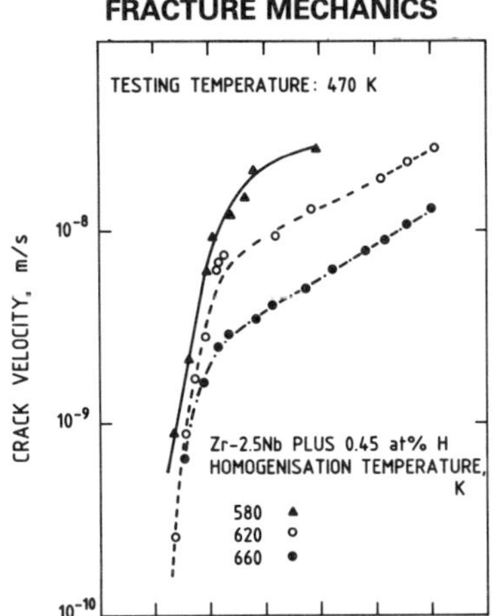

Fig. 8.   Effect of homogenization on V-K_I curve of Zr-2.5 Nb.

Isothermal tests on the as-received material showed no crack initiation after 350 h at 370 K and K_I of approximately 20 MPa√m either in longitudinal or transverse directions. The specimens were broken apart and metallographic examination of the fracture surface confirmed that no DHC had occurred. Under similar conditions, initiation times of 20-100 h have been observed in Zr-2.5 Nb material containing no yttrium.

After 130 thermal cycles under K_I of 20 MPa√m in air, the as-received specimens suddenly failed without any indication of previous cracking. These specimens were covered with a thick white oxide and cracking was probably by oxide wedging. Subsequent testing was done in a protective atmosphere of argon.

DHC was observed in material with a hydrogen concentration of 0.9 at%. Cracking was more difficult in the transverse direction than in the longitudinal direction. Even in the latter direction DHC in these materials generally required very high stresses (K_{IH} = 15 MPa√m), Fig. 9. Incubation times of about 200 h were required to initiate cracking in both furnace cooled and water quenched specimens; incubation times of a few minutes are observed in water quenched yttrium-free material. The V-K_I curve obtained for a specimen containing 1.34 at% hydrogen is included in Fig. 9. The curve is shifted toward a lower K_{IH}. Also the incubation time was reduced to 40 h. The behaviour of the yttrium alloy with 1.34 at% hydrogen was similar to that of the yttrium free material.

The room temperature fracture toughness was 41 MPa√m for as-received material and reduced to 29 MPa√m when 1 at% hydrogen was added.

Crystallographic Texture

Specimens stressed in the thickness direction of the plate (short transverse specimens) cracked faster than specimens stressed in the rolling directions (longitudinal specimens), Fig. 10. Both sets of data follow the standard Arrhenius-type relationship. The threshold stress intensity factor, K_{IH}, was measured at 373, 423, 473, and 523 K. K_{IH} in longitudinal specimens was 13.7 to 20.5 MPa√m and in short transverse specimens it was 9.1 to 10.3 MPa√m. No temperature dependency of K_{IH} was found in the temperature range used.

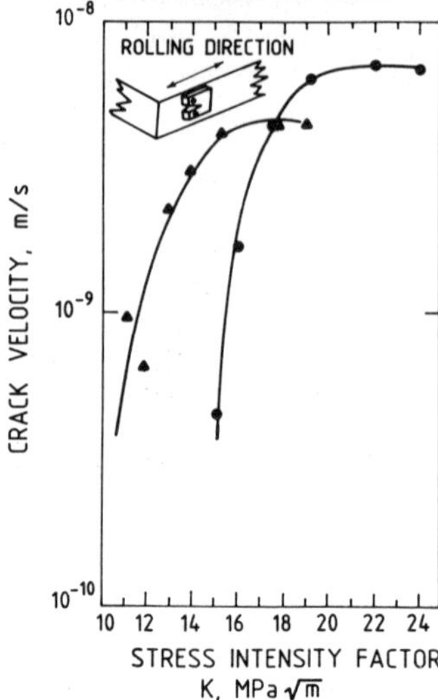

Fig. 9.   V-K_I curves of yttrium alloy at 470 K:
with 0.9 at% hydrogen addition ($\bullet$)
with 1.34 at% hydrogen addition ($\blacktriangle$).

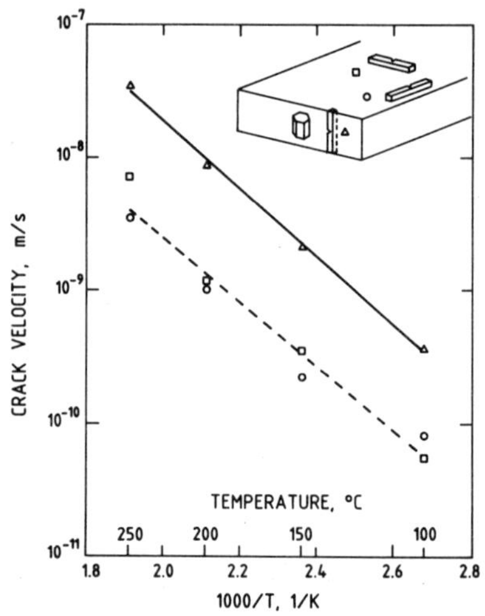

Fig. 10.   Effect of testing direction on crack velocity in warm rolled plate of
Zr-2.5 Nb.

DISCUSSION

The Effect of Heat-treatment on DHC

The experiments have shown that heating Zr-2.5 Nb material at moderate
temperatures reduces the velocity of DHC, Figs. 7 and 8, but has little effect on
K_{IH}, Fig. 8.   The slight decrease in tensile strength (Table 1) is unlikely to
contribute much to the change in crack velocity.   The main change in the
microstructure after heating is the breaking up of the β-phase ligaments, Fig. 4.
This decomposition of the β-phase is probably responsible for the decrease in
crack velocity (Simpson and Cann, 1984).   The extent of this decomposition

increases with increasing homogenization temperature and time at this temperature. Hence the specimens homogenized at higher temperature and time cracked more slowly. The contributing factors are a reduction in diffusivity by reducing the component of diffusion through the β-phase by as much as a factor of 5, and partitioning of the hydrogen between the two phases with concentrations being as much as twice as high in the β-phase as in the α-phase (Sawatzky and co-workers, 1982). Although the flux of hydrogen to the crack tip is affected by β-phase distribution, the conditions for cracking the hydrides are apparently not changed because K_{IH} is not changed.

Yttrium Getter

The results support the notion that hydrogen can be immobilized by yttrium acting as a getter. With low hydrogen concentrations, less than 0.1 at%, initiation of DHC was not achieved in times sufficient to initiate cracking in yttrium free material. At moderate hydrogen concentration, 0.9 at%, DHC was observed but only at high K_I values, but when the hydrogen concentration (1.35 at%) was similar to the yttrium concentration (1.44 at%), the DHC behaviour was similar to yttrium free material, Fig. 9. If the yttrium hydride is YH_2 the lack of complete gettering may be attributed to some of the yttrium combining with oxygen; if the hydride is YH_3 then the latter effect is even larger. The effect of gettering can be thought of as an increase in the solubility limit. Previous work showed that an yttrium alloy was the only one of several alloys of zirconium with other elements (Nb, Ti, Hf, In, Pb, Sb, and O) to show any increase in solubility limit over that of pure zirconium (Erickson and Hardie, 1964). Thus once the solubility limit is exceeded DHC is possible just as in yttrium free material.

Crystallographic Texture

The crack velocity in short transverse specimens was 5 to 8 times faster than the crack velocity in longitudinal specimens, Fig. 10. The decrease in crack velocity in the rolling directions can be attributed to:

- mode of hydride reprecipitation, and
- tensile strength.

Hydride reorientation is difficult if few basal plane normals are parallel to the tensile stress. If the hydrides are forced to precipitate away from their habit planes by adjusting the texture in such a way that the basal plane normals are perpendicular to the direction of the applied stress, the crack velocity will be reduced. In short transverse specimens when hydrides reprecipitate at the crack tip, little reorientation is necessary because initially the hydrides are almost parallel with the crack and the texture is favourable. The converse situation applies with longitudinal specimens.

The strength of the material also affects crack velocity. By increasing the strength, the plastic zone at the crack tip is reduced and the stress concentration is thus increased providing more driving force for the stress induced diffusion of hydrogen to the crack tip. An application of the theory of DHC (Dutton and co-workers, 1977) suggests that in the current example, strength and reprecipitation are contributing similarly to the crack velocity.

APPLICATIONS

To minimize DHC velocity in components made from extruded and cold worked Zr-2.5 Nb alloy, they should be heat-treated to agglomerate the β-phase. For example CANDU pressure tubes are heated to 670 K for 24 h in steam. This treatment provides a wear resistant oxide and appropriate microstructure with little loss in strength.

Yttrium additions to zirconium alloys improve their tolerance to hydrogen but fracture toughness and corrosion resistance (because yttrium is also an effective getter for oxygen) are reduced. The rapid oxidation during thermal cycling tests suggests that this material would need protective cladding, thus direct application of such alloys will require further development.

In Zr-2.5 Nb components, DHC velocity can be reduced and the threshold stress intensity factor increased if the fraction of basal plane normals parallel to the direction of the applied stress is minimized. In CANDU pressure tubes, the majority of the basal plane normals are oriented near the circumferential direction which is also the direction of the principal applied stress - hoop stress. Alternative manufacturing routes of the pressure tubes are being investigated to reduce the number of basal plane normals in this direction.

ACKNOWLEDGEMENTS

We would like to thank L.J. Clegg, B. Ellis, D.E. Foote, R.W. Gilbert, G.W. Newell, E.E. Sexton and J.E. Winegar for technical assistance. Valuable discussion with C.D. Cann, B.A. Cheadle, R. Dutton and C.E. Ells is gratefully acknowledged.

REFERENCES

Brown, B.F. (1966). _Material Research and Standards_, 6, 129-133.
Cheadle, B.A., C.E. Coleman, and M. Ipohorski (1984). Orientation of Hydrides in Zirconium Alloy Tubes. _ASTM STP 824_, 210-221.
Cheadle, B.A., C.E. Coleman, and J.F.R. Ambler (1987). Prevention of Delayed Hydride Cracking in Zirconium Alloys. _ASTM STP 939_, 224.
Coleman, C.E. (1982). Effect of Texture on Hydride Reorientation and Delayed Hydrogen Cracking in Cold Worked Zr-2.5 Nb. _ASTM STP 754_, 393-411.
Coleman, C.E., B.A. Cheadle, J.F.R. Ambler, P.C. Lichtenberger, and R.L. Eadie (1985). Minimizing Hydride Cracking in Zirconium Alloys. _Can. Met. Quart._, 24, 245-250.
Dutton, R., K. Nuttall, M.P. Puls, and L.A. Simpson (1977). Mechanisms of Hydrogen Induced Delayed Cracking in Hydride Forming Materials. _Met. Trans._, 8A, 1553-1562.
Elliot, R.P. (1965). _Constitution of Binary Alloys, First Supplement_, McGraw-Hill, New York, 863-865.
Erickson, W.H., and D. Hardie (1964). The Influence of Alloying Elements on the Terminal Solubility of Hydrogen in α-zirconium. _J. Nucl. Mat._, 13, 254-262.
Mueller, W.M., J.P. Blackledge, and G.G. Libowitz (1968). _Metal Hydrides_. Academic Press, New York, 286 and 451.
Sagat, S., J.F.R. Ambler, and C.E. Coleman (1986). Application of Acoustic Emission to Hydride Cracking. Atomic Energy of Canada Report, _AECL-9258_.
Sawatzky, A., G.A. Ledoux, R.L. Tough, and C.D. Cann (1982). Hydrogen Diffusion in Zirconium-Niobium Alloys. In T.N. Veziroglu (Ed.), _Proc. Miami Intern. Symp. on Metal-Hydrogen Systems_, Pergamon Press, Oxford. pp. 109-120.
Simpson, L.A., and C.D. Cann (1984). The Effect of Microstructure on Rates of Delayed Hydride Cracking in Zr-2.5 Nb Alloy. _J. Nucl. Mat._, 126, 70-73.
Spalthoff, W., and H. Wilhelm (1969). The Use of Hydrogen Getters for Prevention of Hydrogen Embrittlement in Zirconium-Alloy Fuel Cans. _ASTM STP 458_, 338-344. US Patent 1248 184 (1971) Westinghouse Electric Corporation.

SUBZERO TEMPERATURE TENSILE AND FATIGUE BEHAVIOUR OF SELECTED HSLA STEELS

W.J. Bratina and S. Yue*

Department of Metallurgy and Materials Science, University of Toronto,
Toronto, Ontario, Canada M5S 1A4
*Now with the Department of Mining and Metallurgical Engineering,
McGill University, Montreal, Quebec H3A 2T5

ABSTRACT

The tensile and fatigue behaviour of 2 experimental and 1 commercial HSLA steel was investigated over the temperature range 298 to 77 K, and discussed in terms of qualitative observations of the subsequent dislocation configurations. In the static tests, the homogeneity of plastic deformation appeared to be related to dislocation interaction between both coarse and fine precipitates. In fatigue, the absence of quasi-PSB's and well defined dislocation cell walls was mainly attributed to the presence of fine, non-coherent nitrides and carbonitrides.

KEYWORDS

Deformation, static, fatigue, cryogenic, dislocation-precipitation interaction, HSLA steels.

INTRODUCTION

The role of dislocation-interstitial interactions on various mechanical characteristics of a ferritic matrix are well known. The phenomena of yielding, the Cottrell condensed atmosphere, strain aging, the dependence of fatigue strength on temperature and on cyclic frequency, the response to under-stressing, 'coaxing', the mechanism of fatigue limit itself and the trend of ΔK_{th} values versus temperature (Paris and co-workers, 1972) all seem to be satisfactorily explained by the interaction of dislocations with the interstitial carbon and nitrogen in solid solution. However, the interactions of dislocations with very fine precipitates in a bcc lattice has received much less attention (Keh, Leslie and Sponseller, 1965) particularly in cyclically stressed material (McGrath and Bratina, 1965, 1967, 1969, 1970) or at cryogenic temperatures (Yue and Bratina, 1985; Bratina and co-workers, 1986).

Cryogenic tensile characteristics of various *high purity irons* have been investigated extensively (Low, 1963; Warren and Reid, 1963; Matuyama, Meshi and Oikawa, 1985; Hashimoto and Meshi, 1985). For iron single crystals, plastic strains to fracture of up to 0.27 have been reported at 77 K (Hashimoto and Meshi, 1985) and strains of 0.015 at 4.2 K for polycrystalline zone refined iron (Warren and Reid, 1963). In ferritic polycrystalline low carbon steels, the dependence of Luders strain and plastic strain to fracture with subzero temperatures has also been the subject of numerous investigations (Fisher and Rogers, 1956; Conrad, 1963; Love, Bratina and McGrath, 1965). There are always some differences in Luders band propagation characteristics associated with geometric factors affecting triaxiality and therefore with plastic constraint. Even so, in most cases, deformation is preceeded by Luders band localization, as is illustrated by the Nb microalloyed HSLA steel specimen strained in tension at 77 K shown in Fig. 1 On the

305

basis of these earlier studies, the large strains to fracture of about 0.30 and Luders strains of 0.09 which were observed at 77 K in some experimental and commercial HSLA steels, e.g. Fig. 2, were not expected, (Yue and Bratina, 1985; Krishnadev and Ghosh, 1979; Cutler and Krishnadev, 1982; Wilshynsky, 1982; Moon, 1983; Bratina and co-workers, 1986; Tseng, Goel and Tangri, 1986). It is the purpose of this paper to attempt an interpretation for the observed cryogenic behaviour of these steels in terms of dislocation mechanisms, particularly in terms of dislocation interactions with the second phase precipitates that are essential to HSLA steels, i.e. the carbonitrides of niobium, titanium, vanadium. Further, in those HSLA steels which exhibited large fracture strains at 77 K, SEM revealed, predominantly, quasi-cleavage facets with river patterns, i.e. essentially brittle fracture (Yue and Bratina, 1985). The apparent contradiction of large fracture strains in the order of 0.30 being associated with brittle fracture surfaces is addressed in this paper. Finally, fatigue dislocation structure and crack propagation of HSLA steels at cryogenic temperatures are investigated in the present work.

MATERIALS AND EXPERIMENTAL PROCEDURE

Three groups of HSLA steels were investigated. The first two groups, designated as A and B, were comprised of experimental polygonal ferrite steels with Nb as the main microalloying element; the third group consisted of commercial ferrite-pearlite HSLA steels. The low carbon (0.05) niobium (0.022-0.042) containing steels (Group A) were received as cold rolled light gauge ($\sim$ 0.7 to 2.3 mm) strip, the full chemical analysis being: 0.28 Mn, 0.065 Al, 0.002 P, 0.014 S, 0.012 Si, 0.052 As, 0.008 Cr, 0.073 Ni, 0.009 Mo, 0.006 W, 0.002 Ta and 0.023 Co, all in wt. pct. The second group of steels (Group B) contained nominally higher carbon (0.13) levels, and a niobium content of 0.023 and was also received as cold rolled light gauge ($\sim$ 1.2 mm) strip. The other main alloying elements were 0.46 Mn and 0.63 Cr, hence these two groups differed largely in C and Cr content. The commercial HSLA steels contained $\sim$ 0.07 C and were received as a hot rolled plate (finishing temperature about 1073 K) of 12.7 mm thickness. The nominal composition of this material was: 0.007 N, 1.43 Mn, 0.003 P, 0.004 S, 0.29 Si, 0.25 Cr, 0.25 Ni, 0.075 V, 0.020 Nb and 0.011 Ti. All materials were fully stress relieved in a Centorr furnace, in a vacuum better than 10^{-5} Torr, at 940 to 970 K for 2 h followed by furnace cooling. The purpose of this heat treatment was to remove the rolled in dislocation density.

Specimens were subjected to monotonic tensile stresses in an Instron electro-mechanical apparatus, in most cases at a strain rate of $\sim$ 10^{-3} s^{-1}, over the temperature range from 77 to 373 K. Subzero temperatures were obtained by means of a liquid nitrogen cryostat. Cyclic loading was performed on Instron or MTS servohydraulic equipment, generally at R = 0.1 and a frequency of 10 to 20 Hz. Tests at 298 K were performed mostly in air, those at cryogenic temperatures in a nitrogen atmosphere, and at 77 K by immersion in liquid nitrogen. TEM was performed on a Phillips EM 300G microscope at 100 kV and on a Hitachi-H-800 at 150 kV. Fracture surfaces were examined using a Hitachi S520 SEM at 20 kV.

RESULTS

Static Stress-Strain Curves

The full stress-strain curves for one of the steels from group A have been published earlier (Yue and Bratina, 1985), although subsequent work has revealed the strain to fracture at 298 K to be more commonly of the order of 0.30, as opposed to the value of 0.15 reported in the aforementioned paper. Further work on strain rate and temperature effects on the Luders strain magnitude is summarized in Fig. 3. This group of steels, in the annealed condition and with a light gauge is plastically unstable even in the neighbourhood of 273 K and at relatively low strain rates. The yield stress at 298 K was in most cases comparable to or even higher than the corresponding UTS. At 77 K, all these steels were characterized by severe Luders band localization, Fig. 1, and a low strain to fracture. By contrast, in the as-received cold rolled condition the values at 293 K were : UTS = 800 MPa, and strain to fracture 0.011; at 77 K: UTS = 1200 MPa and strain to fracture 0.032. Note the increase in strain to fracture with the lower temperature.

The stress-strain curves for the Group B steels exhibited very large fracture strains at all temperatures. This is illustrated by the curves for the high and low temperature extremes seen in Fig. 2. The Luders strain data for this steel, in the form of an engineering stress-strain curve (Yue and Bratina, 1985), is reproduced in Fig. 4. A reasonably smooth line can be drawn through the points indicating the onset of work hardening. Thus, extrapolating the data to temperatures below 77 K indicates that still larger Luders strains are feasible, although twinning may be activated at such temperatures and could reduce the fracture strain drastically.

The commercial HSLA steels behaved in a manner similar to the above grade. Large elongations were observed at all temperatures, including 77 K and these steels fractured at 77 K with little or no necking. The relevant data is as follows :

	293 K		77 K	
	As-Received	Annealed	As-Received	Annealed
yield (MPa)	440	510	920	980
UTS (MPa)	540	525	920	980
fracture strain	0.37	0.38	0.31	0.29

Microstructural Features of Annealed Material

The main features of the steels with low fracture strains at 77 K (Group A) were a somewhat uneven dispersion of spheroidal $Nb(C,N)$ of up to 40 nm in diameter and an extremely low dislocation density. These annealed-in dislocations appeared to be strongly pinned by $Nb(C,N)$. A few isolated, coarse Fe_3C precipitates of up to 0.5 µm in diameter, approximately one precipitate per grain, were also observed. The average grain size varied from 3-5 µm to 5-10 µm for the lighter (0.7 mm) and the heavier (2.3 mm) gauge material respectively. Since there was no possibility to apply the well known Snoek maximum technique, a crude method of yield point return was employed to generally assess of the amounts of C + N in solid solution. To summarise the results of these experiments, very weak yield point returns were observed after aging at 373 K, indicating some C and N in solution.

A significant feature of the HSLA steels exhibiting large fracture strains at 77 K (Group B) was a high volume fraction of coarse ($\sim$ 0.5 µm) Cr containing Fe_3C type carbides in polygonal ferrite grains of 7-10 µm in diameter. A few annealed-in dislocations were observed which were associated largely with coarse precipitates and pinned by fine $Nb(C,N)$.

In the commercial HSLA steels studied, a ferrite-pearlite structure was observed. The pearlite was present in a banded distribution and ferrite grains were elongated in the rolling direction with the length to width ratio of 1.39. The mean grain size was 6-7 µm. TEM revealed a variety of second phase particles, such as spheroidal $Nb(C,N)$ and cuboidal TiN.

TEM of Statically Deformed Material

Dislocation structures were studied after stressing to the upper yield point, at 50 per cent into Luders strain and at fracture, over the whole temperature range from 77 to 298 K.

In the steel characterized by low fracture strains at 77 K (Group A) the dislocations appeared to be associated with fine Nb(C,N) precipitates, grain boundaries and occasionally with a few widely spaced Fe_3C precipitates. At 298 K the dislocation structure was composed of short segments pinned by fine Nb(C,N) resulting in a very jogged pattern with considerable tangling and numerous dislocation loops. With decreasing deformation temperature, this configuration was gradually replaced by arrays of straight, parallel dislocations, of up to several μm long, typical of the cryogenic temperature deformation of a bcc lattice (Fig. 5). The configuration of these arrays appears to indicate that they are largely imperturbed by the presence of the numerous, fine Nb(C,N) precipitates. However, a heavily jogged dislocation structure resembling the 298 K structure was also present as can be seen in Fig. 5, i.e. 189 K, and even at 77 K, forming a two dimensional grid, a characteristic commonly associated with a latter stage of deformation. In general, at a constant temperature, an increase in dislocation density with increasing strain was noted, and, for the same strain, the dislocation density associated with the low temperature deformation appeared to be considerably higher compared with that at 298 K. TEM of the structure of specimens taken from regions close to the fracture surface was similar for both temperature extremes and revealed very dense dislocation clouds.

The most notable feature of the HSLA steel exhibiting large fracture strains at 77 K (Group B) was an additional strong dislocation activity in the proximity of coarse precipitate groups observed over the whole temperature range investigated. In the Luders region at 298 K a largely radial configuration resulted with the dislocations apparently emanating from the coarse spherical precipitates (Yue and Bratina, 1985), and at 77 K, the otherwise straight, low temperature dislocations appeared to be significantly perturbed in the vicinity of these precipitates with considerable tangling as a result, Fig. 6. Again there was little difference in the dense dislocation structures observed in areas close to the fracture surface of specimens tested at both 77 K and 298 K.

The dislocation pattern of the commercial HSLA steels has not been extensively examined to date. Preliminary observations of specimens deformed at 298 K revealed jogged, tangled dislocations pinned by a variety of second phase precipitates. TEM data of specimens deformed at 77 K is not available yet.

Finally, twins were not observed in this work in any of the HSLA steels investigated at any test temperature.

<u>SEM Fractography</u>

The HSLA steels with low strains to fracture at 77 K (Group A) exhibited ductile fracture surface characteristics, with relatively uniform dimple sizes of up to 2-3 μm in diameter, throughout the temperature range from 77 to 298 K. However, at 77 K, depending on the gauge of the material, a considerable amount of quasi-cleavage facetting was also observed.

In the HSLA steels with large fracture strains at 77 K (Group B) fractography revealed small, ductile dimples of up to 2-3 μm in size and some very large dimples of up to 7-10 μm in diameter for specimens fractured at 298 K, Fig. 7. At 77 K, quasi-cleavage facets with typical river patterns largely prevailed, although fine dimples proliferated at the boundaries of facets together with numerous secondary cracks running perpendicular to the fracture surface, Fig. 8. This gradual transition from a ductile to a quasi-cleavage, faceted fracture surface is apparently contrary to the fracture strain behaviour which indicates comparable or even larger elongations at cryogenic temperatures (Fig. 2).

The fracture behaviour of commercial HSLA steels was almost identical to Group B. As the temperature was decreased from 298 to 77 K, the ductile dimples were gradually replaced with quasi-cleavage facets and secondary cracks, although there was still a surprising amount of deformation via ductile dimples. As in the previous case, a transition from ductile to brittle fracture surface characteristics occurred in spite of comparable total elongations at the two corresponding temperatures.

Fatigue Stressing

Specimens of the HSLA steels with low fracture strains at 77 K (Group A), of geometry similar to that in Fig. 1, but with a shorter gauge length, were cyclically stressed in tension under load control at 10 Hz and at R = 0.1. Prior to fatiguing, the specimens were prestrained at 298 K so as to propagate the Luders band along the entire gauge length, i.e. upto the work hardening stage, thus homogenising the deformation. Without prestraining, fracture tended to occur just prior to or immediately after the initial Luders band had traversed the gauge length. This was thought to be related to the pronounced yield point and very low work hardening characteristics of the annealed material (Yue and Bratina, 1985), combined with the fact that the tests were performed under load control. Since prestraining can be considered merely as 1/4 of the first cycle of fatigue, it is argued that it would not influence the subsequent fatigue dislocation structure. The results at 298 K in air and at 77 K in liquid nitrogen are reported here. Note that the main purpose of this work was the characterization of the fatigue dislocation structures and the fracture surface topography, primarily at cryogenic temperatures. Less attention was therefore paid to the actual stress levels and the number of cycles to which the specimens were subjected.

Since this material exhibited ductile dimples on the fracture surfaces after uniaxial tensile testing over the entire temperature range investigated, i.e. from 77 to 298 K, fatigue striations were expected in the same temperature range. At 298 K, SEM revealed well defined ductile fatigue striations with spacings of about 0.5 µm. Due to the very small specimen dimensions, of the order of millimeters, there was little variation in striation spacings across the fracture surface. At 77 K, the region of fatigue fracture, as indicated by the extent of the striated region, was only a few hundred µm. In this region, SEM revealed fatigue striations, Fig. 9, again with a spacing of about 0.5 µm. The final fracture featured many of the characteristics of brittle fracture, including quasi-cleavage facets, river patterns, secondary cracks, along with ductile dimples bordering the facets, Fig. 10. This is in contrast to the tensile dimpled fracture of this material at 77 K even though the final fracture in fatigue nominally occurs in the tensile mode.

TEM at 298 K revealed some characteristics in common with those observed in monotonic static deformation. A heavily jogged dislocation configuration and the absence of well defined fatigue dislocation cells were evident although there were high density dislocation aggregates resembling the cell walls, but the size of these cells appeared to be too large. Tangles of dislocations and dislocation loops appeared to be associated with fine Nb(C,N) precipitates. Higher dislocation densities were almost invariably found close to grain boundaries but to a greater degree than in static stressing. This is in agreement with previous work concerning fatigued iron-carbon alloys (McGrath and Bratina, 1965). The structure due to fatiguing at 77 K revealed a mixture of dislocation structures. In some areas long, straight, parallel dislocations were observed. In regions of higher dislocation density, however, heavily jogged dislocations and dislocation loops apparently associated with fine Nb(C,N) precipitates were visible, Fig. 11. These areas greatly resembled the 298 K fatigue dislocation structure. Again, the absence of a well defined fatigue cell structure was noted, although dense dislocation structures resembling cell walls were visible.

No work on cyclic stressing the Group B HSLA steels was performed.

To date, the commercial HSLA steels have been employed only in fatigue crack propagation studies. Conventional da/dN vs. ΔK curves were determined at the two temperature extremes of 77 and 298 K, using standard fracture mechanics specimens and techniques. Testing was performed in an atmosphere of dry argon at 298 K and liquid nitrogen at 77K, at a frequency of 20 Hz and at R = 0.1. SEM fractography revealed at 298 K the usual fatigue striations with a spacing from 0.12 to 0.23 µm. The fracture surface at 77 K was essentially brittle, and included quasi-cleavage facets, river patterns, secondary cracks and ductile dimples bordering the facets . A complete absence of fatigue striations was noted.

DISCUSSION

<u>Tensile Stressing</u>

The HSLA steels in the annealed condition exhibited extremely low numbers of dislocation that were strongly pinned by fine carbonitrides. Furthermore, since depinning from a condensed Cottrell interstitial atmosphere was never demonstrated, it may therefore be concluded that dislocation generation must depend on the stress concentrations which are associated with the fine carbonitrides, grain boundaries and with coarse Fe_3C precipitates.

On the basis of the TEM studies in HSLA steels of Group A, fine Nb(C,N) precipitates appear to act primarily as obstacles to moving dislocations and not as generators. This can partly explain the high yield stress ($>$ 400 MPa), and the high flow stress, ($\sim$ 400 MPa) at 298 K. Dislocation generation assisted by the fine precipitates would tend to *lower* the respective yield stress and to *increase* the flow stress as these precipitates would start acting as obstacles to moving dislocations. The very low work hardening and the observed plastic instability in these steels is difficult to explain. It may be that the interaction of dislocations with fine precipitates prevents dislocation-dislocation interactions required for an effective work hardening mechanism. It is likely that the Nb(C,N) precipitates assist in forming the observed jogged and tangled structure. However, the strain field of fine precipitates extends over a very limited range and the interaction between a dislocation and a precipitate can occur only when a dislocation is brought within the precipitate strain field range either by applied stress or by thermally activated motion. Further proof of thermally assisted dislocation-fine precipitate interaction is the configuration at cryogenic temperatures, Fig. 5. The observation that long, straight dislocations appear to be largely unperturbed by the presence of fine precipitates may be attributable to the absence of thermally activated motion. Considerably jogged dislocation networks were, however, also observed at 189 K and even at 77 K, particularly in the regions of high dislocation densities, i.e. in the areas of higher deformation, where stress assisted dislocation motion cannot be ignored. In general, with a decrease in temperature, fewer jogs and tangles can be expected as the dislocation configuration gradually transforms into long, straight, parallel dislocations. Therefore, work hardening, via dislocation/dislocation interaction, should decrease accordingly, and Luders band propagation would become increasingly more difficult, resulting in a strong localization of deformation, as in Fig. 1.

SEM data for these HSLA steels indicates void initiation and coalescence behaviour. Numerous dimples are associated with fine Nb(C,N) i.e. the mechanism of dimple formation is greatly assisted by the presence of fine precipitates. The tearing mechanism via dimples is as efficient at 298 K as at 77 K, although quasi-cleavage fracture processes are also operative at the lower temperature.

The behaviour of HSLA steels with large fracture strains at 77 K (Group B), as well as that of the commercial HSLA steels, is difficult to rationalise. Large strains to fracture can be accounted for by slip, by grain boundary sliding and by crack formation and growth. Since no cracks were observed, all the strain is due to plastic deformation. Although, as seen in Figs. 2 and 4, a part of fracture strain at 77 K can be accounted for partly by an increase in Luders strain, from 0.3 at 298 K to 0.9 at 77 K, and partly by necking, there still remains a very large nearly uniform elongation. The deformation by slip was also confirmed for commercial HSLA steels as an average length to width ratio of individual grains increased from $\sim$ 1.39 for undeformed material to $\sim$ 1.61 for deformed material, i.e. an increase of $\sim$ 16 per cent. Furthermore, slip bands were observed on the surface of specimens strained at 77 K. A phenomenological interpretation of these observations has been advanced recently based on strain rate sensitivity (Tseng, Goel and Tangri, 1986) and an analysis has been performed by Enomoto and Furubayashi (1979) employing the constitutive relationship by

Hart which can at least explain the increasing Luders strain with decreasing temperature. With respect to the influence of microstructure on this phenomenon, there appears to be no correlation with grain size. It was proposed earlier by the present authors (Yue and Bratina, 1985) that the additional dislocation interactions associated with the dispersion of fine and coarse precipitates must be necessary in order to sustain the work hardening process required for Luders band propagation over the entire temperature range from 77 to 298 K. Certainly, the configuration of radially emanating dislocations associated with groups of coarse Fe_3C type precipitates at 298 K (Yue and Bratina, 1985) may be indicative of dislocation generation by these precipitates. Also, in Fig. 6 the low temperature dislocations are seen to be greatly perturbed by groups of medium size Fe_3C type precipitates resulting in dense dislocation tangles. However, the exact relevance of these interactions to the fracture strain behaviour has not been satisfactorily explained.

At 298 K, void initiation and coalescence behaviour is again the fracture mechanism for these steels (Fig. 7). Small dimples are presumably associated with fine Nb(C,N) and large with coarse Fe_3C type precipitates. However, the quasi-cleavage fracture (Fig. 8) associated with large fracture strains at 77 K (Figs. 2 and 4) appear to be contradictory. Noting the presence of secondary intergranular cracks in the fracture surface (Fig. 8), it is thus proposed that quasi-cleavage facets are not necessarily transgranular, but mixed intergranular, i.e. associated with grain boundaries in a manner similar to the secondary cracks. The mechanism of final fracture would be then mixed intergranular fracture assisted by slip as evidenced by the ductile dimples and the slip band traces that are present even at 77 K, Fig. 8. This could result in a combination of large strains to fracture and a quasi-cleavage, nearly brittle fracture surface appearance.

The arguments advanced above regarding the work hardening mechanisms and brittle fracture at cryogenic temperatures could as well apply to commercial HSLA steels. A variety of second phase particles in these steels may account for the work hardening mechanisms required for Luders band propagation at cryogenic temperatures. However, more extensive TEM work on these steels is necessary to define the mechanisms.

Fatigue Stressing

Fatigue striations in HSLA steels, Group A, at both 77 K (Fig. 9) and at 298 K were noted. The brittle final fracture appearance at 77 K (Fig. 10) can be explained in terms of fracture mechanics considerations and K_{Ic} values. The two limiting values, ΔK_{th} and K_{Ic} approximately define the exponent in Paris equation. According to the collection of data by Gerberich and Moody (1979), the exponent varies for different bcc iron base alloys from ~ 4 at 300 K to ~ 24 at 77 K. By comparing the striation pattern and ΔK values at 298 and at 77 K, it appears that the initial portion of da/dN vs. ΔK slope in the Paris region at 77 K does not differ from the 298 K slope, therefore the exponent ~ 4 would apply for early crack propagation stage. However, after reaching the critical crack size of ~ 200 μm for this particular steel and specimen geometry the resulting catastrophic fracture at 77 K, Fig. 10, would correspond to a high exponent in the order of 24 or even more. It is obvious that at 77 K triaxial stress and strain conditions must be taken into consideration and notch assisted cryogenic embrittlement occurs. This is in agreement with the strong temperature dependence of crack growth exponent, elegantly rationalized in terms of ductile and brittle components by Gerberich and Peterson, 1982.

There are still unresolved questions regarding the presence of quasi-PSB's and fatigue dislocation cell structures in HSLA steels subjected to cyclic stressing. Dislocation cell structure formation should be strongly inhibited by the presence of numerous fine second phase precipitates, at higher dislocation densities, i.e. after continued fatiguing,and in the present work only the aggregates of high density dislocations resembling the cell walls and not fatigue dislocation cells were observed. Furthermore, quasi-PSB's were not detected either at 298 or at 77 K. The formation of quasi-PSB's in precipitation hardened bcc (McGrath and Bratina, 1965, 1967) and fcc (Vogel, Wilhelm and Gerold, 1982) alloys seems to be conditional on shearing of precipitates by to and fro moving dislocations. The second phase precipitates in HSLA steels, although some of them possessing a certain degree of coherency with ferrite matrix, are in general very stable, and shearing via moving dislocations is difficult to imagine. At 77 K for example, Orowan type loops associated with precipitates of ~ 10 nm in diameter were observed. However, it must be noted that both dislocation cells and quasi-PSB's have been reported in a commercial HSLA steel fatigued at ambient temperature (Gonzales and Laird, 1983).

In the commercial HSLA steel investigated in this study, macroscopic (da/dN) and microscopic (fatigue striation spacing) values did not coincide. The tortuosity of the actual fatigue crack path is responsible for this, sometimes up to one order of magnitude disagreement as discussed previously in detail (Bratina and Yue, 1985). The da/dN vs. ΔK curves in the vicinity of 77 K should be viewed with caution. Crack propagation may be catastrophic, i.e. static or nearly static, as indicated by the fracture surface topography. Quasi-cleavage facets, river pattern, secondary cracks and some ductile dimples were observed with a complete absence of resolvable fatigue striations. Normally, in HSLA steels, striations with spacings smaller than 0.05 µm or larger than 10 µm are difficult to resolve or recognize. Ductile dimples can be an indication of static stressing or of a high fatigue crack propagation rate. SEM fractography in this work was mainly used in order to study the ductile to brittle transition under cyclic conditions, similarly to the work of Gerberich and Peterson (1982), in an anology with FATT (Fracture Appearance Transition Temperature).

CONCLUSIONS

In the HSLA steels examined in this work, with decreasing temperature, the dislocation configuration in both static and fatigue deformation transformed gradually from tangles of heavily jogged dislocations to arrays of occasionally jogged, parallel dislocations, several microns in length. The diminishing of the jogged nature of the dislocations was thought to be connected with the decreased thermal activation of dislocation movement, leading to decreased dislocation / precipitation interaction.

In static tensile testing, the onset of deformation localisation with decreasing temperature appeared to be dependent on precipitate characteristics. The grade with a distribution of fine carbonitrides only, exhibited strong localisation during the Luders extension at temperatures slightly lower than 293 K, for specimens of the geometry chosen for this work. By contrast, steels with a distribution of clusters of coarse precipipitates, in addition to the usual distribution of fine carbonitrides, demonstrated little tendency for localisation behaviour, thus leading to relatively high tensile fracture strains, even at 77 K. The commercial ferrite / pearlite steel exhibited similar behaviour, suggesting that the variety of precipitates present in this structure was equivalent to the coarse precipitate clusters. Heavy dislocation activity was observed at the clusters of large precipitates at all temperatures, perhaps offsetting localisation at low temperatures, although the exact mechanisms are unclear.

In fatigue, the lack of a dislocation cell structure was attributed to the presence of the fine carbonitrides acting as barriers to dislocation movement, although the subsequent development of a cell structure with increased cycling and/or higher stress levels is not ruled out. The absence of PSB's was thought to be due to the high resistance to shearing, by moving dislocations, of these stable precipitates.

Finally, the steel posessing a high tensile fracture strain at 77 K, exhibited a brittle fracture surface both in tensile testing and in the final stage of fatigue failure. This apparently contradictory behaviour was rationalized, in the former case, by postulating a mechanism of intergranular fracture assisted by slip.

ACKNOWLEDGEMENTS

The continuous financial support of the Natural Sciences and Engineering Research Council of Canada is gratefully acknowledged. Thanks are also due to the Algoma Steel Corp. (Sault Ste. Marie) for technical assistance, and M. Aziza and A. C. Rinella who assisted in the laboratory studies.

REFERENCES

Bratina, W.J., and S. Yue (1985). Fatigue crack growth - A metallurgist's point of view. In A.S. Krausz (Ed.), Time Dependent Fracture, Martinus Nijhoff Publishers, Hague, pp. 27-42.

Bratina, W.J., S. Yue, M. Aziza and A.C. Rinella (1986). HSLA Steels. Proceedings HSLA 85 Steels Conference, Beijing, China, ASM International, Metals Park, pp. 763-770.

Conrad, H. (1963). Yielding and flow of iron. In C.W. Spencer and F.E. Werner (Eds.), Iron and its Dilute Solid Solutions, Interscience Publishers, New York, pp. 315-339.

Cutler, L.R., and M.R. Krishnadev (1982). Proceedings 6th ICSMA, Vol. 1, Pergamon Press, Oxford, pp. 161-166.

Enomoto, M., and E. Furubayashi (1979). Scripta Metall.,13, 113-117.

Fisher, J.C., and H.C. Rogers (1956). Acta Metall., 4, 180.

Gerberich, W.W., and N.R. Moody (1979). A review of fatigue fracture topology effects on threshold and growth mechanisms. In Fatigue Mechanisms, ASTM STP675, pp. 292-341.

Gerberich, W.W., and K.A. Peterson (1982). Micro and macro mechanics aspects of time dependent crack growth. In Micro and Macro Mechanics of Crack Growth, TMS AIME, pp. 1-17.

Gonzales, G., and C. Laird (1983). Metall. Trans., 14A, 2507.

Hashimoto, K., and M. Meshi (1985). Proceedings 7th ICSMA, Vol. 1, Pergamon Press, Oxford, pp. 379-384.

Keh, A.S., W.C. Leslie, and D.L. Sponseller (1965). In Precipitation from Iron-Base Alloys, Gordon & Breach, New York, p. 281.

Krishnadev, M.R., and R. Ghosh (1979). Metall. Trans., 10A, 1941.

Love, R.E., W.J. Bratina, and J.T. McGrath (1965). Proceedings 5th Internat. Conf. Acoustics, Liège, p. D42.

Low, J.R. (1963). The deformation and fracture of iron. In C.W. Spencer and F.E. Werner (Eds.) Iron and Its Dilute Solid Solutions, Interscience Publishers, New York, pp. 217-269.

Matuyama, K., M. Meshi and H. Oikawa (1985). Proceedings 7th ICMSA, Vol. 1, Pergamon Press, Oxford, pp. 373-378.

McGrath, J.T., and W.J. Bratina (1965). Phil. Mag., 12, 1293-1305.

McGrath, J.T., and W.J. Bratina (1967). Acta Metall., 15, 329-339.

McGrath, J.T., and W.J. Bratina (1969). Czech. J. Phys., B19, 284-293.

McGrath, J.T., and W.J. Bratina (1970). Phil. Mag., 21, 1087-1091.

Moon, D.S. (1983). B.A.Sc. Thesis, University of Toronto, Toronto, Ontario.

Paris, P.C., R.J. Bucci, E.T. Wessel, W.G. Clark, and T.R. Mager (1972). ASTM STP 513. After Ritchie, R.O. (1979). Intern. Metals Reviews, Review 245, p. 219.

Stonsifer, F.R. (1976). After Application of Fracture Mechanics (1982). ASM International, Metals Park, Ohio, p. 83.

Tseng, D., N.C. Goel, and K. Tangri (1986). Metall. Trans., 17A, 697-702.

Vogel, W., H. Wilhelm, and V. Gerold (1982). Acta Metall., 30, 21.

Warren, K.A., and R.P. Reed (1963). NBS Monograph, 63, Washington.

Wilshynsky, D.O. (1982). B.A.Sc. Thesis, University of Toronto, Toronto, Ontario.

Yue, S., and W.J. Bratina (1985). Proceedings 7th ICSMA, Vol. 1, Pergamon Press, Oxford, pp. 201-206.

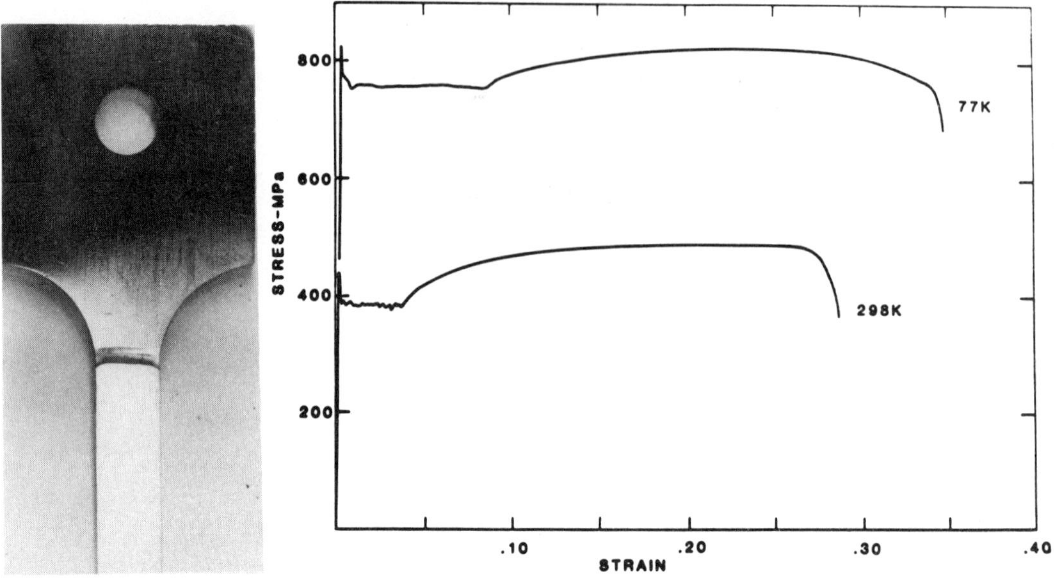

Fig. 1 Lüders band
localization,
group A
steels.

Fig. 2 Stress/strain curves, group B steels. Note increased
Lüders and total strain to fracture at 77 K.

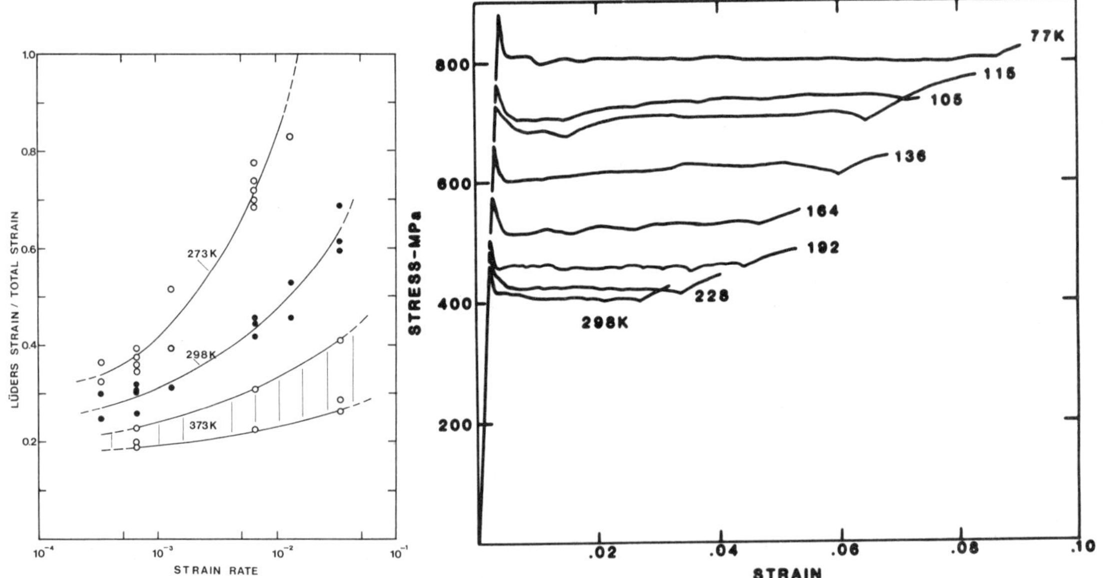

Fig. 3 Variation of Lüders
strain:total strain ratio
with temp and strain
rate, group A steels.

Fig. 4 Increasing Lüders strain with decreasing
temperature, group B steels.

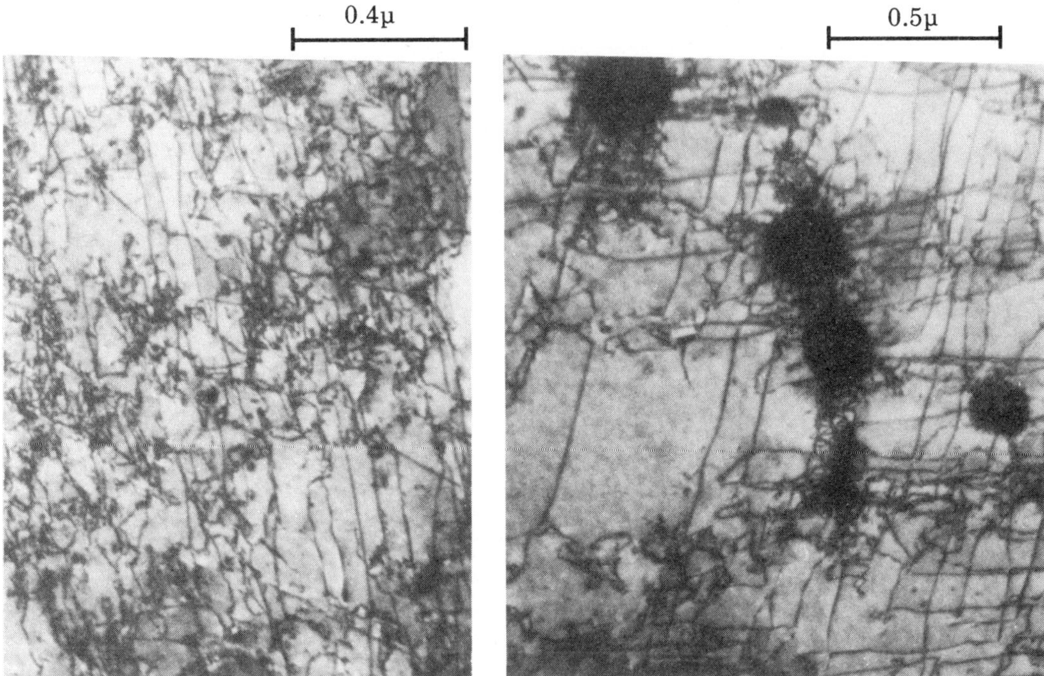

Fig. 5 TEM group A steels, 189 K.

Fig. 6 Note dislocation activity at large precipitates in group B steels, 77 K.

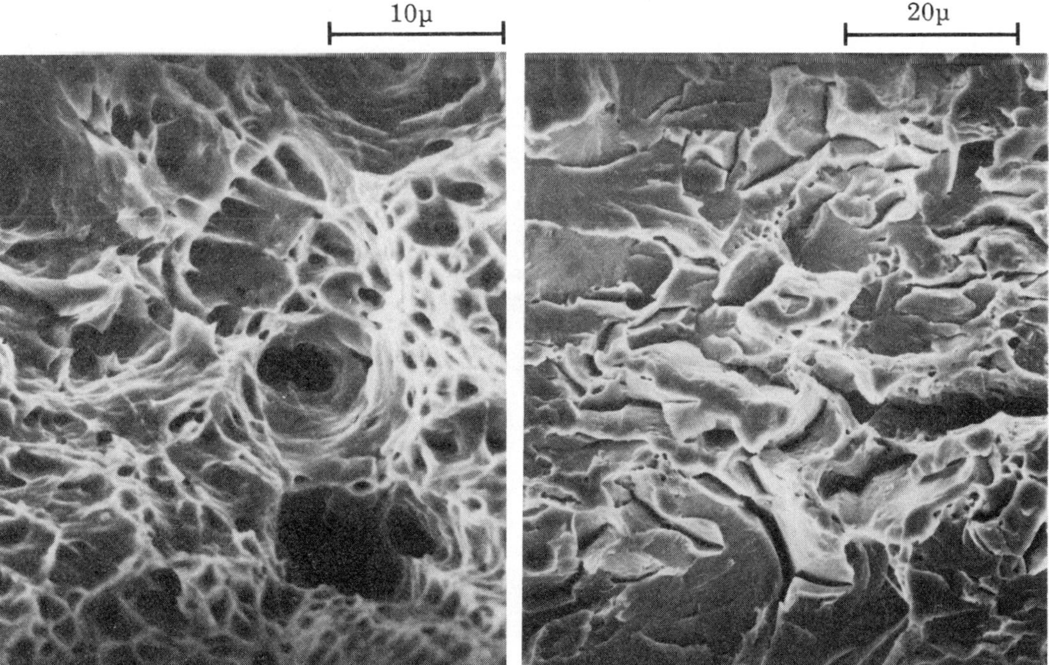

Fig. 7 Tensile fracture surface, group B, 298 K.

Fig. 8 Tensile fracture surface, group B, 77 K.

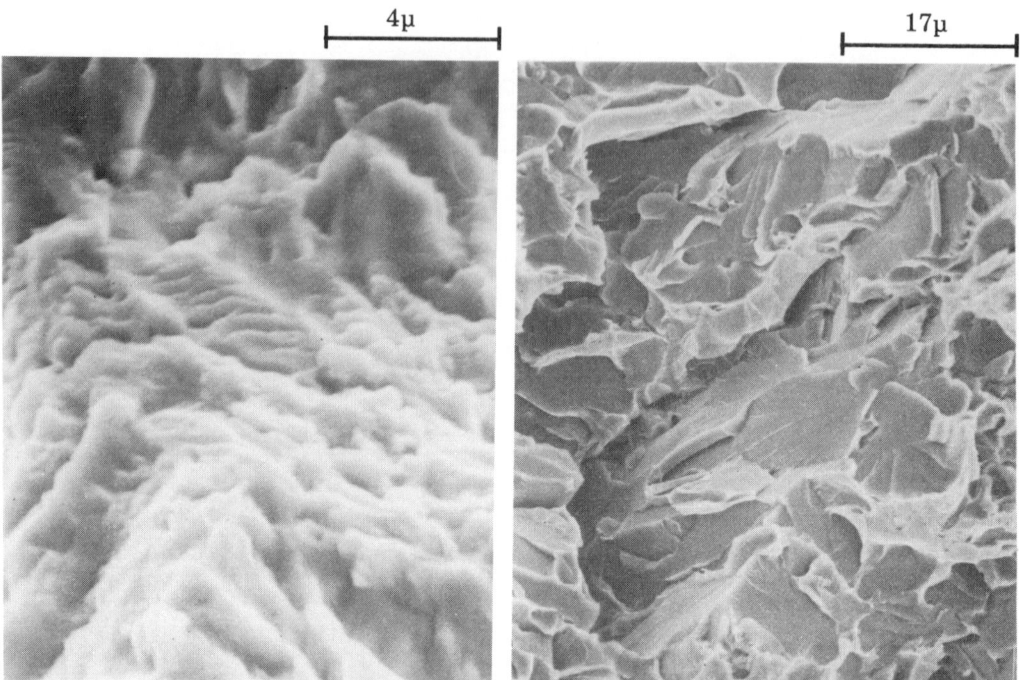

Fig. 9 Fatigue striations, group A, 77 K.   Fig. 10   Final fracture of fatigued specimen, group A, 77 K.

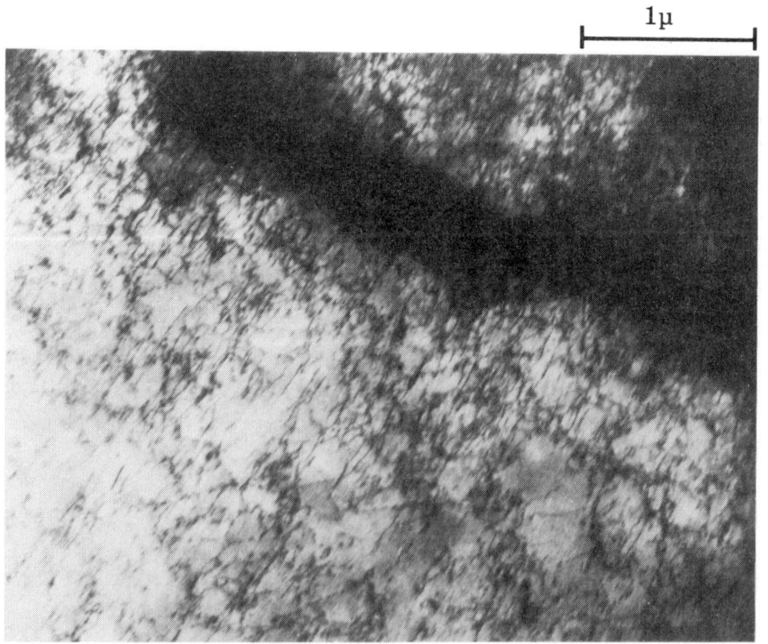

Fig. 11   Fatigue dislocation configuration, group A, 77 K.

HEAT AFFECTED ZONE CRACKING IN SUPER ALLOYS

K. J. ARMSTRONG and N.L. RICHARDS
BRISTOL AEROSPACE LIMITED, P.O. BOX 874
WINNIPEG, MANITOBA, CANADA R3C 2S4

ABSTRACT

Whilst Electron Beam Welding [E.B.W.] is a viable production route for welding the Super Alloys, metallurgical observations of these welds have frequently shown both weld metal and heat affected zone [H.A.Z.] defects. This investigation examined the H.A.Z. cracking of cast nickel based Super Alloy Inconel 718 and a wrought iron based alloy, Incoloy 903.

In both alloys, H.A.Z. cracking was observed after Electron Beam Welding, the defects being associated with grain boundary liquation of matrix phases. In Alloy 718, the liquation/cracking reaction was associated with Laves phases, whilst in the Incoloy 903 alloy, cracking was due to liquation of primary carbides of the MC type.

KEYWORDS

Electron Beam Welding, Heat Affected Zone Cracking, Grain Boundary Liquation, Metallography, Super Alloys.

INTRODUCTION

The nickel based Super Alloys are generally considered difficult to weld, especially those alloys utilizing a precipitation hardening reaction involving titanium and/or aluminum (Betteridge and Heslop, 1974). Cracking can occur during welding in both the weld metal and heat affected zone [H.A.Z.] and during subsequent post weld heat treatment [P.W.H.T.].

Heat affected zone [H.A.Z.] cracking has been attributed to (Baker, 1976; Owczarski and co-workers 1966; Thamburaj and co-workers, 1983);

(a) liquation cracking - where, due to local compositional variations arising from segregation, melting of the grain boundaries is possible.

(b) constitutional liquation - of primary carbides or carbo-nitrides.

(c) sub-solidus ductility dip.

The liquation cracking mechanism is related to the formation in intergranular regions in the H.A.Z. of a liquid phase which is too weak to withstand local thermal
stresses. Not all grain boundaries are affected however, and cracks that do form
are typically from one grain diameter (several tens of microns) up to 1.2mm long.

H.A.Z. cracking of Inconel 718 has been observed by several authors. Duval and
Owczarski (1967) observed liquation reactions occurring on heating during simu-
lated H.A.Z. experiments using the Gleeble, with cracking being associated with
MC carbides. Bologna (1969) using high voltage Electron Beam Welding (150 KV)
found the alloy susceptible to H.A.Z. cracking when welded in the solution treated
and in the aged conditions. Grain size was, however, viewed as a significant parameter.

Early work by Thompson (1969) on Inconel 718, related cracking to constitutional
liquation from Laves phase or possibly carbides and silicides. After gas tungsten
arc welding [G.T.A.W.] and H.A.Z. simulation, grain boundary liquation was confirmed at 1150-1200°C. Lucas and Jackson (1970) using low (22-26 KV) and high
(100-150 KV) E.B.W. and G.T.A. welding on Inconel 718, observed cracking to be
associated with niobium carbide films. The high KV E.B. welds showed less cracking
tendency than low KV welds, due primarily to weld geometry.

Gordine (1971) observed cracking in G.T.A. welds of Inconel 718, but was unable to
relate cracking to any compositional factors. Recent work by Thompson and Genculu
(1985) using H.A.Z. simulation, found constitutional liquation from MC particles
in Inconel 718, and a significant effect of heat treatment. This in contrast to
the work of Bologna (1969). Reference for welding problems in Incoloy 913 were not
observed in the literature.

The present work is concerned with the dynamic cracking that occurs in the H.A.Z.
during the welding process and not with weld metal or P.W.H.T. cracking. Specifically, an investigation was made of the H.A.Z. cracking of cast and solution
treated Inconel 718 [INCO] and forged and solution treated Incoloy 903, an iron
based precipitation hardening alloy [INCO].

EXPERIMENTAL

Electron Beam Welding was carried out using a 7 KW Electron Research Welder and a
30 KW Sciaky Mark VII Welder, both having a maximum available voltage of 60 KV.
Individual welding parameters are given for each alloy combination in the 'Results'
section.

Scanning Electron Microscopy [S.E.M.] and Energy Dispersive X-Ray analysis [E.D.X.]
were performed on a JEOL JXA 850 Microscope equipped with a Tracor Northern solid
state x-ray analyzer. Quantitative analysis was carried out on selected phases
using a standardless ZAF correction procedure.

Standard metallographic procedures were used for mechanically polishing samples
using silicon carbide papers and diamond to a one micron finish. Final polishing
was achieved with 0.05 micron alumina. Kalling's reagent was used for etching the
Inconel 718 and, 92 HCl, $5H_2SO_4$, 3 NHO_3 for the Incoloy 903.

The following joint combinations were welded: -

   (1) Cast Inconel 718 - butt weld (.300 ins., 0.76 cms thick) - Sciaky

(2)   Forged Incoloy 903 (0.340 ins., 0.86 cms thick) to sheet Inconel 718
       (0.063 ins., 0.16 cms thick) - butt weld - Electron Research.

RESULTS AND DISCUSSION

<u>E.B.W. of Cast and Solution Treated (954°C) Inconel 718</u>.   Examination of the
microstructure showed a coarse dendritic morphology with an austenitic matrix and
interdendritic areas containing a light-grey phase and an acicular phase; additionally, primary carbides were present being located both at grain boundaries
and within the grains, Fig. 1.   SEM/EDX analysis, Table 1, showed the grey phase
to be a Laves phase, A_2B (Fe+Cr+Ni, 64%; Ti+Nb+Mo, 35%), whilst the acicular phase
was too small to identify.   Inspection of a T-T-T diagram for Inconel 718 and the
work of Muzyka and Maniar (1969), clearly indicates that the acicular phase is the
orthorhombic δ-Ni$_3$Nb compound.

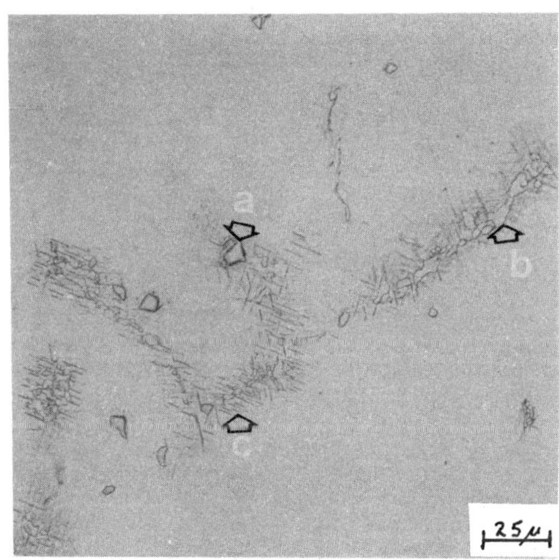

Fig. 1. Micrograph of Parent Material in Cast Inconel 718.
          a = NbCN,   b = Laves,   c = acicular phase

Electron Beam Welding the alloy in the solution treated condition at 5.8 kj/inch
(40 KV, 120 ma, 50 ipm) showed no cracking in the weld metal, but several micro-
fissures, up to .15 mm long in the heat affected zone, Fig. 2(a).   In this region,
the acicular Ni$_3$Nb phase had re-solutioned, whilst the Laves phase morphology was
altered to a more spheroidized structure due to dissolution and interaction with
the matrix.

S.E.M. examination of the H.A.Z. cracks show them to be associated with a lamellar
structure, Fig. 2(b), typical of an eutectic type reaction.   E.D.X. analysis,
Table 1, shows an analysis similar to the Laves phase, but with additional Ni from
the Laves/matrix eutectic interaction.   This constitutional liquation leads to a
small amount of liquid phase on the grain boundary which would not support local
thermal stresses.

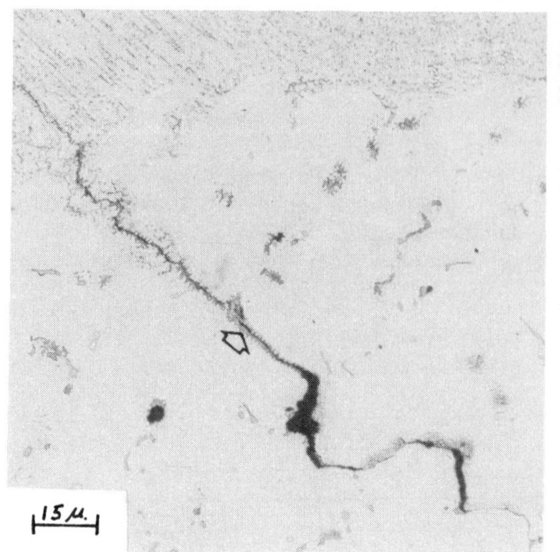

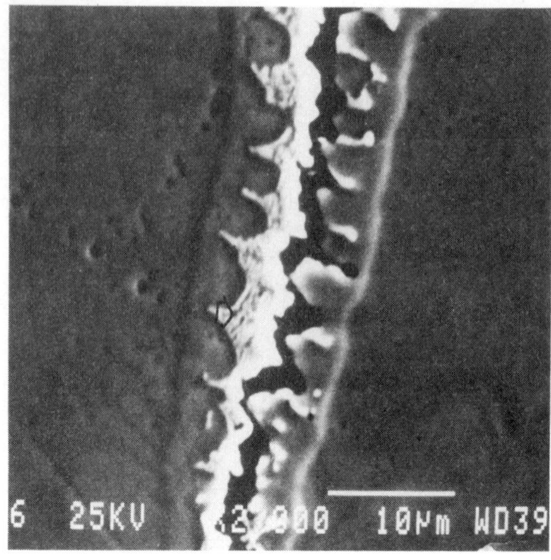

Fig. 2(a). Optical   Fig. 2(b). S.E.M.

Fig. 2. Micrographs of H.A.Z. Crack in Cast Inconel 718.
          Arrow indicates H.A.Z. phase analyzed in Table 1.

TABLE 1. SEM/EDX Analysis of Cast and Solution Treated Inconel 718
                      "As Welded" Wt. Per Cent

Phase	Ni	Cr	Fe	Nb	Mo	Si	Ti	Al
Matrix	51.8	20.1	20.7	3.0	3.0	0.07	0.85	0.27
Primary C-N	0.77	N.D.*	0.98	80.4	N.D.*	1.2	15.2	0.93
Laves	35.9	14.1	14.1	18.8	15.3	0.97	0.62	0.24
HAZ Crack	45.0	14.3	12.1	25.3	N.D.*	0.85	1.5	1.1
Base Metal Analysis	52.5	18.7	Bal.	4.95	2.95	0.14	0.90	0.46

* N.D. Not Determined   0.003B/0.06C/0.003S

Removal of the Laves phase is an obvious way of reducing the cracking. This would
necessitate solution treating above 1040°C to re-solution the Laves phase which
would also lead to re-solution of the δ - Ni_3Nb phase. In Cast material, it is unlikely that this would affect the properties of the alloy since the grain size is
very coarse. In wrought materials, the δ-Ni_3Nb controls the grain size and the
Laves phase is generally not observed, thus, solution treatment temperatures above
the solvus of the δ-Ni_3Nb are unnecessary and detrimental to fine grain size
requirements ($\geqslant$ ASTM 6/7).

<u>E.B.W. of Forged Incoloy 903/Sheet Inconel 718</u>. Optical microscopy of the Incoloy
903 alloy solution treated at 843°C showed a mixed grain structure with small
grains 'decorating' larger grains, (Fig. 3); additionally, due to warm working the

grain size was elongated being approximately ASTM 3 by ASTM 5. The Inconel 718
sheet was of ASTM 7 grain size after solution treatment at 954°C. Primary particles were observed in the Incoloy 903 alloy located on grain boundaries and within grains. Most of the primary carbides were mauve in appearance and were identified by EDX analysis (Table 2) as niobium rich, Titanium containing precipitates
being carbide or carbo-nitride in nature. Some titanium carbide or carbo-nitride
having a light orange colour optically were also observed but the majority of the
primary particles were niobium rich. A discrete precipitate was also observed on
grain boundaries in the matrix. EDX analysis showed 2-3 times the matrix value
for niobium but the analysis is considered tenuous due to the fine size of the precipitate and the influence of the matrix. Comparison with Inconel 718 would suggest a Ni_3Nb type precipitate for the Incoloy 903, though further work is necessary to substantiate this fact.

E.B. welding the alloy in the solution treated condition at 1.9kj/inch (50 KV, 22
ma, 35 ipm) resulted in HAZ microfissuring in the Incoloy 903 material, Fig. 4(a)
- (b), but not in the Inconel 718 sheet material. Optical microscopy showed that
the niobium rich primary particles were generally associated with the cracking,
Fig. 5 (a)-(c). SEM/EDX analysis, Table 2, and Fig. 6 (a)-(b) show a niobium rich
phase present on the grain boundaries. Again, the matrix probably affects the
analysis due to the fineness of the precipitate, probably resulting in a lower
than actual niobium level.

TABLE 2.   SEM/EDX Analysis of Forged Incoloy 903, Matrix and Weld,
Wt. Per Cent

Phase	Fe	Ni	Co	Nb	Ti	Al	Si	Cr
Matrix	42.7	36.3	15.8	2.5	1.4	0.80	0.16	0.22
Primary C-N	0.98	0.8	0.50	80.4	15.2	0.9	1.2	0
HAZ GB Precipitate	27.7	41.1	15.8	6.4/10.7	3.4	0.86	0.1	0.10
Weld [White Phase]	36.3	39.2	14.8	6.5	1.8	0.96	0.34	0.02
Base Metal Analysis Bal.	38.8	14.9	2.93	1.41	0.95	0.07	0.11	

0.007B/0.028C/0.001S

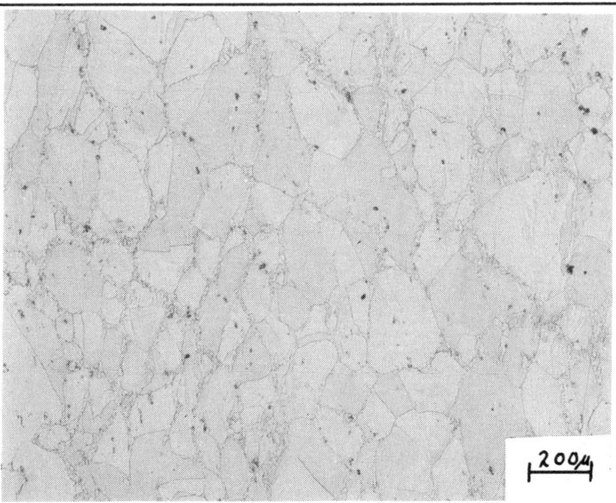

Fig. 3. Optical Micrograph of Forged Incoloy 903.

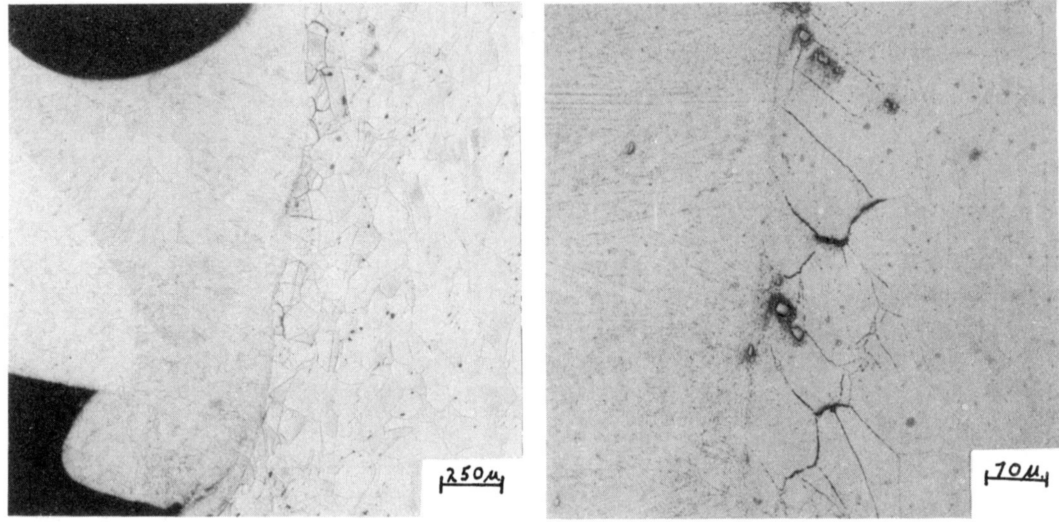

Fig. 4(a). General View of Weld.   Fig. 4(b). HAZ Cracks in Incoloy 903.

Fig. 4. Weld in Forged Incoloy 903/Sheet Inconel 718.

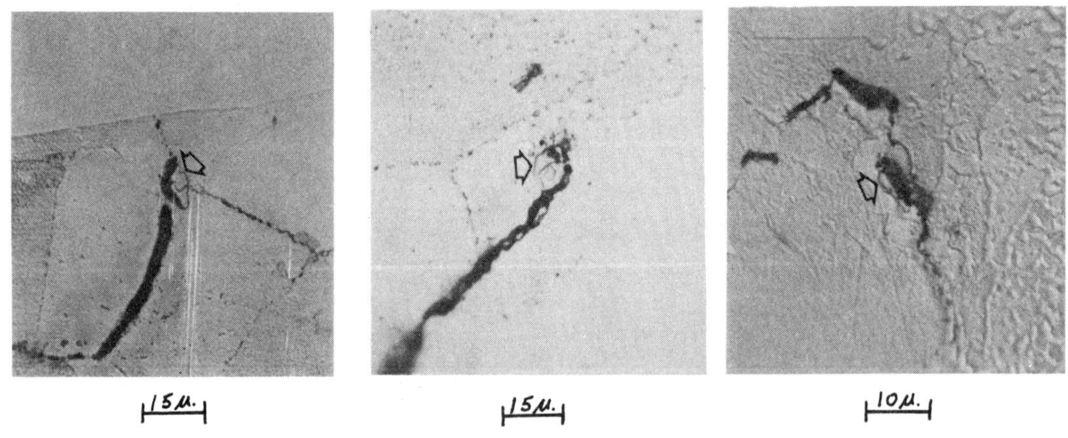

Fig. 5(a)   Fig. 5(b)   Fig. 5(c)

Fig. 5 (a) - (c). Partly Liquated Primary C-N (Arrows) with
HAZ Cracks in Incoloy 903.

The cause of the cracking is probably due to partial liquation during welding of large primary niobium rich particles and complete liquation of smaller particles, Figs. 5 and 6. Furthermore, the presence of a niobium enriched grain boundary precipitate present in the matrix would also contribute to the presence of a niobium rich liquid phase which would be susceptible to cracking under the influence of a thermal stress gradient.

An additional phenomenon observed under Nomarski contrast, Fig. 7, was intrusions of weld metal along cracks in the grain boundaries of the HAZ. Analysis of the weld phase showing white in Figure 7. Table 2, again showed enrichment of niobium by a factor of two to three times that of the matrix value. A similar analysis was also obtained in the backfilled HAZ region.

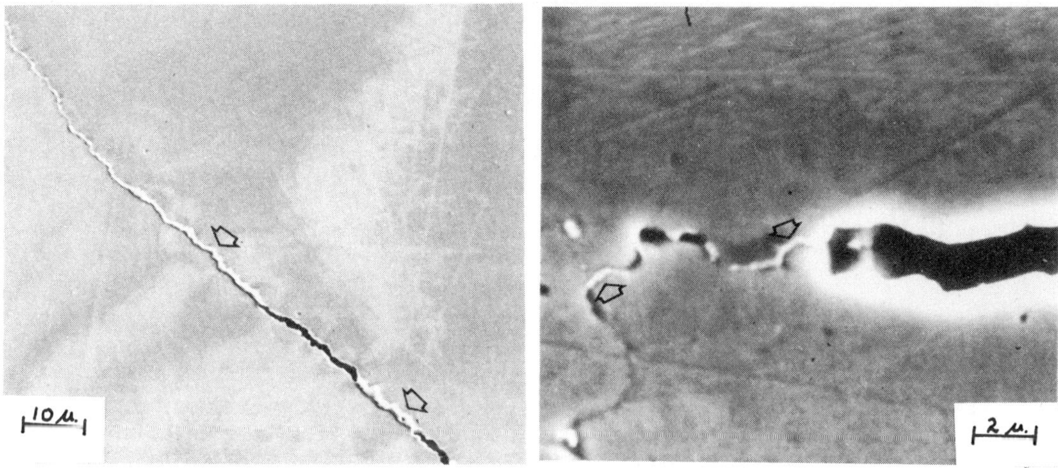

Fig. 6 (a).   Fig. 6 (b).

Fig. 6 (a) & (b).   SEM Micrograph of G.B. Precipitates (Arrows) in H.A.Z. of Incoloy 903.

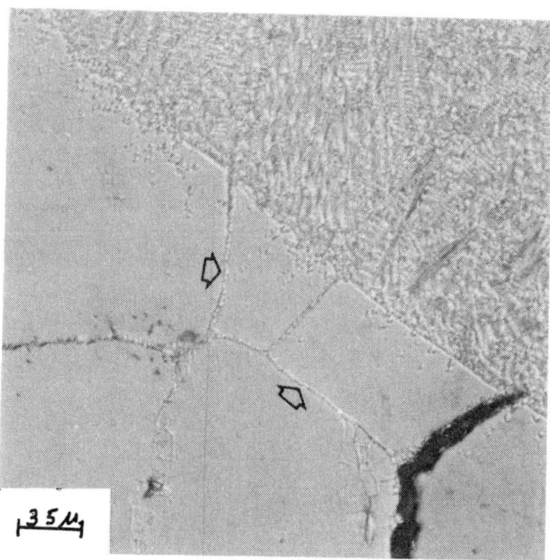

Fig. 7.   Intrusions (Arrows) into H.A.Z. in Incoloy 903.

GENERAL DISCUSSION

The dynamic cracking that occurs in the heat affected zone of Super Alloys requires
a combination of two factors to cause failure: -

(1)   a level of stress which triggers cracking in:
(2)   a microstructure with a propensity to failure, i.e., containing one
       or several of: low melting point grain boundary phases, brittle
       phases(s), large grain size, elemental segregation.

The main stress component is generated by thermal stresses arising from welding,
where the high temperature of the beam coupled with rapid heating and cooling
leads to high stress levels. Koren and co-workers (1982) have correlated the
number of cracks per unit length with welding speed and current using multiple
linear regression analysis. Minimization of cracking was achieved by using the
slowest speed and highest current when other variables were kept constant. Further
analysis showed that reducing the welding speed resulted in a reduction of the
thermal gradient in the heat affected zone leading to reduced strain on the weld.

Cracking due to grain boundary segregation has been verified by Auger Electron
Spectroscopy in nickel and nickel alloys, Docherty (1974), with sulphur being one
of the main segregating elements. Hafnium, zirconium, and lanthanum were found
effective in controlling segregation and improving ductility.

The presence of a brittle phase occurring after welding is not easily controlled.
The brittle phase in the HAZ of the Cast 718 alloy in the present work was directly
caused by the presence of a parent material Laves phase which reacted with the
austenite matrix. Similarly, the niobium rich HAZ grain boundary phase observed
in the Incoloy 903 alloy was probably formed from liquation of niobium rich carbonitrides with supplementary niobium from matrix grain boundary precipitates.

Fine grain sizes have been shown in many investigations to be beneficial to minimizing HAZ cracking. Unfortunately, high temperatures in the HAZ can lead to rapid
grain coarsening with increased crack susceptibility, unless precipitates are present to control the HAZ grain size.

Consideration of the factors controlling heat affected zone cracking in Electron
Beam welds leads to the following:

(a)   weld slower to reduce thermal stresses
(b)   reduce KV and increase ma to reduce thermal stress
       (for low voltage welding machines)
(c)   preheat to reduce thermal stresses (if necessary)
(d)   weld in solution treated condition to reduce grain
       boundary segregation of alloying elements and other
       elements such as sulphur
(e)   ensure fine grain size ($\gtrless$ ASTM #5) in HAZ by using
       grain boundary precipitates to control grain size
       and solution treating temperature below the grain
       coarsening temperature.

CONCLUSIONS

(1)   Heat Affected Zone cracking in Cast and solution treated Inconel 718
       was related to liquation of Laves phases, leading to grain boundary
       rupture under the influence of welding thermal stresses.

(2)   Similar cracking in forged and solution treated Incoloy 903 was attributed
       to liquation of niobium rich primary carbides (MC) with possible assist-
       ance from matrix niobium rich phases.

(3)   Intergranular HAZ cracking is not amenable to detection by non-destructive
       means.

(4)   HAZ grain boundary cracking is a complex function of welding and material
       parameters.

REFERENCES

Baker, R.G. (1976). _Philos. Trans. R. Soc. A287_, 207.
Betteridge, W., and J. Heslop (1974). _The Nimonic Alloys_. 2nd ed.
  E. Arnold, p.371.
Bologna, D.J. (1969). _Metals Eng. Quart._, _11_, 37-43.
Docherty, J.E. (1974). In J.L. Walter (ED.), _Grain Boundaries in Engineering
  Materials_, Fourth Bolton Landing Conference, p.619.
Duval, D.S., and W.A. Owczarski (1967). _Welding Journal_, 46(9), 423-S-431-S.
Gordine, J. (1971), _Welding Journal, 50(11)_, 480-S-484-S.
Koren, A., M. Roman, I. Weisshaus and A. Kaufman (1982). _Welding Journal,
  61(11)_, 348-S-351-S.
Lucas, M.J., and C.E. Jackson (1970). _Welding Journal, 49(2)_, 46-S-54-S.
Muzyka, D.R., and G.N. Maniar (1969). _Metals Eng. Quart., 9(4)_, 23-27.
Owczarski, W.A., D.S. Duval, and C.P. Sullivan (1966). _Welding Journal, 45(4)_,
  145-S-155-S.
Thamburaj, R., W. Wallace, and J.A. Goldak (1983). _Int. Met. Rev. 28(1)_, 1-22.
Thompson, R.G. (1969). _Welding Journal, 48(2)_, 70-S-79-S.
Thompson, R.G., and S. Genculu (1985). _Welding Journal, 64(12)_, 337-S-345-S.

THE INFLUENCE OF MICROSTRUCTURE ON THE FRACTURE TOUGHNESS
AND YIELD STRENGTH OF Mn-Ni-Al SHIP PROPELLER BRONZES

J.I. Dickson[*], L. Handfield[*] J Hallen-Lopez[*],
Y. Blanchette[*] and M Sahoo[**]

*Departement de génie métallurgique, Ecole Polytechnique
Montréal (Québec) Canada H3C 3A7
**PMRL-CANMET, 568 Booth Street, Ottawa, Canada K1A 0G1

ABSTRACT

The fracture toughness of 8.4-14.2 wt % Mn, 2.0-2.2% Ni, 6.7-8.5% Al was shown to
increase with decreasing Mn composition and with decreasing yield strength. In
slowly-cooled alloys which fractured partly by cleavage, the fracture toughness
was determined primarily by the volume fraction and mean free path of b.c.c. β
phase. An anneal of 24 hours at 720°C increased the resistance to cleavage, as a
result of fine precipitation within the β phase. Laboratory as-cast alloys fractured by microvoid formation and coalescence and their fracture toughness was
governed by the volume fractions of β phase and K_I particles, through their influence on the localization of the crack tip plasticity within the soft f.c.c. α
phase and on microvoid nucleation. The yield strength of these bronzes was determined primarily by the volume fraction and mean free path of β phase.

KEYWORDS

Fracture toughness, Mn-Ni-Al bronzes, cleavage, microstructure, yield strength.

INTRODUCTION

Mn-Ni-Al bronzes and Ni-Al bronzes are materials often employed for large ship
propellers. One of the important mechanical properties of propellers for ice-
breakers and for arctic navigation is the fracture toughness, since the probability of impact with ice cannot be neglected. While some studies have been carried
out on the Charpy impact properties of propeller bronzes (Thomson, 1968; Sahoo,
Edwards and Thomson, 1980. Sahoo and Couture, 1984), a detailed study of their
fracture toughness does not appear to have been previously reported.

The objective of the present study was to determine the influence of Mn content
and of microstructure on the fracture toughness of Mn-Ni-Al bronzes having compositions similar to or containing somewhat lower Mn and Al than those of Cu alloy
C95700 which is covered by ASTM specification B148-86 and which permits 11-14 wt %
Mn and 7-9 wt % Al.

EXPERIMENTAL

The bronzes studied were copper-based alloys containing 8.4-14.2% Mn, 6.7-8.5% Al, 2.0-2.2% Ni and 2.9-3.4% Fe for which the Al and Mn contents tended to increase simultaneously. The composition of the alloys studied are reported in Table 1. These alloys will be referred to by their Mn compositions and by their material conditions.

<u>TABLE 1   Chemical Analyses (wt %)</u>

Mn	Al	Ni	Fe
8.4	7.0	2.0	2.9
8.7	6.8	2.1	3.3
8.8	6.7	2.0	3.2
10.5	7.1	2.1	3.2
10.7	6.7	2.0	3.0
10.9	6.9	2.1	3.2
12.2	8.3	2.1	3.0
12.3	7.9	2.1	3.3
14.0	8.5	2.1	3.4
14.2	8.0	2.2	3.4

The bronzes were cast at PMRL-CANMET into plates 295 x 148 x 25.4 mm, employing sand moulds and a central riser. Some of the as-cast (AC) plates were annealed for 2 hours at 900°C and cooled at 10°C/hour to 250°C to obtain normalized and slowly cooled (NSC) material. Some of this NSC material was heat treated in an attempt to improve the toughness. These heat treatments were either a two hour anneal at 690°C followed by water quenching or a 24 hour anneal at 720°C followed by air cooling to produce the HT1 and HT2 material conditions, respectively.

The fracture toughness was determined at room temperature following ASTM-E813-86 recommended procedure for J_{Ic} testing. The CT specimen geometry was employed, with specimens of thickness B = 22.2 mm and width W = 50.8 mm. All materials were tested employing specimens with side grooves resulting in a net thickness of 17.8 mm. Some of the materials were also tested employing specimens without side grooves.

The specimens were fatigue precracked to a crack length a/W $\simeq$ 0.55, employing an R-ratio = 0.1, a cycling frequency of 20 Hz and an initial stress intensity range ΔK of 14 MPa$\sqrt{m}$. The ΔK was automatically adjusted to maintain an approximate cracking rate of 3 x 10^{-5} mm/cycle. An elastic modulus $E/(1-\nu^2)$ of 125,000 MPa was found to permit to monitor satisfactorily the crack length from the elastic compliance and was also employed to calculate K_{Ic} values from the measured J_{Ic} values.

During J_{Ic} testing, a crack opening displacement rate of 0.001 mm/s was applied at the load line. A partial unloading of 10% was performed at every 0.1 mm of load line displacement to obtain the J-Δa curve. The critical value of J was taken as the intersection of this line with the crack blunting line, J = $2\sigma_f\Delta a$ where σ_f is the mean between the yield and ultimate tensile strengths.

Dynamic fracture toughness tests were performed on a Tinius Olsen Charpy impact tester, instrumented with Dynatup equipment. A digital oscilloscope was employed to record the load vs time curve. The specimens were Charpy type specimens 50 mm (span 40 mm) x 10 mm x 10 mm which were fatigue precracked at $\Delta K \simeq 14$ MPa(m)$^{1/2}$ and R = 0.1 to a crack length a/W $\simeq$ 0.5. The load vs time curve recorded was converted into a load vs load-line displacement curve, employing the apparent mass and apparent length of the impact tester. The J_{ID} values were then computed from the strain energy to the maximum load obtained, employing the equations recommended by ASTM E-813-86.

RESULTS AND DISCUSSION

<u>Fracture Toughness</u>. Table 2 reports the J_{IC} results (average of 3 tests on specimens with side grooves) as well as the corresponding calculated K_{IC} values. Except for the 8.4% Mn AC material, all materials with less than 9% Mn did not satisfy the thickness requirement that $B_{net} > 25$ J_{IC}/σ_y. Several procedures were compared to estimate the J_{IC} values of these higher toughness materials. The most conservative estimates were obtained from the stretch zone widths, δ_s, measured in the mid-thickness region and from the relationship obtained for the other materials between J_{IC}/σ_f and δ_s (Figure 1), where σ_f is the mean between σ_y and the tensile strength. The δ_s values correspond to measurements made by scanning electron microscopy on specimens tilted 45° with respect to both the tensile axis and the fracture plane. The toughness values thus obtained are identified by asterisks in Table 2 and are approximately 30% higher than the limiting J_{IC} values required to satisfy ASTM E-813-86. Values of J_{IC}, valid according to ASTM E-813-86, were also obtained on specimens without side-grooves for the 12.2% Mn AC, 12.3% NSC, 14.0% AC and 14.2% NSC materials. On the average, these values were 2.1 times higher (K_{IC} 1.45 times higher) than those for specimens with side grooves.

<u>TABLE 2   Static Fracture Toughness</u>

Material	J_{IC} KJ/m^2	K_{IC} MPa$\sqrt{m}$	Material	J_{IC} KJ/m^2	K_{IC} MPa$\sqrt{m}$
8.4% Mn AC	225*	167*	12.3% Mn AC	97	110
8.7% Mn AC	654*	286*	12.3% Mn NSC	104	113
8.7% Mn NSC	600*	274*	12.3% Mn HT1	93	108
8.8% Mn HT1	546*	261*	12.3% Mn HT2	87	104
8.8% Mn HT2	594	272*	14.2% Mn AC	42	71
10.7% Mn AC	254	178	14.2% Mn NSC	55	82
10.9% Mn AC	273	184	14.2% Mn HT1	64	89
10.9% Mn NSC	332	204	14.2% Mn HT2	93	105
10.5% Mn HT1	200	158			
10.5% Mn HT2	238	172			

* Value estimated from δ_s

The J_{Id} (average of three tests) and corresponding K_{Id} values obtained at temperatures of 24°C, 4°C and -19°C or 24°C and -8°C are reported in Table 3. These show similar trends but are higher than the static toughness values of Table 2. The average difference in toughness ($J_{Id} = 1.66$ J_{IC} and $K_{Id} = 1.29$ K_{IC}) is less than the average decrease in the measured static toughness caused by the use of side grooves. This suggests that the dynamic toughness which would be obtained on specimens with side grooves would be somewhat lower than the static toughness. The test temperature had very little influence on the dynamic toughness, although a tendency for a slight decrease with decreasing temperature is suggested by the results.

The K_{Ic} values of table 2 are plotted as a function of the amount of Mn and of σ_y in Figures 2 and 3, respectively. It is seen that K_{Ic} decreases with increasing σ_y and Mn content, with the AC and HT2 conditions giving the highest K_{Ic} for a given yield stress at higher Mn levels. The 8.4% Mn AC alloy gives a particularly low K_{Ic} value for its Mn composition.

<u>TABLE 3   Dynamic Fracture Toughness, K_{Id} (MPa$\sqrt{m}$)</u>

Material	24°C	4°C	-8°C	-19°C
8.4% Mn AC	235	238	- - -	225
8.7% Mn AC	302	295	- - -	283
8.7% Mn NSC	307	- - -	292	- - -
8.8% Mn HT1	306	- - -	308	- - -
8.8% Mn HT2	300	- - -	285	- - -
10.7% Mn AC	249	245	- - -	241
10.9% Mn AC	255	215	- - -	218
10.9% Mn NSC	218	- - -	198	- - -
10.5% Mn HT1	276	- - -	261	- - -
10.5% Mn HT2	237	- - -	239	- - -
12.2% Mn AC	157	152	- - -	143
12.3% Mn NSC	110	102	- - -	101
12.3% Mn HT1	113	- - -	105	- - -
12.3% Mn HT2	147	- - -	149	- - -
14.0% Mn AC	105	90	- - -	87
14.2% Mn NSC	90	82	- - -	88
14.2% Mn HT1	92	- - -	76	- - -
14.2% Mn HT2	143	- - -	133	- - -

<u>Microstructure and Fractographic Features</u>. The microstructure of the bronzes with 8-9% Mn consisted of relatively soft, f.c.c. α grains, some portions of which were surrounded by a harder phase (Figure 4). For the three slowly-cooled alloys, this phase was richer in Ni, Al and Fe than the b.c.c. β phase in the AC alloys or in all higher Mn alloys other than the 10.9% Mn NSC bronze and appeared to be similar to one of the T phases described for Cu-Mn-Al alloys (Köster and Gödecke, 1966). Different Fe-Mn-Al intermetallic particles are present which are similar in shape, in distribution and in characteristics to the K_I, K_{II} and K_{IV} precipitates described for Ni-Al bronzes (Hasan and others, 1982). The largest of these are the K_I particles, which have an interdendritic shape in the AC alloys. These are near the centers of α grains, since they precipitate first and then act as nucleation sites for α precipitation within the β phase.

As the Mn composition increases, the β and K_I volume fractions, V_β and V_K, increase and the α grain size decreases. The β phase starts to completely surround the α grains (Figure 5), permitting to observe that numerous α grains are present within each β grain. Except for the AC material, the widths of the β layers also increase with increasing Mn. Both the 14.2% Mn AC (Figure 5) and 14.2% Mn HT2 alloys contain almost continuous layers of α grains lining β grain boundaries, with the K_I particles in the former tending to be concentrated near the β grain boundaries. Both the HT1 and HT2 (Figure 6) heat treatments precipitate fine secondary α grains within the β phase of the higher Mn alloys, with the HT2 treatment (Figure 6) also precipitating numerous very fine ($\simeq$ 0.1-0.2 μm) particles (possibly K_{IV}) within the β phase.

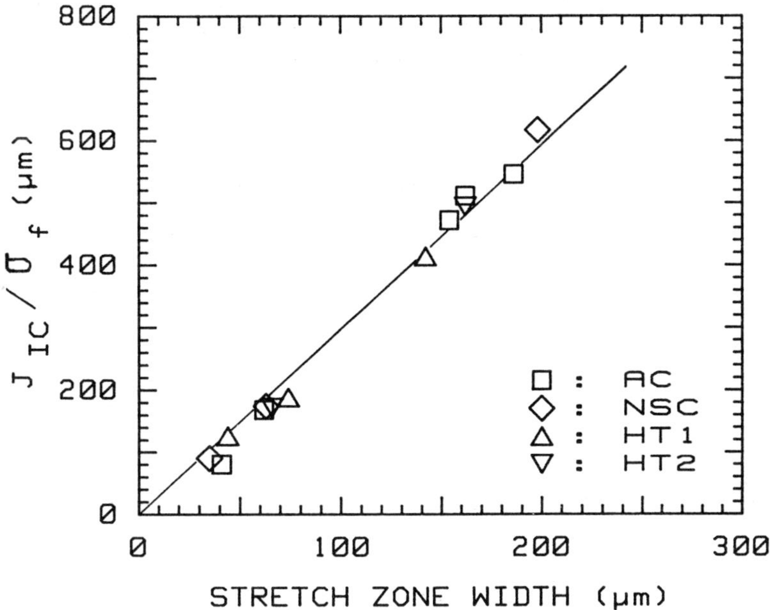

Fig. 1 Relationship between J_{IC}/σ_f and stretch zone width.

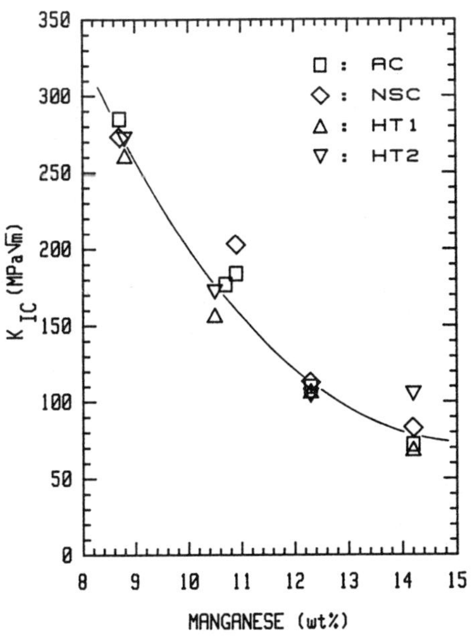

Fig. 2 Fracture toughness vs Mn composition.

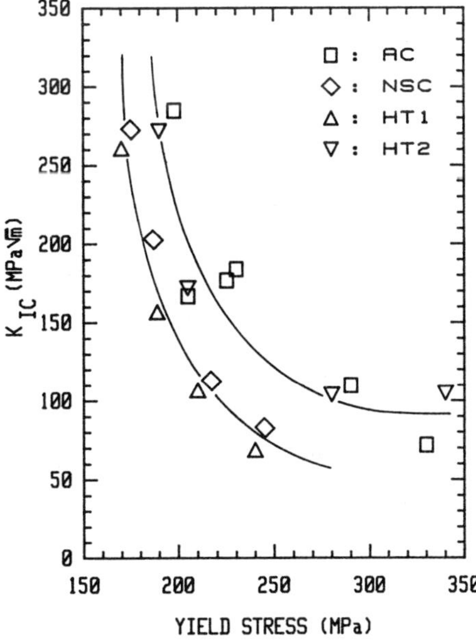

Fig. 3 Fracture toughness vs yield strength.

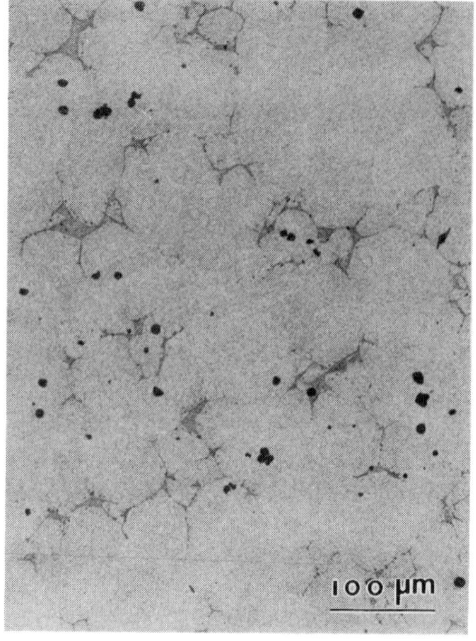

Fig. 4   Microstructure of the
8.7% Mn AC alloy.

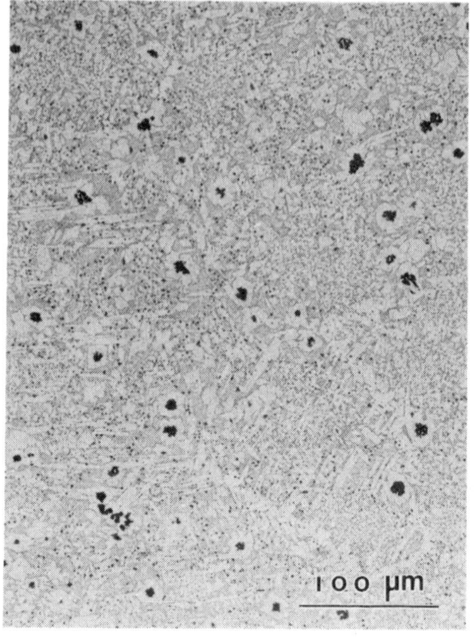

Fig. 5   Microstructure of the
14.2% Mn AC alloy.

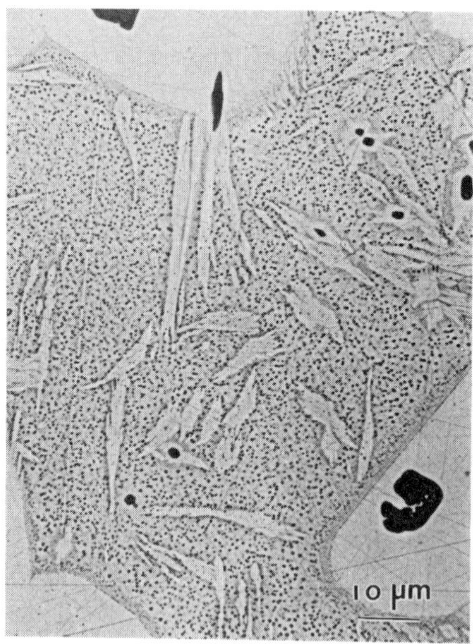

Fig. 6   Precipitates in the β phase
of the 12.3% Mn HT2 bronze.

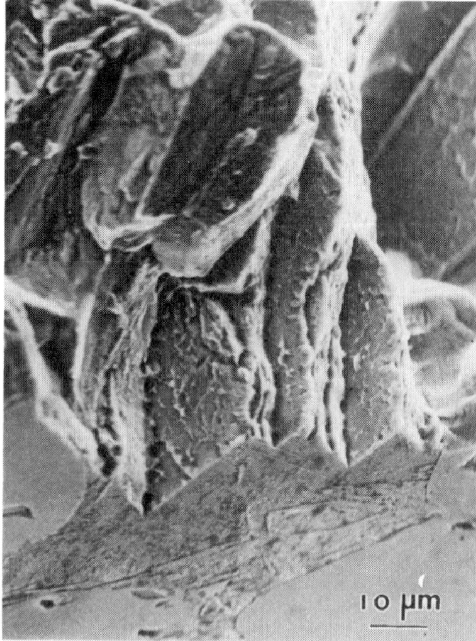

Fig. 7   Finely dimpled crystallographic facets in β phase
of 12.3% Mn HT2 bronze.

The fractographic observations showed that fracture in the AC alloys occurred by formation and coalescence of microvoids, with the 14.2% Mn AC alloy presenting largely a ductile intergranular aspect, associated with the concentration of K_I particles in the region of the β grain boundaries, and the 12.3% Mn AC alloy a considerably less intergranular aspect. The NSC and HT1 alloys presented increasing amounts of cleavage with increasing Mn. For the HT2 alloys with more than 10% Mn, the β phase appeared finely dimpled on the fracture surface. The preparation of mid-thickness profile specimens (Dickson and others, 1987) as well as stereographic observations showed that these finely dimpled regions corresponded to crystallographic facets (Figure 7) formed by a combination of cleavage and of nucleation of small microvoids at the very fine precipitates produced by this heat treatment. For the 14.2% Mn HT2 material, the crack path had some tendency to follow almost continuous lines of α grains along β boundaries (Figure 8).

For the alloys with less than 11% Mn, and especially for the 8.4% Mn AC bronze, a number of large dimples presented near their centers finely-cracked layers (Figure 9), rich in Fe, Mn and at times P. It was verified from energy dispersive X-ray analysis and from their presence on freshly produced fracture surfaces that these layers did not correspond to mud-cracks in corrosion product formed after fracture. These brittle layers which were sufficiently thin to be at times semitransparent in the scanning electron microscope were shown (Figures 9 and 10) to be along some α grain boundaries.

<u>Relationships Between K_{Ic}, σ_y and the Microstructural Parameters</u>. Relationships between the fracture toughness and different microstructural parameters, evaluated employing an IBAS2 quantitative image analyzer, were investigated. The relative resistance to cleavage of the slowly-cooled alloys was determined by comparing the fraction of cleavage f_c to V_β (Figure 11). For the NSC material f_c was considerably greater than V_β indicating a high susceptibility to cleavage. For the HT1 material, f_c was closer to V_β, indicating some increase in the resistance to cleavage. For the HT2 material, f_c was considerably less than V_β, indicating considerably improved resistance to cleavage as a result of the formation of microvoids at the fine precipitates in the β phase produced by this heat treatment. For the 14.2% Mn alloy the HT2 treatment increased K_{Ic} by 28% with respect to the NSC material and this heat treatment also reduced f_c to approximately half of V_β. This increase in the resistance to cleavage results in the HT2 condition giving, for the higher Mn alloys, the highest K_{Ic} value for a given V_β (Figure 12). For a given β volume fraction, K_{Ic} of the AC alloys, none of which presented any cleavage on the fracture surfaces, is only slightly higher than that of the HT1 and NSC materials. It is also seen that the low toughness of the 8.4% Mn AC alloy is largely explained by its relatively high β volume fraction associated in part with a relatively high equivalent Al (Sahoo and Couture, 1984) and possibly with some imprecision in the chemical analysis. Nevertheless the evidence suggested that the presence of the finely-cracked brittle layers observed in some of the large dimples on the fracture surfaces of the 8.4% Mn AC and had a moderate effect in decreasing K_{Ic}. The detailed results also indicate that while the HT2 and, to a considerably lesser extent, the HT1 heat treatments increase the relative resistance to cleavage. this increase often tends to be offset by an accompanying increase in the β volume fraction. The HT2 heat treatment only clearly increased K_{Ic} of the 14.2% Mn alloy, for which the presence of an alternate fracture path in the α-phase permitted f_c to be reduced particularly strongly with respect to V_β The results. however, indicated that this treatment also significantly increased K_{Id} of the 12.3% Mn alloy.

The relative yield strengths of these bronzes is largely determined by the β volume fraction (Figure 13) and β mean free path. defined as the average thickness of β layers between primary α grains (Figure 14). The higher yield strengths obtained in Figure 13 at a given V_β for the AC alloys can be explained by these

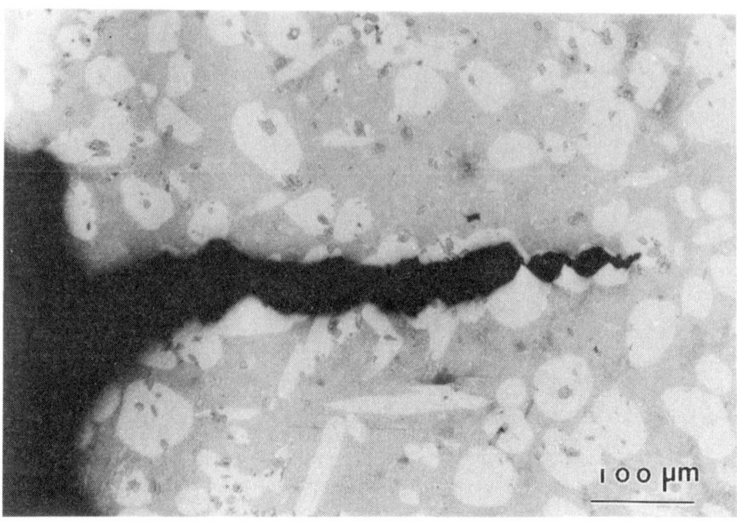

Fig. 8   Secondary crack following α grains along β grain boundary, 14.2% Mn HT2 bronze.

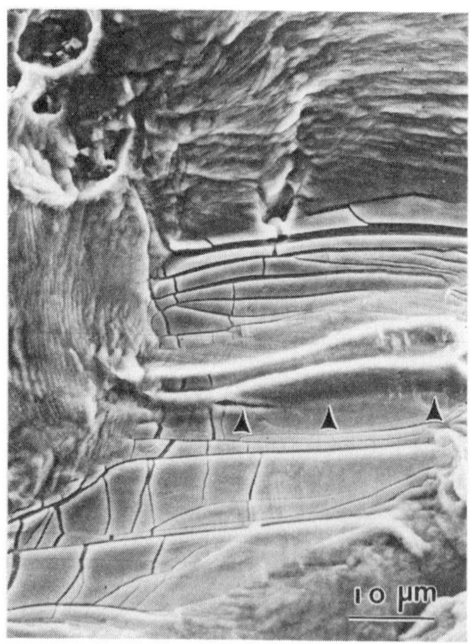

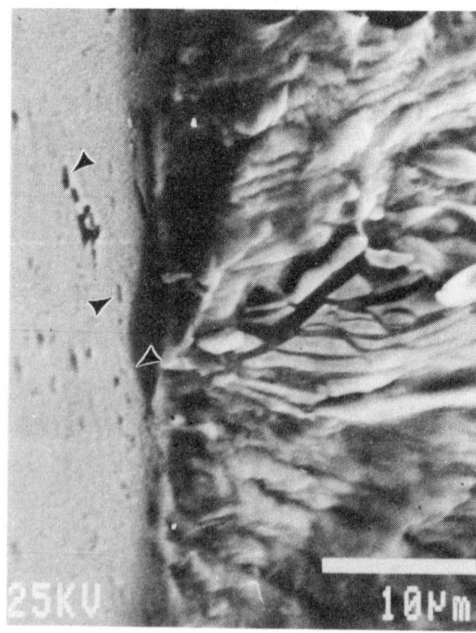

Fig. 9   Brittle layer near α grain boundary (arrows) in 8.4% AC bronze.

Fig. 10   Brittle layer near α grain boundary (arrows) in 8.8% Mn HT1 bronze.

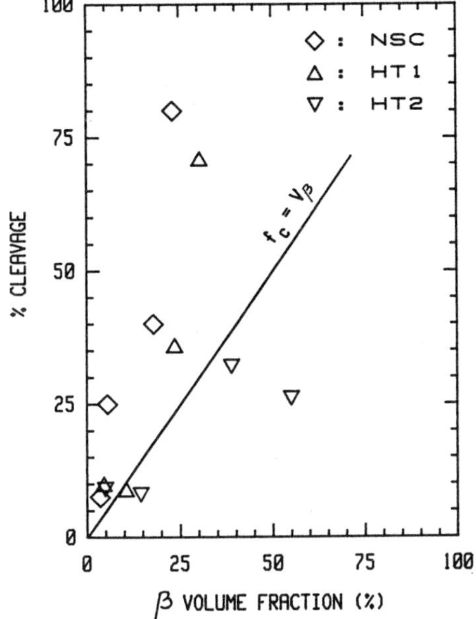

Fig. 11   Cleavage fraction vs
          β volume fraction.

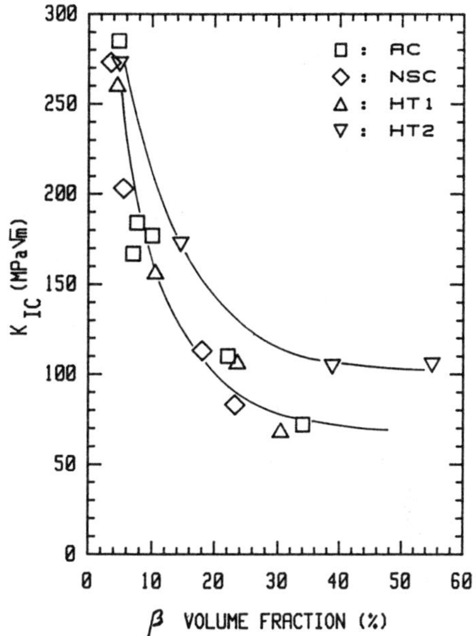

Fig. 12   Fracture toughness vs
          β volume fraction.

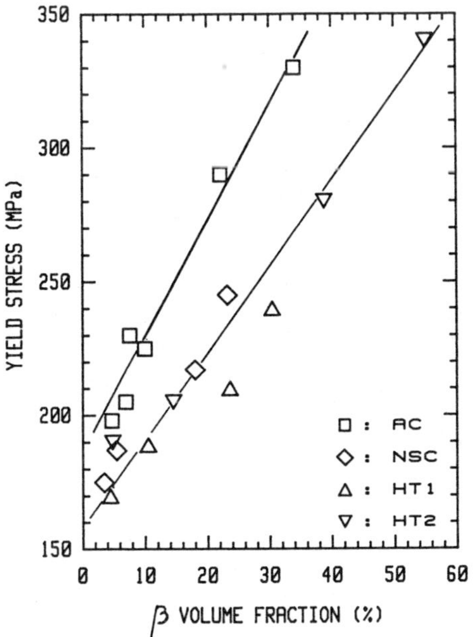

Fig. 13   Yield strength vs
          β volume fraction.

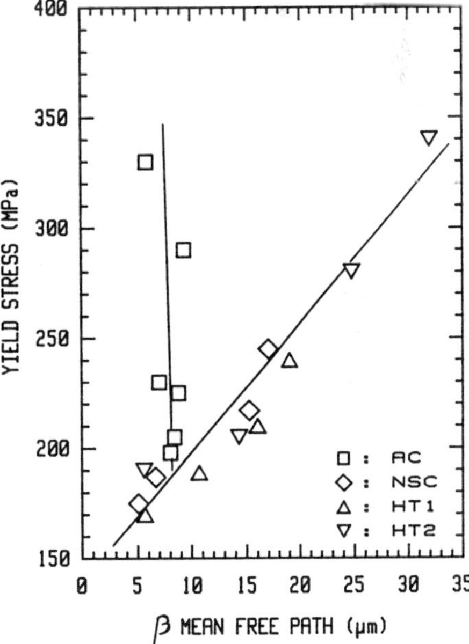

Fig. 14   Yield strength vs β mean
          free path.

laboratory cast alloys having a small β mean free path for all compositions. The increase in V_β and in cleavage resistance produced by the HT2 heat treatment in the higher Mn alloys resulted in the highest K_{Ic} for a given yield strength for the slowly-cooled conditions investigated.

CONCLUSIONS

From the present study, it can be concluded that the fracture toughness of the bronzes studied decreased with increasing Mn composition and with increasing yield strength. The fracture toughness of the slowly-cooled alloys was primarily controlled by the β volume fraction and mean free path, through their influence on the occurrence of cleavage. The fracture toughness of the AC alloys was controlled by the β and K_I volume fractions through the influence of the former on the localization of the crack tip plasticity within the soft α phase and of the latter on the nucleation of large microvoids The yield strength of these bronzes was controlled primarily by the volume fraction and mean free path of the β phase. The HT2 heat treatment produced fine precipitation within the β phase which increased the resistance to cleavage. An accompanying increase in β volume fraction resulted in a relatively high toughness for a given yield strength, in comparison to the other slowly-cooled alloys.

ACKNOWLEDGMENTS

The present work was performed under DSS contract 24ST.23440-5-9054 for PMRL-CANMET. The authors are indebted to Nicole Roy for typing the manuscript.

REFERENCES

Dickson, J.I., J. Hallen-Lopez, L. Handfield and M. Sahoo (1987). "The Relationship Between the Microstucture and the Fracture Toughness of Mn-Ni-Al Ship Propeller Bronzes". Submitted to Microstructural Science.

Hasan, F., A. Jahanafrooz, G.W. Lorimer and N. Ridley (1982). "The Morphology, Crystallography and Chemistry of Phases in As-Cast Nickel-Aluminum Bronze". Metall. Trans. A, 13A, 1337-1345.

Köster W. and T. Gödecke (1966). "Das Dreistoffsystem Kupfer-Mangan-Aluminium". Zeitschrift für Metallkunde. 57. 889-901.

Sahoo. M. (1982). "Structure and Mechanical Properties of Slow-Cooled Nickel-Aluminum Bronze Alloy C95800". AFS Transactions. 90. 913-926.

Sahoo. M. J.O Edwards and R. Thomson (1980). "Influence of Corrosion Inhibiting Heat Treatment on the Microstructure, Impact and Fatigue Properties of Nickel-Aluminum Bronze Alloy C95800'. AFS Transactions. 85. 769-776.

Sahoo, M. and A. Couture (1984). "Microstructure and Mechanical Properties of Continuously Cast Mn-Ni-Al Bronzes". AFS Transactions, 92, 217-226.

Thomson, R. (1968). "Charpy Impact Properties of Bronze Propeller Alloys". AFS Transactions, 76, 99-109.

SUBJECT INDEX

337